T0146217

Cancer Virus Hunters

Cancer Virus Hunters

A History of Tumor Virology

Gregory J. Morgan

JOHNS HOPKINS UNIVERSITY PRESS BALTIMORE

Additional images in this book were made possible through the generous support of the Stevens Institute of Technology.

Printed in the United States of America on acid-free paper
9 8 7 6 5 4 3 2 1

Johns Hopkins University Press
2715 North Charles Street
Baltimore, Maryland 21218-4363
www.press.jhu.edu

Library of Congress Cataloging-in-Publication Data

Names: Morgan, Gregory J., author.
Title: Cancer virus hunters : a history of tumor virology / Gregory J. Morgan.
Description: Baltimore : Johns Hopkins University Press, 2022. | Includes bibliographical references and index.
Identifiers: LCCN 2021039028 | ISBN 9781421444017 (hardcover) | ISBN 9781421444024 (ebook)
Subjects: LCSH: Oncogenic viruses—Research. | Cancer—Etiology—Research. | Virology—Research.
Classification: LCC QR372.O6 M67 2022 | DDC 616.99/4019—dc23/eng/20211014
LC record available at https://lccn.loc.gov/2021039028

A catalog record for this book is available from the British Library.

The text for this book was set in Gotham Book & Chaparral.

For my COVID-19 bubble: Stacey, Edward, and Cam

It could be said further that this work we are doing is all very well, but what all of us are really interested in is people—not chickens. I can only repeat the story of the English gentleman leaving the opera one night who passed a man under a lamp-post, looking for something in the gutter. On being informed that the searcher had lost his watch, the first man got down to help, and looked and looked. Finally he asked the searcher if he had dropped his watch right here under the lamppost. "No," the man said, "I really dropped it around the corner—but I'm looking for it here because there's so much more light."

We are in the position of the man under the lamp-post. What we are primarily interested in is not chicken cancer, but since we have so much light there, we look. Maybe we will find something better than a watch.

Howard Temin, *Cancer and Viruses*, 1960

Contents

Glossary and Abbreviations

ACS	American Cancer Society
ALV	avian leukosis virus
antigen	A regular molecular structure that can be bound by a specific antibody.
ATP	adenosine triphosphate. An energy-rich molecule that provides energy for a large number of processes in the cell.
Au	Australia antigen. A surface antigen of the hepatitis B virus.
bacteriophage (phage)	A virus that infects bacteria.
capsid	The outer protein shell of the virus that protects the genetic material.
CDC	Centers for Disease Control and Prevention
cell line	A population of cells with uniform genetic characteristics derived from a single cell that will continue to proliferate as long as the cells have fresh medium and space to grow.
CSHL	Cold Spring Harbor Laboratory
dalton	A unit of mass equal to one-twelfth the mass of a carbon atom or roughly the mass of a hydrogen atom. One thousand daltons is called a kilodalton. Commonly used to measure the mass of proteins.
DNA	deoxyribonucleic acid. The primary molecule that encodes genetic information.
EBV	Epstein-Barr virus
FeLV	feline leukemia virus
GTP	guanosine triphosphate. An energy-rich nucleotide used in a variety of cellular processes.
HBV	hepatitis B virus

HIV	human immunodeficiency virus
HPV	human papillomavirus
HTLV	human T-cell leukemia virus
ICRF	Imperial Cancer Research Fund
IL-2	interleukin 2
IM	infectious mononucleosis
immortalized cell line	A population of cells from a multicellular organism that would normally stop proliferating but, because of mutations in the cells causing them to evade normal cessation mechanisms, can proliferate indefinitely.
JAMA	*Journal of the American Medical Association*
JAX	Jackson Memorial Laboratory
LMB	MRC Laboratory of Molecular Biology
lymphocyte	A type of white blood cell. B-cells and T-cells are types of lymphocytes.
lymphoma	A blood cancer that develops from lymphocytes.
MMTV	mouse mammary tumor virus
MRC	Medical Research Council (UK)
MSKCC	Memorial Sloan Kettering Cancer Center
NCI	National Cancer Institute
NFIP	National Foundation for Infantile Paralysis
NIH	National Institutes of Health
p53	A tumor-suppressing protein with a molecular weight close to 53 kilodaltons.
plaque	A visible circular structure in cell culture due to either the growth or death of cells.
PNAS	*Proceedings of the National Academy of Science of the United States of America*
provirus	A virus genome integrated into the genome of a host cell.
ras	A proto-oncogene mutated in many human cancers. Mutated versions of *ras* were discovered in the genomes of Harvey sarcoma virus and Kirsten sarcoma virus.

Ras	The protein product of the *ras* gene involved in cell signaling.
RIMR	Rockefeller Institute of Medical Research, later renamed Rockefeller University
RNA	ribonucleic acid. Present in all cells and a messenger between DNA and ribosomes where protein synthesis occurs. Retroviruses have RNA genomes.
RSV	Rous sarcoma virus
sarcoma	A cancer of the connective tissue.
serial passage	The process of repeatedly growing a virus in a given host.
src	An oncogene first discovered in the genome of Rous sarcoma virus. Later the cellular proto-oncogene *c-src* was distinguished from the viral oncogene *v-src*.
Src	The protein product of the *src* gene with the cellular function of phosphorylating other proteins.
STS	Science and technology studies (alternatively, science, technology, and society)
supernatant	The fluid that does not sink to the bottom of a centrifuge tube.
TMV	tobacco mosaic virus
transfection	The introduction of nucleic acid into eukaryotic cells.
transformed cell line	A cell line that acquired the potential for infinite growth after insertion of viral gene components into the cell's genome.
SV40	simian virus 40
SVLP	Special Virus Leukemia Program, later called Special Virus Cancer Program
UCSF	University of California San Francisco
virion	A viral particle consisting of a protein coat, either DNA or RNA, and sometimes a membrane envelope.
zoonosis	The process whereby a virus jumps from one host species to another.

Cancer Virus Hunters

Introduction

The Untold Story of How a Century of Tumor Virology Changed Biomedicine

The rise of molecular biology counts alongside the rise of both modern physics and computing as one of the most influential technoscientific developments of the twentieth century.[1] As such it is central to any account of how and why the developed world underwent its dramatic technoscientific transformation over the course of the century. And yet, for many in the communities that draw actionable lessons from the past—the educated public, biologists, medical professionals, and historians of science, to name a few—the full story of molecular biology's incredible jumps in understanding remains a mystery.[2] Or rather, it represents an *unknown unknown*, a story that many don't realize they don't know.[3] A large missing piece of this puzzle is the rise of *tumor virology*, the study of viruses that can cause cancer. One of the first articles written by a historian of science on this topic was published in a volume appropriately titled *Hidden Histories of Science*.[4] The absence of tumor virology from the story of molecular biology prevents us from a complete understanding of how molecular biology became the technoscientific force that it did. Furthermore, the story of the impact of tumor virology on the development of contemporary biomedicine brings together many biomedical discoveries that have not been gathered before.

I wrote this book to elevate tumor virology from this state of relative obscurity. By better understanding this subfield in its own right, we will better understand how contemporary molecular biology and biomedicine came to be. Insights gleaned from the study of these viruses provided insight that extended beyond tumors into other diseases, treatments, and fundamental biological mechanisms. I show how discoveries made with tumor viruses came to dominate the contemporary biomedical understanding of the nature of cancer, both virally caused and not. Furthermore, experiments performed with tumor viruses informed and enriched biologists' understanding of *normal* cellular regulation at the molecular level. Without these insights, many significant theoretical advancements made over the past four decades in cellular and molecular biology would have not happened or would have been significantly delayed.

More practically, the study of tumor viruses saved thousands, if not millions, of lives. First, it led to vaccines—for hepatitis B virus (HBV) and human papillomavirus (HPV)—against two forms of human cancer. It is estimated that about 20% of human cancers worldwide are virally caused, with HBV and HPV being the leading culprits. Vaccination has saved many from these cancers and has the potential to wipe them out permanently. Second, the discovery of the hepatitis C virus and the invention of a test for these viruses in the blood has alleviated the risk of chronic hepatitis, which sometimes results in liver cancer, among people who receive blood transfusions. Third, tumor virology accelerated the scientific community's response to the AIDS pandemic. The well-studied discovery of HIV falls along a longer historical arc that began in the search for a human virus that causes cancer. Although HIV is not technically considered a tumor virus—it is indirectly associated with Kaposi's sarcoma and lymphoma—the knowledge and techniques developed while investigating RNA tumor viruses translated relatively quickly into a test to detect HIV, an RNA retrovirus, in infected people.

Appreciating how tumor virology has infused twentieth-century molecular biology also enhances an understanding of the overall development of modern biology itself. Starting in the 1970s, biologists unraveled key molecules in cellular regulation. It became clear that enzymes called kinases modify various proteins to change their activity and control key cellular regulatory pathways. These pathways control numerous cellular processes including nuclear transcription, cell proliferation, signal transduction, and endocrine homeostasis. That viruses infecting cells directly affect fundamental cellular pathways provided a clue to cellular gene and protein regulation, the cornerstone in biologists' understanding of how cells keep themselves alive.

In addition to the *products* of science—the theories, the experiments, the techniques, the inventions—this volume also illustrates the *process* of science, how the products of science are created. The word "science" is ambiguous: it can refer to either.[5] What can the history of tumor virology tell us about the process of science? Five themes stand out:

1. The value of creating good model systems of complex phenomena. Cancer in animals is a complex process. Studying cultured cell lines rather than organisms simplified the research problem and allowed experiments to be more easily quantified. Focusing research on model organisms like the Rous sarcoma virus (RSV) or SV40 and their associated host cells allowed laboratories around the world to access and contribute to a common knowledge pool. Relatedly, the development of repeatable, accurate assays to measure the "malignant transformation" (often called just "transformation") of normal cells into "cancerous cells" took cancer research out of the clinic and into the laboratory, a change that allowed the tools of molecular biology to be brought to bear and in turn further developed.
2. Sexism in science.[6] In several episodes, women scientists faced more challenges getting recognition for their work than did men. In the most prominent example, the electron microscopist Louise Chow did not receive the 1993 Nobel Prize despite performing some of the key experiments in the discovery of RNA splicing. There are several other lesser-known cases as well: as discussed in chapters 2, 3, and 5, virologists Sarah Stewart and Marguerite Vogt received less recognition than they might have had they been men, and Yvonne Barr dropped out of research partly because of her experience of sexism. Sexism in the development of molecular biology is more pervasive than many current practitioners think.
3. The difficulty in achieving consensus about assignment of credit to individual scientists for major discoveries. In many of the following episodes, the central players did not agree about how to attribute relative credit. This disagreement can be exacerbated when a Nobel Prize is at stake, because the ultimate scientific prize can be given to at most three people, yet often discoveries require many more than three researchers. In some cases, notably the HIV test and the HPV vaccine, the issue of credit entered the legal sphere when valuable patents were contested by teams of scientists, scientific institutions, and their lawyers. Legal rulings did not settle disputes outside the legal realm, however.
4. The disproportionate influence of certain institutions in supporting the development of tumor virology.[7] The Rockefeller Institute for Medical Research (RIMR, later Rockefeller University) and Cold Spring

Harbor Laboratory (CSHL) in the United States, and the Imperial Cancer Research Fund (ICRF, now Cancer Research UK) in England all feature prominently in this history. The US federal government provided millions of dollars of research funds for tumor virology in the 1960s and 1970s in the hope of discovering human tumor viruses.[8] Journals like *Proceedings of the National Academy of Sciences* (*PNAS*) quickly disseminated new results and established priority for various discoveries. Mary Lasker and the Albert and Mary Lasker Foundation also supported tumor virology. Many of the tumor virologists featured in this volume won the foundation's prestigious Lasker Award, which often serves as the precursor to a Nobel Prize.

5. The international nature of twentieth-century biomedical research. Research developments in tumor virology came from biologists working in the United States, western Europe, eastern Europe, Australia, and Asia. Scientists moved locations and research foci because of international military conflicts, especially World War II and the Vietnam War. In times of peace, international conferences allowed researchers to exchange results and approaches and learn from one another.[9] Researchers, students, expertise, ideas, techniques, and model systems circulated among laboratories around the world. In some cases, such as Jan Svoboda's in chapter 6, being cut off from international contacts attenuated a scientist's career.

For each of these five themes, historians, sociologists, and philosophers of science, as well as scholars working in science and technology studies (STS) have generated a large literature. Rather than engage with the secondary literature extensively in the text, footnotes and the appendix point interested readers to additional scholarly resources.

Relatedly, driven as it is by the nature of the available primary sources in these episodes and the constraints of a single volume, I am more concerned with the experiments performed and the concepts and theories proposed by virologists to generate and account for their data, and the virologists themselves. I am less concerned with the social and political context of twentieth-century biology and medicine, interesting as it can be.[10] One hope is that this first account of a century of tumor virology might prove a useful resource to future historians of science investigating social and political themes in twentieth-century biomedicine in more detail.

Overview of the Book

An unfashionable area of research for half a century, tumor virology began in the early twentieth century with the study of tumors in chickens, rabbits, and mice. At a time when other possible causes of cancer were prioritized, tumor virology proved a fertile field in which devoted biologists, through careful observation, improved the understanding of the causes of cancer and the nature of viruses. The early pioneers Peyton Rous and Richard Shope worked at the Rockefeller Institute in New York City and Princeton, respectively, while John Joseph Bittner worked at the Jackson Laboratory in Maine. The work of these three researchers and their collaborators is the core of the early phase of tumor virology, from 1909 to the 1940s, and the subject of chapter 1.

The second phase of the history of tumor virology began in the 1950s with two significant developments. First, Ludwik Gross, Sarah Stewart, and Bernice Eddy characterized polyoma virus, a small DNA tumor virus that reliably causes a variety of tumors in laboratory animals. Second, animal virology took a quantitative turn when approaches from the phage school of genetics were imported into tumor virology. In particular, Renato Dulbecco developed an influential assay for accurately measuring the number of viruses infecting a plate of lab-cultured animal cells,[11] and Howard Temin and Harry Rubin invented an assay to quantify the cells transformed into cancer by a given tumor virus. This quantitative turn allowed tumor virology to advance in much the same way as results from quantitative phage biology drove bacterial genetics in the preceding two decades.[12] The development of these assays is discussed in chapter 3. By the end of the 1950s, tumor virology began to gain the respectability that it lacked in the preceding decades. Tools from the nascent field of molecular biology were increasingly used to analyze viruses. For example, the electron microscope allowed biologists to visualize virus particles in the 1940s and 1950s. A new understanding of viruses came into focus: they were particles made up of pieces of nucleic acid surrounded by a protective protein coat.

By the late 1960s and early 1970s—the third phase of tumor virology—the field had partially split into two scientific communities investigating tumor viruses: those investigating viruses with DNA genomes and those investigating viruses with RNA genomes. (Both DNA and RNA are long molecules that contain genes, sequences of bases that code for proteins; they differ in which sugar—ribose or deoxyribose—makes up their backbones.) The split was partly due to differences in how two types of viruses reproduced. Viruses of the latter type were at first called "RNA tumor viruses" but were later renamed "retroviruses" following Howard Temin and David Baltimore's 1970 discovery of an enzyme called reverse transcriptase used by RNA tumor viruses to replicate by making DNA copies of their RNA genomes.[13] Prior to the discovery of reverse transcriptase, only the opposite direction (DNA copied into RNA) was known. The surprising discovery of this enzyme revealed a new tool for virologists and molecular biologists: just like an RNA tumor virus, scientists could use the enzyme to make more stable DNA copies of less stable RNA, thus opening up a wide range of new experiments.

Both groups of scientists had regular meetings at Cold Spring Harbor Laboratory, which functioned as the mecca for molecular biology in this period. The summer pilgrimage to conferences at CSHL on Long Island allowed presentation of the latest results to the field and fruitful discussion among biologists, both formally at the talks and informally afterward at the bar on campus.[14] James Watson, the director of CSHL from 1968 to 1994, had turned the laboratory into a world center for tumor virology by hiring a critical mass of highly talented tumor virologists, including three scientists who had previously worked in Renato Dulbecco's laboratory. One of them, Joseph Sambrook, showed that SV40, a virus closely related to polyoma virus, has DNA that can integrate into a host cell's genome during carcinogenesis. Many of the techniques developed or refined to study SV40 and adenoviruses, two types of tumor virus, were included in *Molecular Cloning: A Laboratory Manual*, an influential handbook of molecular biology techniques published by CSHL.[15]

Funded by the Special Virus Leukemia Program (later called the Special Virus Cancer Program) of the US government, researchers during this period looked for human retroviruses that caused cancer.[16] They were inspired by Scottish veterinary scientist Bill Jarrett's 1960s discovery of feline leukemia virus (FeLV), a retrovirus that causes leukemia in cats, along with the conjecture over the existence of endogenous tumor viruses, that is, cancer-causing viruses embedded into the genomes of human cells. However, after some reports were revealed to be false alarms, many biologists became skeptical that a human cancer-causing retrovirus existed.[17] As discussed in chapter 14, Robert Gallo finally succeeded in isolating a human cancer virus that he called human T-cell leukemia virus (HTLV) in 1980. There were earlier discoveries regarding DNA viruses, too: Epstein-Barr virus (EBV), a type of herpes virus that causes Burkitt's lymphoma, an aggressive facial cancer found in African children, was discovered by Anthony Epstein in 1964. And hepatitis B virus was shown to cause liver cancer. These discoveries are the subjects of chapters 5 and 7.

Before the invention and adoption of recombinant DNA technology, in the 1960s and 1970s, because viral genomes are so small, biologists were able to study and manipulate a small number of oncogenic genes. During this time, much of the growth of molecular biology was driven by a study of virus-host interactions. Tumor viruses interact with cellular machinery to reproduce themselves, and uncovering these interactions helped researchers to understand the mechanisms behind the normal biochemical pathways of cell regulation and cell division, in addition to abnormal cancerous processes.[18] Research performed on the West Coast of the United States on the retrovirus Rous sarcoma virus yielded the first oncogene, named *v-src*. It turned out to be a mutated version of a normal cellular gene, *c-src*, that the virus had acquired from its chicken host in the distant past. If *v-src* were present in a cell, the oncogene could transform it into a cancer cell. Harold Varmus and Michael Bishop from the University of California San Francisco (UCSF) won the Nobel Prize for this discovery, although the core experimental work was performed by the visiting French scientist Dominique Stehelin, who had inherited a project created by Bishop and Varmus. Additional oncogenes were discovered in other retroviruses; chapter 12 follows the four laboratories racing to discover the *ras* oncogene in Harvey sarcoma virus.

Work on DNA tumor viruses also yielded important molecular biology knowledge beyond virology. Chapter 11 tells the story of the discovery of RNA splicing in adenovirus. Teams of researchers at Cold Spring Harbor Laboratory and MIT used electron microscopy to see that pieces of the adenovirus messenger RNA were spliced out and not translated into protein. Their results, which revealed a new form of gene regulation, had implications well beyond virology, leading to a Nobel Prize along with some controversy about who deserved primary credit.

Work on DNA tumor viruses on both sides of the Atlantic also revealed an important regulatory protein called p53.[19] After some false starts, it was determined to be a tumor-suppressor protein—it functioned as a brake in the cell cycle and in cellular replication. Oncogenes like *v-src* and *ras* code for proteins that function as an accelerator. Work on p53 led to a simple model of cancer: mutations turn the accelerator on, in concert with further mutations in the same cell that disable the brake, leading to uncontrolled cellular growth, or cancer. Chapter 13 describes the work of David Lane and Arnold Levine, two pioneers in p53 research.

Chapter 14 explains how the speed of developing an accurate test for the retrovirus HIV was made possible by earlier work on RNA tumor viruses. Thus, RNA tumor virus research had an important impact on the biomedical community's response to the AIDS epidemic. Tumor virology also drove the development of two anticancer vaccines: the hepatitis B virus vaccine for liver cancer and the human papillomavirus vaccine for cervical and oropharyngeal cancer. The work of Harald zur Hausen, Ian Frazer, and others on HPV is summarized in chapter 15. Like many other episodes, the HPV story involved researchers from around the globe: British, German, Australian, American, and Chinese scientists made contributions that finally led to an effective vaccine.

It is difficult to quantify how important a field like tumor virology is, but one crude measure of the impact of tumor virology on biology and medicine is the number of Nobel Prizes in medicine and physiology given for work stemming from tumor viruses. There have been seven: Peyton Rous (1966); David Baltimore, Howard Temin, and Renato Dulbecco (1975); Baruch Blumberg (1976); Michael Bishop and Harold Varmus (1989); Richard Roberts and Phillip Sharp (1993); Harald zur Hausen (2008); Harvey Alter, Michael Houghton, and Charles Rice (2020). The work leading to all of these prizes is discussed in this book.

Given my century-long scope, for reasons of space and to aid in understanding the narratives, I have focused more on primary figures and less on secondary ones in these episodes. A more encyclopedic history of early tumor virology can be found in Ludwik Gross's *Oncogenic Viruses*.[20] Additionally, by structuring each chapter around the careers of prominent biologists and their model systems, I hope that these episodes, which are arranged chronologically, will be of interest and use to my primary audience: biologists, medical researchers, virologists, oncologists, and others working or aspiring to work in the biomedical sciences.[21] Understanding the history of tumor virology preceding the development of the HPV vaccine should also enrich the current popular debate over its health benefits and scientific validity.

Tumors and Virology before Tumor Virology

Before examining the beginnings of tumor virology in chapter 1, with the work of Peyton Rous, some context about the general state of virology at the end of the nineteenth century and beginning of the twentieth is useful. The 1890s is often considered the founding decade of virology.[22] In this decade, the Russian microbiologist Dmitri Ivanovski tried to isolate the cause of tobacco mosaic disease, an agriculturally important disease that can cause significant economic losses in infected plants.[23] Infected tobacco plants have mottled mosaic leaves, hence the name. Ivanovski ground up infected leaves and ran the extracts through a fine porcelain filter that was used to remove bacteria. Finding that the filtered extract could still transmit the disease to healthy plants, Ivanovski thought that an especially small bacterium or perhaps a poison created by bacteria was causing the disease. Martinus Beijerinck, a Dutch microbiologist, obtained similar results that he published in 1898. He famously called the infectious agent *contagium vivum fluidum* (contagious living fluid or "virus") and claimed that it was not a microbe.[24] Later researchers would question whether it was a fluid and whether it was living. It survives being dried out but not boiling, which destroys its ability to infect plants.[25] Tobacco mosaic virus (TMV) is not alone; other pathogens, such as the agent that causes foot and mouth disease, pass through fine filters.[26]

In 1908, Vilhelm Ellermann and Oluf Bang, working in Copenhagen, were able to transmit a form of chicken leukemia.[27] They used cell-free filtrates from infected chickens, but at the time their results did not garner much attention. Their work was not viewed as an advance in the understanding of cancer largely because leukemia was poorly understood and thus was not considered a form of cancer. Furthermore, the dominant theory of cancer did not include a viral cause. Ellermann, a physician, classified eight strains of what today would be called avian leukosis viruses.[28] These viruses, later revealed to be retroviruses, would feature in the search for the first oncogenes, as described in chapter 9. A year after the Danes' work on chickens, the American Peyton Rous took a similar approach, but his work received more recognition, as his chicken virus caused solid tumors. His work begins the first chapter.

While the causes of cancer were largely unknown at the beginning of the twentieth century, some things were known. In retrospect, oncologists have identified certain cases in the history of medicine as revealing significant clues about the nature and causes of cancer.[29] Hippocrates (c. 460–370 BCE) is often credited for naming certain tumors "carcinomas," which when Latinized by Celsus (25 BCE–50 CE) and others became "cancer."[30] In both Latin and Greek, the term had a connection with crabs—tumors, much like crabs, often have "claws" projecting outward.[31] Georgius Agricola, in his *De Re Metallica* of 1556, reported on miners who suffered lung diseases and tumors at a high rate, caused presumably by the working conditions in the mines.[32] Many of those miners probably suffered from lung cancer.[33] In the late eighteenth century, heavy use of tobacco snuff was documented to cause cancer in some cases.[34] However, it took until 1912 before good evidence that tobacco smoking can cause tumor growths in the lungs was found.[35] In 1775, the English surgeon Percival Pott noticed that chimney sweeps had high rates of scrotal cancer.[36] Chimney sweeps in eighteenth-century England spent their lives in contact with black chimney soot, so it was a natural to infer that something in the soot caused the cancer. This cancer became known as chimney sweep carcinoma. Domenico Rigoni-Stern, an Italian surgeon, looked in 1842 at decades of mortality statistics in Verona from the end of the eighteenth and beginning of the nineteenth century and discovered that cervical cancer was significantly more prevalent in married women than in nuns and virgins.[37] The natural inference from the data presented was that cervical cancer was somehow related to sexual activity.

In addition to various reports of cases of cancer in particular environments, physicians have a long history of speculation about more proximate causes of cancer. In the ancient world, Galen's ideas were influential. Like Hippocrates before him, Galen (130–210 CE) believed that sickness was caused by an imbalance of the four humors—black bile, blood, phlegm, and yellow bile. Cancer was thought to be caused by an excess of black bile.[38] This theory was influential for more than 1,000 years. In the seventeenth and eighteenth centuries, other causes of cancer were postulated: lymph dysfunction, trauma, irritation, and contagion. The historian Alan I. Marcus identifies two schools of thought by the mid-nineteenth century. "Constitutionalists" thought cancer was a systemic disease occurring when an individual's bodily system was disrupted. Cancer could occur in different places in the body because the overall system was out of balance. On the other hand, "cellularists," following German pathologist Rudolf Virchow (1821–1902), thought cancer was an internal disease of cells already in the body. These cells distributed around the body were undifferentiated and dormant unless disturbed by an external mechanical or chemical agent. Cancer that metastasized could be explained by renegade cells traveling to different places in the body. The success of the germ theory of disease also influenced theorizing about cancer in the late nineteenth century. Some thought that, like other microbial diseases, a particular cancer would be found to be caused by a single type of microbe.

Let me end this brief introduction with a few remarks on the title of the book: *Cancer Virus Hunters*. It is inspired by the wildly successful 1926 book by Paul de Kruif, *Microbe Hunters*, which devoted a chapter each to some of the founding figures of bacteriology and microbiology: Louis Pasteur, Robert Koch, Antonie van Leeuwenhoek, and many others.[39] *Microbe Hunters* was a popular book, selling half a million copies by 1936, and it inspired many young minds to begin careers in medicine and biology.[40] While *Cancer Virus Hunters* is a more scholarly book than *Microbe Hunters*, it shares the aim of being accessible to readers beyond historians of biology and medicine. I hope that researchers, physicians, students, and others aspiring to work in the biomedical sciences will find the history of tumor virology an enlightening perspective from which to consider contemporary biology and medicine. Knowledge of the leading figures of biology and what they and their collaborators did to earn their reputations can only enrich an understanding of biomedicine and the development of twentieth-century science.

Cancer Virus Hunters does not dramatize possible conversations that might have taken place, as *Microbe Hunters* did, and is more grounded in archival sources, but like *Microbe Hunters*, it is structured around the most productive career phases of the most important figures in the field.[41] Many researchers featured have had long, productive research careers. For reasons of space, I have consequently had to make judgments about which parts of their careers have contributed the most to driving the field forward. Additionally, most episodes in the history of tumor virology are too complex to be reduced to the work of a single individual. As is apparent in the following chapters, it normally takes teams of people and collaborations among researchers to make major discoveries.[42] This growth in the size of research teams became especially pronounced as the field matured.[43]

1

The Beginnings

Peyton Rous and Chickens, Richard Shope and Rabbits, and John J. Bittner and Mice

The Early Career of Peyton Rous

Peyton Rous (1879–1970) was raised in Maryland and was drawn from a young age to the study of nature. He also had a flair for crafting prose. In 1900, at the age of 21, he wrote a brief article on Maryland flowers for the *Baltimore Sun* that proved so successful he was asked to write a monthly column. As Eva Becsei-Kilborn points out, Rous first and foremost valued the careful observation and description of nature.[1]

After completing public high school in Baltimore, Rous won a scholarship to attend Johns Hopkins University, the first graduate research university in the United States. After completing undergraduate studies, he stayed at Hopkins for his medical studies, which he began in 1900. He received thorough training in pathology from accomplished teachers including William H. Welch, who like many of the Hopkins faculty was trained at a major university in Europe. Experimental pathology, the challenge to determine what caused a particular human pathology, interested Rous. His secular rationalism motivated his empiricism and his suspicion of what he called a priori reasoning. In 1906 Rous decided against a career as a physician and instead began teaching pathology at the University of Michigan. His mentor at Michigan helped him arrange to spend some time in Germany to study human cadavers, a practice that was uncommon and not generally accepted in the United States. Rous's work at Michigan was funded by the Rockefeller Institute for Medical Research,[2] and by 1908 he was publishing his results on the nature of lymphatic cells in RIMR's *Journal of Experimental Medicine*.[3]

Simon Flexner, the director of the RIMR and also a graduate of Johns Hopkins School of Medicine, was impressed with Rous's work and offered

1.1. Peyton Rous in 1923. Image courtesy of Rockefeller Archive Center

him a position at the RIMR in New York City in 1908 (figure 1.1). Flexner convinced Rous to begin work on "the tumor problem"—despite Rous's earlier mentors warning him to avoid what they thought was an intractable problem.[4] To investigate "the tumor problem," Rous and Flexner meant to bring physical, chemical, and biological knowledge to bear on the nature and formation of neoplasms.[5] Flexner argued that this was a more significant research subject than Rous's other research interests.[6] It was also in line with the medical mission of the Rockefeller Institute. Rous was convinced. Despite knowing little about cancer, he took the position. Like others experimenting on cancer at the time, Rous started transplanting solid tumors from sick mice into healthy mice, sometimes grinding up the tumor first.[7] He was interested in how tissues could change from one type to another, a research question that spanned oncology and embryology.[8] He would observe the effect in mice of the injected or transplanted tumor on the host, while varying conditions such as the age of the mouse, its nutrition levels, and the size of the tumor. He was interested in finding general laws regulating transplanted tissue. As his biographer Becsei-Kilborn puts it, "He viewed the pathology of a disease as some kind of secret script presented by nature that needed to be deciphered."[9]

The Rockefeller Institute was modeled on the Institut Pasteur in Paris and the Robert Koch-Institut in Berlin, but with an American twist—it was not consolidated around a single figure, and it was interdisciplinary by design. This approach resonated with John D. Rockefeller's business ideology: collaborative efforts in pursuit of a common goal.[10] The RIMR was not affiliated with a university, so its scientists could devote themselves fully to research without the obligation of teaching.[11] It aimed to employ "pathfinder" scientists who would venture into new unexplored areas, and it succeeded in this goal when Flexner hired Peyton Rous.

In 1909, in what Rous later called a "fortunate accident," a local chicken breeder brought a chicken with a tumor to the RIMR (figure 1.2). It was a purebred, light-colored, barred Plymouth Rock hen. The chicken had a tumor in the right breast for two months before it was brought to Rous. The growth looked like a sarcoma—a malignant tumor of the connective tissue. Under the microscope, it appeared to consist of loose bundles of spindle cells. Rous took pieces of the tumor and injected them into the chicken's left breast and peritoneal cavity.[12] He also injected them into other chickens from the same clutch of eggs. A month later the chicken was dead from the growth, and the other chickens had growths at the injection sites. It turned out that the tumor could be repeatedly transplanted into closely related purebred chickens. Rous tried to transplant the tumor into different breeds of chicken, into pigeons, and into guinea pigs as well, but to no avail. In his research, Rous found that younger chickens were better hosts for the tumor than older ones, and that after inoculation with the tumor preparation, a chicken appeared to be protected from further inoculations of the tumor. He submitted these

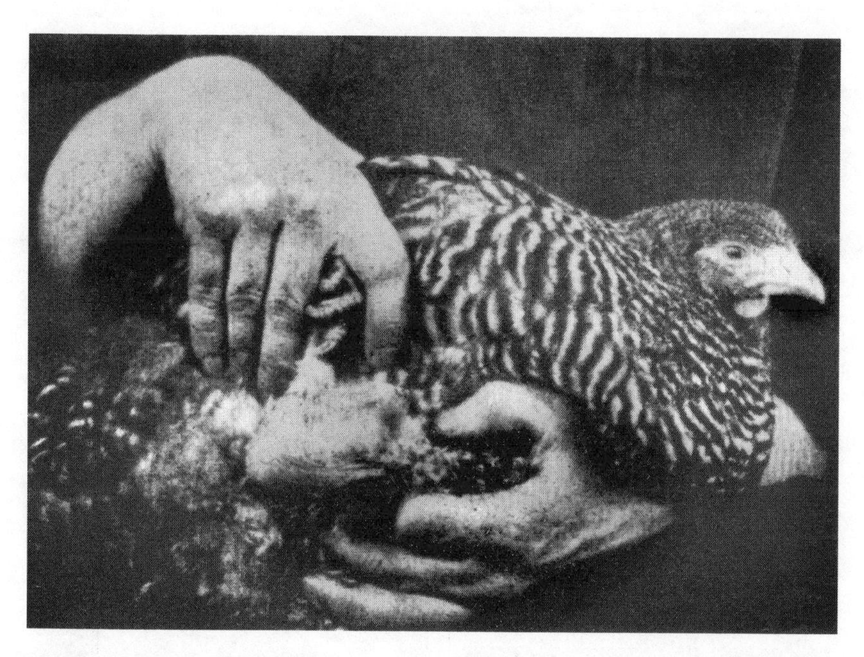

1.2. The chicken that started it all. From Peyton Rous, "A Transmissible Avian Neoplasm (Sarcoma of the Common Fowl)." ***Journal of Experimental Medicine*** **12, no. 5 (1910): 696–705.**

results to the *Journal of Experimental Medicine* in June 1910, in a paper that described the first avian tumor that proved transplantable in other individuals.[13]

In his next paper on the subject, Rous tried to identify the cause of the tumors.[14] His strategy was to test cell-free filtrates for their ability to cause tumors. He ground up tumors and suspended the cells in a fluid. At first, he used ordinary filter paper. He then injected 2 milliliters of the filtrate into purebred chickens. To his surprise, the filtrate caused new tumors to grow. Next, Rous repeated the experiment, this time using centrifugation to remove practically all the cells from the fluid. This method also successfully transmitted tumors into previously tumor-free chickens. To be sure that some cells did not get past centrifugation, Rous then used Berkefeld filters to filter out material as small as a bacterium. Before passing through the Berkefeld filter, the fluid was yellow and cloudy; afterward, it was faintly yellow. These experiments, refined with more sensitive filtering, reinforced the conclusion that the cell-free filtrate caused tumors, leading Rous to surmise that something in the filtrate—not the tumor cells themselves—caused tumors. He tried injecting the filtrate into pigeons, ducks, rats, mice, and guinea pigs, but again found that it caused the tumors to grow only in young chickens. He also found, interestingly, that the tumors caused by filtrate injections seemed to become more malignant after passing through a number of hosts, an observation that suggested that perhaps the "agent" could travel through the blood or perhaps the lymph system. Using a dark-field microscope, Rous tried to find a parasitic organism but could not see anything. Reluctant to use the word "virus," he instead noted that "the first tendency will be to regard the self-perpetuating agent active in the sarcoma of the fowl as a minute parasitic organism," but it was conceivable that it was instead a chemical stimulant produced by the neoplastic cells. Although in private writings he considered the possibility that it could be a virus, in print he was cautious. What was needed was in vitro propagation of the agent to demonstrate it was a living thing, and such propagation was elusive.[15]

Like much work in the sciences, Rous's advances, though spectacular in hindsight, were constrained by the technology and the scientific interests of his time. Further experimentation along this path would have required isolating the tumor-causing agent from the filtrates, but given the equipment available to him in 1911, Rous was unable to move forward. The ultracentrifuge, in particular, would be the key piece of equipment needed to isolate virus particles, and in the 1930s it would be used to investigate viral proteins.[16] Although some members of the medical community expressed interest in the chicken sarcoma findings, the medical community at large did not change its approach to cancer in response to his research, and in 1915, Rous abandoned this experimental work.

He did not completely abandon thinking about the tumor agent, however, as private correspondence demonstrates. In a 1922 letter to the pathologist T. Mitchell Prudden, Rous listed four questions of "prime importance" about the growth:

1. Is the growth a tumor?
2. Is the agent associated with the growth . . . merely a surviving tissue element or a distinct entity?
3. Is the filterable agent living, or some strange chemical thing?
4. Have the findings with the growth any bearing on the problem of mammalian tumors?[17]

Rous thought that pathologists generally believed the growth was a genuine tumor. However, in the beginning years some scientists still agreed with a pathologist, probably Ernest Bashford, director of research at the Imperial Cancer Research Fund, who reportedly said to Rous, "Look here, young man, that can't be a cancer if you found the cause!"[18] Nonetheless, Rous continued to get requests for samples of the dry tumor to study. Regarding the second question he posed, Rous believed that the ability of the agent to pass through Berkefeld filters, its ability to withstand higher temperatures than tumor cells could, as well as the fact that it had retained its activity for seven years in its dry state all indicated that it was indeed a distinct agent. One could potentially claim that some unknown tissue could do this, but this position was "camel-swallowing," that is, taking a much more difficult route when an easier one was available.[19] The agent was its own thing. Further, Rous viewed the agent as a living thing and part of the animal kingdom. Tuberculosis and syphilis could induce cell proliferation, so why not the chicken sarcoma agent also? Similarly, Rous was optimistic about the existence of mammalian tumor agents. As he wrote, "I believe mammalian growths are due to like agents, but to agents conditioned in their activity by accessory factors that are exceedingly difficult to bring into play experimentally. It must be confessed that the belief is wholly the product of a priori reasoning."[20] Rous thought that as the number of factors needed to cause a cancer increased, so too would the difficulty of studying it. Eventually, the view that cancer is infectious led to concrete experiments, and so just like the germ theory of disease, it opened up a fruitful research program.

Despite Rous's hope, the outbreak of World War I prevented further work on this topic. Instead, Rouse contributed to the war effort, figuring out how to safely store blood, a vital technology for treating wounded soldiers. In 1925, interest in the tumor agent was rekindled on the other side of the Atlantic, when William E. Gye and J. E. Barnard published articles in the *Lancet* promoting the idea that cancer was caused by both a chemical and a microbial agent.[21] Gye reached out to Rous, sparking a friendship, though Rous claimed that he wanted to remain on the sidelines of the debate. Although some of Gye's experiments could not be repeated, he nevertheless rose to prominence, becoming the director of the ICRF in 1935.[22] While Rous did publish a single paper on cancer in mice in 1932, the work of one of Rous's colleagues studying rabbits would push the story forward.[23]

The Early Career of Richard Shope

Born and educated in Iowa, Richard "Dick" Shope grew up with animals and had an affinity for pigs, which his family raised each year to sell at the market. As Peyton Rous later said, Shope was not sentimental about animals but nevertheless had compassion for them.[24] Despite being overworked, his father often took him hunting or fishing, which further developed his interest in the animal kingdom.[25] Shope followed his father into medicine, graduating with an MD from the University of Iowa in 1924. After serving as an instructor at his alma mater, he started research at the Rockefeller Institute for Medical Research in Princeton, a New Jersey branch of the RIMR, in 1928, working at first on the biochemistry of cholesterol and also on tuberculosis. After several months at Rockefeller, he began to focus his research on virology. At Princeton, Shope learned from Theobald Smith and Paul Lewis, who both believed that disease should be studied under natural conditions, and that careful observation was as important as laboratory gadgetry. Their views reinforced Shope's approach to research. Tragically, Lewis contracted yellow fever in South America while studying the virus, which he had begun to research after making little progress investigating tuberculosis.[26] He died in 1929 of yellow fever.

After spending some months trying to understand tuberculosis, Shope shifted his attention to diseases of pigs.[27] Studying livestock was relatively easy given the farm-like setting of RIMR-Princeton. He first made his name with work on the etiology of swine influenza and its cause (figure 1.3). Having isolated a bacillus (*Haemophilus influenzae*) that regularly accompanied the swine flu but proved harmless when injected by itself into pigs, Shope finally isolated a virus but found that, alone, it caused only mild symptoms in his subjects. Based on these observations, Shope argued that the bacterium and a virus could be working in concert to cause swine flu.[28]

This research, which he published in 1931, inspired the British team of Wilson Smith, Christopher Andrewes, and Patrick Laidlaw. Two years after Shope's work, they discovered a virus that causes human influenza.[29] Having succeeded at infecting ferrets with a filterable virus isolated from humans, they found that Shope's virus—which he had sent to them—also infected ferrets and caused similar symptoms. Previous infection with swine influenza virus provided the ferrets with protection from the human influenza viruses. Andrewes, by now one of the most eminent British virologists, trav-

1.3. Commander Richard Shope receives the John Scott Award at the American Philosophical Society for his work on swine influenza in 1943. Image courtesy of the American Philosophical Society

eled to RIMR-Princeton to compare notes with Shope, and the two began a lifelong friendship.

This transatlantic collaboration resulted in important insights about influenza. For example, the 1918 flu pandemic that killed 40–60 million people worldwide was likely caused by a virus very similar to the one Shope isolated in pigs, as human survivors of the epidemic were found to have antibodies reactive to the swine virus. Independently, Shope and Laidlaw proposed that swine influenza arose from the transmission of human influenza to a pig in 1918. (In the 1940s, Shope proposed that the virus could be transmitted via earthworms pigs eat after inclement weather. Opinion was divided on the life cycle proposed by Shope. As Andrewes put it, "Some accepted it as gospel, others have been wholly skeptical."[30] Andrewes tried to confirm Shope's hypothesis by taking infected earthworms back to England, but they did not make his test animals sick.

Shope also found that the terrible cattle disease called "mad itch" had a viral cause. This disease caused cattle to bite and tear at themselves repeatedly until they died of self-inflicted injuries. Shope proposed that the cause of this disease was identical to that of pseudorabies found in Hungarian cattle. He proposed that cattle were infected when in close proximity to pigs, who in turn were infected by rats. Shope's career was on a roll, but he was not finished.

Shope and Rabbit Papillomas

While hunting in New Jersey in 1932, Shope discovered fibromas—small lumps beneath the skin—in cottontail rabbits. He subsequently showed that these fibromas were caused by a filterable virus similar to the virus for myxomatosis, an often-fatal disease that afflicts European rabbits. The fibromas often regressed over time. He found that rabbits that already had one of these "tumors," which Shope purposely defined loosely to include these repressible "warts," were immune to further infection by the tumor virus. Moreover, while infectious myxoma virus was often fatal, when injected into rabbits with actively growing tumors, it was not: the "striking" result was that 14 of the 15 injected rabbits lived.[31] He reported that an injection made from fibromas could be used to immunize rabbits against myxomatosis. Shope suspected the virus might be transmitted mechanically by a biting insect like a mosquito. Unlike Rous 20 years earlier, Shope confidently called the tumor agent a virus. It was resistant to glycerol, gave rise to specific immunity, was host specific, and displayed preference for one tissue type. (Other pathogenic microorganisms such as poliovirus were known to be resistant to glycerol.)[32] To attempt to classify it otherwise would seem "artificial," he reasoned.

Later in 1932, Shope was showing an elderly midwestern visitor to Princeton some of his rabbit fibromas when the old man said "disdainfully" that this was nothing compared to the sort of things he had seen in midwestern rabbits. He told Shope that he had shot rabbits with horns that grew out of the side of their heads like Texas steers or out of the front of their heads like rhinoceroses.[33] This disease was probably the cause of the myth of the jackalope—a rabbit with antelope horns popularized in the 1930s in the American West. Shope thought the visitor's description sounded like papillomata. In September, he set out for Cherokee, Iowa, with the gentleman and spent four days hunting rabbits. Unfortunately, none of the rabbits Shope and his party shot had the "horns" he was seeking. He returned to Princeton, but not before giving five dollars and a bottle to a young man in the hunting party, Cliff Peck. He told Peck that he would give him another five dollars if he returned the bottle with some warty rabbit horns. Within a week Peck sent some horns to Shope, who identified them as papillomas.[34]

Upon receiving the rabbit horns, Shope was "immediately interested" in whether they could be transmitted. He shaved the fur and "scarified" the skin of recipient rabbits with a needle, and then rubbed in a suspension of papillomas made by grinding up the warts with sterilized sand. After a week or so, he could see the beginnings of papillomas, which continued to grow but were benign. The largest warts, often black, grew to 3 centimeters long. The causative agent proved to be filterable and capable of prolonged storage in glycerol. It also survived being heated at 65°C for 30 minutes, suggesting it was one of the more stable viruses, fortuitous for Shope as it meant he could be leisurely with his work on the virus without it degrading.

His first experiments did not go as expected. He could transmit the virus from his wild rabbits to domestic rabbits, but could not repurify, or recover, the virus from the new warts that grew on domestic rabbits. Furthermore, Shope suspected that if he gave the domestic rabbit papillomas to tumor specialists, they would not be able to determine the etiology of the tumor. There was some precedent for this lack of transmission in "spontaneous" mammalian tumors. Shope thought that this was evidence that some spontaneous tumors were in fact caused by viruses.

Later work showed that even though the virus could not be shown to be present in the domestic rabbit warts, it was there in a "masked," or noninfectious form.[35] Applying a suspension of rabbit warts to scarified rabbit skin did not cause papillomas, but it did indeed immunize them against the effect of the wild rabbit virus.[36] Furthermore, the serum from immunized rabbits was capable of neutralizing wild rabbit virus. This classic experiment thus provided evidence that biologists could prevent a viral infection by generating a neutralizing antibody.[37]

Shope's successes with transmissible tumor virus in rabbits reinvigorated Peyton Rous, who had been standing on the sidelines of cancer research. With a possible second cancer virus to consider, Rous now reentered the fray. After Shope gave him some rabbit virus, Rous showed that it did cause a tumor, even though the tumor did not metastasize.[38] Additional work showed that the tumors, in combination with chemicals, could develop a malignant character. Rous's letters to his scientific ally Andrewes make his excitement clear: "Will you believe that an old-stager like me can lie awake at night and quake at these findings?—yes, be deliciously happy and quake alternately. . . . The age of adventure is not done, is it, when one can sail the seas of viruses?"[39] Despite Rous's optimism, many in the medical community were skeptical of cancer viruses.[40] One early and persistent detractor was Rous's colleague James B. Murphy, who, like Rous, was trained at Johns Hopkins Medical School and who also worked at RIMR, studying chicken sarcomas. He argued that other approaches to the cause of cancer—parasitic, chemical, and genetic—were supported by more evidence than the alleged viral cause.[41]

Richard Shope's Later Career

Shope's research focus eventually moved on to other topics. During World War II, he was given an assignment by the secretary of war to find a vaccine against rinderpest, a fatal viral cattle disease—considered "the worst of all cattle plagues" at the time.[42] The War Department feared that the virus might be introduced as a biological weapon, and the Americans needed a defense. Work on the project, a joint American-Canadian enterprise, was done in secrecy and isolation on Grosse Isle, Quebec. Over the course of 18 months, Shope and his team adapted the rinderpest virus by repeatedly growing it in chicken's eggs, eventually attenuating the virus so that it could be used as a live vaccine. Shope's team was successful—after the war, the vaccine was used commercially in agriculture—but this was not the end of his contributions to the war effort. He asked to be transferred to the navy and was subsequently ordered to build a laboratory unit on Guam. He was then ordered to scout Okinawa for diseases that might affect soldiers in combat after the American invasion. The time in Okinawa was hellish; attacks came without warning—at one point, Shope was forced to jump into a foxhole in such a hurry that he had to leave behind his helmet, rifle, and pants. Despite the dangers of being part of the military's push into Japan, the results of Shope's investigation were anticlimactic: his team determined that the diseases on Okinawa were mild.

In 1947, the Princeton branch of the Rockefeller Institute was closed, and Shope decided to move to Merck & Co. While he could have joined Peyton Rous at the Rockefeller Institute in Manhattan, Shope was a country boy at heart and preferred not to uproot to the big city. However, he soon found that the position of assistant director of research at a commercial company did not harmonize with his temperament, and he ultimately moved to the Rockefeller Institute in New York in 1952. There, he was able to interact with his friend Peyton Rous while continuing to spend weekends "in the country" at his home in Princeton.

In 1959, the father of American virology, the University of California virologist Wendell Stanley,[43] nominated Shope for the Nobel Prize for his "unusual research ability" and research of "the first magnitude."[44] In addition to his work with tumor viruses, Stanley highlighted Shope's work with swine influenza and his idea that a virus could have intermediate host reservoirs, but the nomination was unsuccessful. Nonetheless, Shope and Stanley remained close, exchanging letters for years. The two, in fact, had jointly nominated Peyton Rous for a Nobel Prize in 1953. But while Rous eventually received the Nobel Prize in 1966 for the work he did 55 years earlier, Shope never received the award. Richard Shope died in 1966 at the age of 64.

John J. Bittner, Mice, and the "Milk Factor"

In 1933, the staff of the Roscoe B. Jackson Memorial Laboratory (JAX) of Bar Harbor, Maine, published a two-page note in *Science* summarizing three years of research on inbred mice.[45] The note reported a surprising result: the existence of an "extrachromosomal influence" on the incidence of "spontaneous" mammary tumors—or breast cancer—that extended for many generations. Female mice from seven different strains exhibited varying probabilities of developing spontaneous mammary cancer. (Male mice very rarely develop mammary tumors.) Some inbred mouse strains—such as high-tumor strains C3H—had many tumors, while other, low-tumor strains had few to none. It had long been known that some families of mice developed spontaneous mammary cancer more frequently than others. To stabilize this tendency, inbred lines of mice were created by mating male and female littermates. Applying classical genetic methodology, researchers crossed low-tumor and high-tumor strains and examined the progeny, termed the "F1 generation" by geneticists. It turned out that high-tumor females crossed with low-tumor males produced high-tumor daughters. Conversely, low-tumor mothers crossed with high-tumor fathers produced low-tumor female progeny. When the granddaughters of the original strains—the F2 generation—were examined, those with high-tumor grandmothers had higher incidences of spontaneous mammary tumors than those with high-tumor grandfathers. "Because of that fact," the researchers concluded, "the simple Mendelian genetic nature of the biological agents influencing the etiology of spontaneous mammary tumors [was] definitely disproved." According to a simple Mendelian model, it should not matter whether an organism inherits a particular gene from its grandmother or its grandfather.[46] Thus the JAX researchers concluded that the best explanation for intergenerational patterns of tumors was a factor that could be inherited yet was not part of the mouse chromosome.

It was ironic that the discovery of an *extrachromosomal* factor would be made at the Roscoe B. Jackson Memorial Laboratory (JAX). As detailed by the historian of biology Karen Rader, this laboratory and mouse breeding program designed to investigate the genetics of cancer was set up by Clarence Cook ("C. C.") Little with funding from Detroit motoring moguls. Roscoe B. Jackson had been one of the owners of the Hudson Motor Car Company. He convinced other wealthy friends, including Edsel Ford, son of Henry Ford, to fund Little's new approach to understanding cancer. Little, president of the University of Michigan at the time, hoped that inbred lines of mice could be used to find patterns of inheritance of certain types of cancer. His plans were announced in the *New York Times* in 1929. His Motor City backers would serve as trustees of the Bar Harbor, Maine, facility. As he wrote in 1928, "The fact that the tendency to certain forms of cancer is hereditary in mice has been established for some years. . . . For a time the medical profession was not very willing to accept this fact."[47] Little's hope was that mice would reveal this better than humans, as they could be carefully bred and had a short lifespan, enabling researchers to see multigenerational patterns. Little's views of cancer research were connected to his eugenic ideals of racial improvement. For him, the key to both was an understanding of the underlying genetics, and this required a standardized research object allowing for repeatable experiments—an inbred mouse line, or many of them.[48] As the historian Jean-Paul Gaudillière has remarked, this approach echoes the practice of industrial standardization, which was one reason the tycoons thought it would be a fruitful approach to the cancer problem.[49] The laboratory opened its doors just weeks before the stock market crash of 1929, and the crash hit its financial backers hard. As a result, JAX was forced to raise revenue by selling mice to labs around the country, and it eventually became the standard source for many widely used mice lines. As Karen Rader has pointed out, this short note in *Science* on inheritance patterns of cancer in mice reframed the JAX assumption that the genetics of inbred mice would be insightful models of cancer research. Instead of abandoning the model, it presented the findings as a problem for further research.

The obvious question arose: What was the source of the extrachromosomal factor? The JAX researchers considered three possibilities: hormones, milk, and blood. They divided up the experiments among themselves. "It chanced that [John J.] Bittner, who had been assigned to study the maternal milk, was the lucky one," Little later reported.[50] Bittner had a background in cancer, having studied transplanted tumors in mice for his PhD in 1930 (figure 1.4).[51] In his initial experiments on the extrachromosomal factor, which started in January 1934, Bittner transformed the daughters of high-tumor strains into low-tumor strains by nursing them only with low-tumor mothers, implying that the extrachromosomal factor was in the milk.[52] He wrote up his preliminary results and published them in *Science* in 1936. Primarily just describing his experiment, he ended his article tentatively: "Should further study demonstrate that the incidence of mammary gland tumors in mice may be affected by nursing an explanation may be offered for the so-called extrachromosomal influence as a cause in the development of this type of neoplasm."[53]

Following the 1936 publication, Bittner continued to work on the "milk influence," thinking that the right approach was to work on inbred mice. He and his boss C. C. Little believed that inbred mice were to biologists what pure chemicals were to chemists and therefore an important basis for further discoveries.[54] By 1942, he knew the milk factor was active in lyophilized (dried-out) tissue. When passed through Seitz filters and ultracentrifuged, it was found in the fat fraction of the supernatant, the liquid that remains after heavier material is pulled out. "It is probably a colloid of high molecular

1.4. John J. Bittner. Image courtesy of his daughter Betsy Loague

weight," he concluded.[55] He proposed that the origin of breast cancer was caused by a tripartite set of influences:

1. Genetic—susceptibility to spontaneous mammary tumors was inherited
2. Hormonal—mammary tissue was stimulated by estrogenic hormones
3. Milk—it was generally transferred from mothers to young by nursing

He considered this to be his greatest contribution to cancer research.[56] Following his reasoning, researchers of the period began to speak of the hormones activating the virus.

Bittner's discovery was recognized as important at the time. *Time* magazine ran an article on him in its June 9, 1941, issue.[57] Given the war in Europe, the cover image depicted an aging dictator, Mussolini, but inside, the widely read magazine devoted a small article to Bittner, calling his "the most important basic medical discovery of a generation." After describing Bittner's experiments, *Time* offered a positive message: some inherited tendency for cancer might be averted by simple precautions. It ended by quoting Bittner:

"It may be that some women with breast cancer in their family should not nurse their daughters." This alarm was tempered somewhat by an earlier footnote stating that Bittner was not claiming that milk causes cancer, and moreover that "top flight" researcher Cornelius Rhoads of Manhattan's Memorial Hospital believed that milk contained a protein that protected rats from a type of cancer caused by chemicals.

That same year, Bittner summarized his understanding of the milk agent in *Biology of the Laboratory Mouse*, a handbook written by staff at JAX that symbolized the ascendancy of the laboratory's inbred mice as biomedical research tools in the mid-twentieth century. As reported in the handbook, if the mice were kept more than 24 hours with their birth mothers, the dramatic effects Bittner had discovered could not be reproduced. The "breast cancer causing influence," as it was termed, was now known to occur in many of the internal organs of the high-tumor strains of mice, but its nature "ha[d] not been determined."[58]

In March 1946, Bittner's research was again profiled in *Time*.[59] The magazine reported that Bittner believed that the milk agent was a virus because it grew in cells, was too small to be seen in a microscope, and caused "immune agents" (antibodies) to be formed when it was injected into animal tissues. By this time, Bittner had worked on the milk agent question for four years with his Minnesota collaborator Robert Green and had used 104,275 mice.

Bittner was reluctant to use the word "virus" in print, and he was not alone in this reticence. W. Ray Bryan, H. B. Andervont, and several colleagues at the National Cancer Institute (NCI) called the active agent "mammary tumor inciter."[60] One reason for their reluctance to use the word "virus" was that they did not want to scare the public with the idea that cancer was infectious.[61] The NCI group confirmed Bittner's findings. Andervont began performing caesarean sections on pregnant mice so that newborn mice had no chance to receive milk from their birth mothers. The NCI group also showed that the milk agent could pass through filters that retained bacteria.

By 1945, Andervont, who like Bittner had devoted his research to the milk factor, was able to assert that filtration and ultracentrifugation experiments "suggest[ed] the action of an agent belonging or related to the viruses."[62] In an NCI symposium devoted to cancer research, he then drew the comparison between this line of research and Rous sarcoma virus.

By 1949, British researchers from the Imperial Cancer Research Fund in

London were using the term "Bittner virus" in the title of an article on how low temperatures (–79° C) did not inactivate the virus but did kill tumor cells.[63] Additional evidence for the milk influence or agent being a virus came from electron micrographs.[64] In 1948, Keith Porter and H. P. Thompson of the RIMR took micrographs of particles they found inside the tumor cells that averaged 75 nanometers in diameter. They called these particles "virus-like bodies" and noted that when found in large numbers, the cells "show[ed] signs of degeneration." Given that these particles were derived from tumors in a high-tumor strain of mice, Porter and Thompson concluded that "it seem[ed] reasonable to assume tentatively that the particles [were] in fact the milk agent."[65] Bittner himself used the word "virus" to describe the milk agent in 1947.[66] The name "Bittner virus" lasted less than a decade. Even in the late as the 1950s and early 1960s, Bittner preferred to call it "mammary tumor agent."[67] In the 1960s, others used the more descriptive "mouse or murine mammary tumor virus" (MMTV), which replaced "Bittner virus" as the standard way to refer to the milk agent.

In the first-ever textbook of virology, *General Virology*, published in 1953, Salvador Luria devoted two pages to the milk factor. Interestingly, he did not explicitly call it a virus but nevertheless stated that the milk factor "ha[d] all the properties of a virus."[68] He included an electron micrograph from Keith Porter and noted that the abnormal particles seen were "presumably viral particles." Consistent with Bittner's tripartite theory of cancer, Luria noted the role of hormones as well as the milk factor: "It is difficult here to decide which is the provocative and which the actuating cause of the cancer."[69] This complication might explain Luria and others' reluctance to use the word "virus." Luria believed that the milk factor could assume a number of different forms, which would explain how there can be low- and high-tumor strains of mice. In the next decade, another textbook for virology would devote another page to the milk factor.[70] During the same period, it was also shown that hormones were important in the formation of tumors by giving estrogens to castrated male mice, who, it turned out, could develop mammary tumors. As the British virologist Anthony Waterson wrote in 1961, "There can be little doubt that the essential factor in the production of this particular cancer of the breast is something which can be called a virus, but it is important to remember that it has not yet been grown in cells in culture, nor adequately characterized structurally or chemically."[71]

In 1942, Bittner left JAX for a position at the University of Minnesota, where he continued to work on the milk agent. Tragically, a huge fire ripped through JAX in 1947, destroying it and all its inbred mice lines. The facility was rebuilt with the help of numerous donations. Researchers from around the world sent back breeding pairs of various lines to reestablish the colony. Bittner contributed to this effort, sending many strains of mice back to his former employer.

J. J. Bittner died in Minneapolis on December 14, 1961, at the young age of 57. He had been the George Christian Professor of Cancer Research and director of the Division of Cancer Biology at the University of Minnesota for nineteen years. Though he received honorary degrees from Bard College in New York and University of Perugia in Italy during his lifetime, along with the 1957 Bertner Award for distinguished research in cancer, he would undoubtedly have received more recognition had he lived longer. At a time when most cancer research was focused on chemicals and other mutagens, Rous, Shope, and Bittner introduced three different tumor viruses into the research community, discovering three different objects for future research. Taken together, they laid the foundation for the next phase of tumor virology.

2

True Believers

Ludwik Gross, Sarah Stewart, Bernice Eddy, and Polyoma Virus

It was September 1939 and the beginning of World War II. German troops were invading Poland, and the Polish army was in retreat. Ludwik Gross was driving a friend's abandoned car, desperately trying to make it to the Romanian border before the German army caught up with him. He had moved his mother and sister from their home in Krakow to Lwów to protect them from bombing and now was making a run for freedom.[1] But he had a serious setback: the car ran out of gasoline.[2] Would he make it?

Ludwik Gross (1904–99) was born into a prominent Jewish family. His father, Adolf Gross (1862–1936), was a lawyer and had represented the Jewish population of Krakow in the Austro-Hungarian parliament. The Grosses were concerned with the plight of others. Adolf was known for passing a bill to support widows and orphans of World War I. Ludwik began medical school at Jagiellonian University in Krakow in 1923, graduating in 1929. While in residency at St. Lazar General Hospital in 1931 and 1932, he operated on patients with cancer of the lip, which often occurred in peasants who smoked pipe tobacco. He noticed that the disease often spread to the lymph nodes, moving in a manner that resembled a spreading infection.[3]

In addition to practicing medicine, Gross was a medical columnist, writing science articles for a widely read Krakow newspaper *Ilustrowany Kuryer Codzienny* (Illustrated Daily Courier).[4] Among his readers was the wife of Marian Dąbrowski, the publishing magnate who owned the newspaper. Mrs. Dąbrowska altered the course of his career when she became sick during a visit to Paris and summoned Gross to consult with her French physician. Gross traveled to Paris, where in addition to looking after her, he took the opportunity to visit the Institut Pasteur (figure 2.1). There he discussed the

2.1. Ludwik Gross in Europe. Image courtesy of Augusta Gross

idea that tumors may be caused by transmissible viruses with immunologist Alexandre Besredka (1870–1940), who proved interested enough that he invited him to join the Institut Pasteur as a temporary guest investigator. Gross would ultimately work at the Institut Pasteur for eight years.

Early Tumor Research in Europe

Gross brought a mouse tumor called Sarcoma 37, a well-studied transplantable sarcoma discovered in 1907, with him to Paris.[5] Following his mentor's interests, Gross spent years trying to immunize mice against transplanted tumors. His work at the Institut Pasteur steeped him in its rigorous research traditions. There, he also fed his fascination with the nature of scientific discovery, speaking with a number of eminent scientists from the previous generation.[6] Gross and Besredka obtained some positive results, which they presented at the Académie des sciences in 1935. They reported that even though implanting tumor cells below the skin of mice proved fatal, implanting them intradermally had the potential to immunize mice against subsequent implants from that tumor. But despite years of effort, Gross and Besredka could not identify a virus that caused cancer.

To publicize his results on tumor implantation and immunization, Gross wrote to Peyton Rous at the Rockefeller Institute.[7] A year later, Gross wrote again, this time looking for a job.[8] Rous thought that Gross tended to make excessive claims without sufficient evidence. He also thought that Gross's reputation would be enhanced if he published less with Besredka and more by himself.[9] For his part, Besredka held Gross in high regard, writing that he was driven by a "feu sacré" (sacred fire) and thought of him as a son.[10] Given the number of American competitors for the position, Gross unsurprisingly was not offered a job at the Rockefeller Institute.

In the fall of 1938, Gross traveled to the United States on the French Line steamship SS *Paris* to look for a position in person. He met with Surgeon General Thomas Parran and inquired at Yale, Columbia, the Memorial Hospital, the Cleveland Clinic, and the Rockefeller Institute but once again came up empty handed. As he put it, "Nobody [I spoke with] believed at the time in the existence of a cancer or leukemia virus."[11] As Creager and Gaudillière point out, in the 1920s and 1930s, the most popular theories of cancer causation focused on the role of chemicals.[12]

Reluctantly, Gross returned to Europe to consider the offer of a position at the Marie Sklodowska-Curie Radium Institute in Warsaw. But before Gross could begin work in Warsaw, Germany attacked Poland. Gross decided to flee to Romania, which had opened its borders to Polish refugees. Luckily for Gross and the future of tumor virology, he beat the invading army to the Romanian border. It is unclear whether he found gasoline or abandoned the car; either way, his escape was successful. While staying in Bucharest, Gross made plans to return to the United States. To raise the necessary funds, Gross sold his Citroen, which he had stored in Paris, and bought a ticket on the SS *Rex* from Genoa to the United States. The next obstacle was to get to Italy and obtain a visa to enter the United States. By using a publishing contact, the *New York Times* correspondent in Bucharest, Gross secured a meeting with the US ambassador, who gave him a visitor's visa. He also wrote to Pietro Rondoni, the director of the National Cancer Institute in Milan, to obtain an official invitation to visit his laboratory so that he, an Eastern European Jew, could obtain an Italian visa. Gross traveled to Paris to close his laboratory and then to Italy to board the *Rex*. While he was crossing the Atlantic, France fell to the German army.

Not all of Gross's family were as lucky. One brother, Felix Gross, escaped to the United States via Japan (Chiune Sugihara, a heroic Japanese ambassador in Lithuania saved thousands of Jews by giving them Japanese visas) and eventually settled in New York City, where he became an accomplished sociologist. Two brothers remained hidden in Europe and survived. Another brother and his family were captured and killed. One sister was sent to a Siberian work camp and somehow survived. His younger sister and mother were sent to a concentration camp, where his mother was killed.[13]

A New Start in the United States

Although there was no Poland to return to, Gross was still allowed to enter the United States on his visitor's visa. He traveled to Cincinnati to take a position at the Jewish Hospital, the first Jewish hospital in the United States, founded in 1847. Despite holding a position in Ohio, Gross took and passed the New York state licensing board exam in medicine, as New York allowed individuals to take the exam before they obtained full US citizenship. Now certified, he applied for a commission in the US Army Medical Corps. For Gross, joining the army was a way to express his gratitude for having been allowed into the United States.[14] He wrote to Peyton Rous for help getting his official status changed from "visitor" to "resident."[15] Rous complied with the request, writing that he was impressed with Gross's "enthusiastic endeavors to discover facts which bear upon the cancer problem."[16] While at Christ Hospital, his second position in Cincinnati, Gross continued to work on tumor immunity in mice. He read case histories describing how retinoblastoma was found in a grandfather, a father, an uncle and a nephew, and speculated—mistakenly, as judged by other biologists—that a virus transmitted from one generation to the next could explain the data. He was impressed with Bittner's discovery of the "milk factor" that could be transmitted from mother mice to offspring that caused mammary cancer (see chapter 1).[17] Because the prevailing climate in cancer research did not favor a viral etiology of cancer, Bittner had avoided using the word "virus." Nonetheless, based on Bittner's work, Gross became more convinced that both solid tumors and leukemia in humans were caused by viruses.[18]

A few weeks later, his orders arrived: now–captain Ludwik Gross was to report to the US Army Medical Field Service School at the Carlisle Barracks in Pennsylvania. Based on what he had seen at the Institut Pasteur, Gross wrote to the US surgeon general during his brief stay in Carlisle with a suggestion to improve the treatment of athlete's foot.[19] Soon afterward he was ordered to report to Atlanta and then posted to Tullahoma, Tennessee. During this period, Gross started to plan experiments that would test his ideas about leukemia and viruses. Because of their short generation time and relative ease of use, Gross planned to continue working with mice. He thus needed a donor strain of inbred mice that would frequently develop "spontaneous" leukemia, along with a recipient strain of inbred mice that would not. Gross planned to remove the embryos from young, healthy female donor mice, grind them up, and inject the supernatant into non-leukemic mice.

As his established colony of C3H mice in Ohio was destroyed after he reported for military duty, Gross wrote to Bittner at the University of Minnesota to obtain some as the non-leukemic recipients, which Bittner mailed to him.[20] With no laboratory space of his own, Gross raised his new colony of mice in coffee cans in his room and in the trunk of his car—an image that would later become part of Gross's mystique as a figure utterly devoted to science despite extremely limited resources. He also wrote to the surgeon general, requesting to be transferred to an Army medical school so that his "experience in research could be utilized."[21] As he noted at the beginning of his long justification of the reassignment, "I have a definite plan for active immunization against spontaneous mammary carcinoma in mice, and sufficient training in research to carry out the plan."[22] Unfortunately for Gross, given the need for medical officers at his station in Tullahoma, his request was "not favorably considered." Two years later, however, Gross got his wish. He was reposted first to the station hospital in Camp Butner, North Carolina, and then to the Veterans Administration Hospital in the Bronx. Gross suspected that Balduin Lucké (1889–1954), the discoverer of a tumor virus in frogs, had played a role in his transfer.[23] Lucké, who worked at the University of Pennsylvania, was a consultant to the surgeon general. He had observed the mice in the trunk of Gross's car and felt the young researcher had something to offer.[24] The VA in the Bronx was recognized as a place where anti-Semitism did not influence hiring decisions and talented Jewish medical researchers could flourish.[25]

Ludwik Gross at the Veterans Administration Hospital in the Bronx

Once in New York, Gross began to search for an inbred line of mice that would develop spontaneous leukemia at high frequency.[26] There were two possible lines: the C58 line, discovered by E. Carleton MacDowell (1887–1973) and Maurice N. Richter (1897–1988), and the Ak line, bred by Jacob Furth (1896–1979). Gross drove to Cold Spring Harbor Laboratory and took MacDowell out to dinner to ask him for some mice. But in stark contrast to Karen Rader's portrayal of MacDowell as someone who "invested significant time and effort in making mouse material more readily available to those in need," he declined Gross's request for reasons lost to history.[27] This incident also illustrates the value of standardized inbred lines and the ways leading researchers could control the field by controlling access to coveted sublines—more generally what the historian of science Lorraine Daston calls the "moral economy" of science.[28] Gross then approached Furth, who had selected mice for leukemia and lymphoma. After 20-plus generations of selection, Furth's mice had a 90% chance of developing leukemia. The Hungarian Furth was generous, giving Gross 11 mice to begin a colony. He gave his mice freely in response to being refused animals earlier in his own career.[29] Gross stored them in coffee cans—and later in wooden boxes—in his makeshift one-room "laboratory" in the basement of the VA hospital.

For the next four years, Gross tried to confirm his hypothesis. He filtered extracts from spleens, thymuses, and lymph nodes of leukemic Ak mice and injected them into his non-leukemic C3H mice, but to no avail. He seemed to be in the same predicament as earlier investigators who had looked in vain for mammalian tumor viruses. Gross attempted to obtain grants to support his work, but with no positive experimental results, he was turned down repeatedly. The viral theory of cancer was so unpopular that the American Association for Cancer Research twice refused his membership application.[30] Gross was nearly ready to give up.

The turning point came during a lecture at the VA hospital by Gilbert Dalldorf, the discoverer of Coxsackievirus. He mentioned that Coxsackieviruses were neurotropic, causing paralysis in mice injected with the virus within 48 hours of birth. Prior to learning this from Dalldorf, Gross and others in the field had injected only adult mice with viruses. Before the lecture was finished, Gross ran to his laboratory. He found a litter of C3H mice born the previous day and injected an extract made from ground-up leukemic organs such as the lymph nodes and spleen. All the mice he injected developed generalized leukemia and died within two weeks. He then performed an experiment that used ground-up embryos from a four-month-old Ak female mouse that was in good health but whose mother and grandmother had developed leukemia. Four of the six C3H mice inoculated with cell suspensions of embryonic tissue developed leukemia at 8.5 months of age.[31] After publishing his results, he was given more money for cages, equipment, and assistants. Positive results also meant that Gross was at last cleared to work on research full time. In October 1952, he was appointed chief of research at the hospital.[32]

To demonstrate that a virus could transmit leukemia, Gross had to remove all the mouse cells from the leukemic extract and show that the purified extract still caused leukemia. To do this, he used a centrifuge and Berkefeld filters. After a long latency period of 10–18 months, a significant number of the inoculated mice developed leukemia. Publishing these results in 1953, Gross also proposed that the transmission of the virus was "vertical," that is, from mother to fetus, one generation to the next.[33] Interestingly, this type of transmission lies between infection and heredity and between endogenous and exogenous viruses, a distinction that Ton van Helvoort has explored to analyze the twentieth-century development of virology.[34]

Most of the reaction to Gross's papers was negative, as he recounted in an unpublished manuscript: "With only a few exceptions, the scientific community was generally critical, and frequently hostile. Most of the oncologists and hematologists did not take my work seriously, and a few questioned my integrity. I was severely, sometimes even viciously criticized. Nobody took me seriously."[35] At one lecture at Memorial Hospital in New York City, a junior pathologist even refused to shake Gross's hand. The idea that viruses could cause cancer in mammals was deeply unfashionable. It was hard to believe that cancer was infectious because oncologists and others who worked with tumors never appeared to "catch" it. It did not help that Gross's results were difficult to replicate, because not all leukemic extracts were leukemogenic. Gross himself prepared 80 before finding one that would reliably produce leukemia. Eventually, Gross isolated a potent virus strain that he called "Passage A," which he could ship to anyone who was interested. He also visited other laboratories to demonstrate his techniques, cultivating a network of researchers by sharing his mice and samples.[36] As Gaudillière and Löwy have pointed out, Gross's purebred mouse line and the leukemia virus can be thought of as an integrated system.[37]

Gross did have some supporters in this early period. Richard Shope, who had worked on rabbit tumor viruses and was featured in chapter 1, offered encouragement and advised caution about drawing conclusions from his experiments. Similarly, Joseph Beard (1906–1983) of Duke University showed interest,[38] as did Cornelius P. Rhoads (1898–1959), director of the Sloan-Kettering Institute for Cancer Research in New York City. W. Ray Bryan from the National Institutes of Health (NIH) wrote, "May I congratulate you on what appears to be a very important discovery."[39] Jacob Furth wrote, "Your findings are exceedingly interesting and if you will prove beyond doubt that there is an agent associated with leukemia similar to that of the Rous or Bittner agent, or different from them but filterable, you [will] certainly have a magnificent contribution."[40] However, the congratulatory letters were relatively few. Moreover, even the most enthusiastic supporters hedged their praise and suggested that further work was needed to be clear exactly what Gross had discovered.

Perhaps Gross's most important ally was Albert Sabin, who became famous for his attenuated poliovirus vaccine in 1955. Gross had known Sabin since the early 1940s, and he and his wife, Dorothy, socialized with Sabin and his wife, Sylvia. Like Gross, Sabin was of Polish Jewish ancestry and had served in the US Army Medical Corps in World War II. In May 1953, referring to Gross's experiments on the mouse leukemia virus, Sabin wrote, "I am personally convinced that you have made one of the most important discoveries in cancer research in recent years. Since I left you last Thursday I have acted as if I were your personal press agent."[41] Sabin teamed up with Shope to lobby for more funds and resources for Gross. Believing that one of Gross's greatest obstacles was not having a colony of mice large enough to perform all the necessary experiments, Sabin wrote to officials in the Veterans Administration to request more space and a better location for Gross.[42]

The Discovery of Parotid Tumor Virus / Polyoma Virus

While Sabin was excited about the discovery of what came to be called Gross murine leukemia virus, he was thrilled when a second tumor virus in the leukemic extracts that could cause tumors of the parotid (major salivary) glands was found. Gross recorded that on November 9, 1951, an injected mouse was noted to have "neck tumors."[43] By February 1952, Gross had the pathologist of his hospital, Theodore Ehrenreich, examine slides of the tumors. He also arranged for B. Gordon, Maurice Richter, and Jacob Furth to study the tumors. Some mice developed additional tumors under the skin of the abdomen, in the mammary glands, and in the adrenal glands. In the resulting paper, which appeared on June 8, 1953, Gross noted, "The results were surprising: while only 9 of the 84 mice that had been inoculated with the fresh filtered extracts developed generalized leukemia, 15 other mice developed at an average age of 3.3 months, bilateral tumors in the neck region. These tumors slowly enlarged and eventually formed enormous confluent masses around their necks."[44] Interestingly, like Bittner, Gross was unwilling to use the term "virus" in print, still preferring the less controversial term "agent." Unlike "virus," the weaker term did not imply any biological multiplication and left open the possibility that his results were caused by a merely chemical phenomenon. Additionally, the concept of the virus was just beginning to crystallize into its modern form.[45]

Gross used several methods to demonstrate that an agent other than the leukemia virus had caused these tumors. If the leukemic extracts were heated at 63°C for 30 minutes, their leukemogenic potency was destroyed, but they could still produce parotid tumors. Presumably, the parotid tumor virus was more resistant to heat. If the leukemic extracts were spun in the Spinco Model L ultracentrifuge, the supernatant was still able to induce parotid tumors. The leukemic potential, however, was found in the pellet, the solid material at the bottom of a centrifuge tube.[46] Thus, the centrifuge could separate the two viruses. The parotid tumor virus also appeared to be less dense and more stable than the leukemic virus.[47]

An Ambitious Competitor from the National Public Health Service Emerges

In December 1952, Gross received a letter from Sarah Stewart (1906–1976), a "vivacious microbiologist" in the Cancer Research Unit of the US Public Health Service (USPHS) hospital in Baltimore, Maryland.[48] "It is with much interest that we have been following your work on the etiological agent of mouse leukemia," she wrote. "I have tried to confirm some of the work which you have reported but with no success" (figure 2.2).[49] Some of her injected mice were 15 months old but still had not developed leukemia. She asked him whether he could send some of his mice with the tumor. Gross invited her to visit his lab in the Bronx so that she could pick them up and observe his procedures, which she did in late January 1953.[50]

Stewart, like Gross, had an enduring interest in the idea that cancer could be caused by a virus. Like Gross, she was determined to work on tumor virology and fought the established prejudices against the view.[51] Her master's degree research was on botulism. After first working in soil bacteriology, she started working at the NIH as a bacteriologist in 1935. During her first year there, she worked without pay, as there was a position but no funding due to the Great Depression.[52] Fortunately for Stewart, her work led to a paid position in the USPHS. She then returned to college, graduating with a PhD in microbiology from the University of Chicago in 1939 before returning to the USPHS. During World War II, she worked on developing toxoids—toxins of pathogenic organisms treated so as to destroy their toxicity—for therapeutic use, but she desired to work on cancer as her family had suffered from the disease.[53] But her requests to change research focus and move into cancer research at the NIH, part of the USPHS, were denied. She was told that she was not qualified to work in the cancer field. Despite this setback, she persisted. Although Georgetown University was not yet coeducational, Stewart used her status as an instructor at the medical school to take classes. As she put it, "My idea was to get my medical degree and go into cancer research. This was 1944. My feeling has always been that certain cancers are virus induced. When I finished my medical degree in 1949, I wrote up a protocol for work on viruses and cancer, to investigate whether human leukemia is cancer induced and I submitted this to Dr. [John] Heller, who was the director [of the National Cancer Institute]."[54] The first woman to graduate from

2.2. Sarah Stewart examining a mouse. Image courtesy of New Mexico State University Library

Georgetown Medical School with an MD, Stewart did not realize that "there was a tremendous feeling against viruses and cancer," and although Heller seemed receptive to her proposal, Stewart was given an assignment in gynecology, which was not her specialty.[55] Finally, in 1951, she was reassigned to the Marine Hospital in Baltimore maintained by the USPHS, allowing her to begin to work on cancer. It was there that she attempted to confirm Gross's work on mouse leukemia virus. But instead of finding leukemia, Stewart found solid tumors: "I got tumors of the parotid gland, which were tumors that had never been observed as spontaneous tumors. There had been some instances of parotid tumors that had been induced by carcinogens, but none with spontaneous parotid tumors. I became very excited about these tumors. I tried to convince the people here that there was something to it, but nobody seemed to think that there was."[56] She first observed parotid tumors in July 1952, presented the results at the American Association for the Advancement of Science meeting in December 1953, and an abstract was published in *Anatomical Record*.[57]

Stewart tried many times to fulfill what bacteriologists call the Koch postulates. In the late nineteenth century, the German Robert Koch, often called the father of modern bacteriology, suggested that to identify the microbial cause of a disease, researchers had to

1. find the agent in all cases of the disease,
2. isolate the agent,
3. grow it in culture, and finally
4. reinfect a healthy host.[58]

Stewart reisolated the virus from parotid tumors and then tried to use the isolated virus to again produce parotid tumors, but she failed in every instance. Here, she might have set the bar too high for herself. As Thomas Rivers wrote in 1937, "In regard to certain diseases, particularly those caused by viruses, the blind adherence to Koch's postulates may act as a hindrance instead of an aid."[59] Since viruses need host cells to replicate, it was very difficult for Stewart to satisfy Koch's postulates, which require the pathogen to be isolated from the host. After these initial failures, it occurred to her to try to culture cells as there might be only a small amount of virus present in mouse extracts.[60] However, not having her own tissue culture lab made the work difficult. She approached several people at NCI and NIH with tissue culture labs, proposing a collaborative study, but nobody was interested.[61]

Stewart was apparently unaware that Gross had already observed parotid tumors, and she consequently thought that the salivary tumor agent was her own unique discovery.[62] She believed that Gross had only learned of parotid tumors in January 1953 when she visited his lab and told him about her work. After they both published subsequent papers, each admonished the other for not citing their respective papers.[63]

A number of scientists were inspired by Gross's successes and began working on newborn mice, including a number of women in addition to Stewart.[64] Gross shared his new results with others in the field, including Anna Dulaney at the University of Tennessee. As she noted, "Dr. Gross is ready to do all he can to have other laboratories confirm his findings and certainly that is the most important thing at this time."[65]

Jeana Levinthal of the University of Michigan also attempted to replicate Gross's results.[66] John Bittner received "so many requests" for C3H mice that he sent some to Jackson Memorial Laboratory so researchers could purchase mice directly from them.[67] Lloyd Law (1910–2002) at the National Cancer Institute, and formerly of JAX, was able to induce parotid tumors in CH3 mice but had more difficulty inducing leukemia. Gross interpreted this as evidence that the substrains of C3H were variably susceptible to leukemia: Bittner's had a higher susceptibility to the cancer, while the NCI's was lower.[68]

With new researchers getting into the field, the time was ripe for a small conference on viruses and cancer. Jacob Furth organized the 1955 Gordon Conference at Colby College in New Hampshire. It included a session on how viruses could cause leukemia and tumors. Furth invited Gross to talk about mice and Wendell Stanley to chair the session. Others, including Lloyd Law, Arthur Kirschbaum (1910–1958), George W. Woolley, Sarah Stewart, and John Bittner would make up a roundtable. Although Stanley was a plant virologist, he was a Nobel laureate and arguably the most eminent American virologist at the time—and the perfect chair for a panel in the newly hot area of virology.[69] Stanley had experience building institutions and was adept with the popular press, and he would be a good spokesperson to push these new ideas.[70]

After three years working on tumor viruses, Stewart found a collaborator with whom she could improve the understanding of the parotid tumor virus, Bernice Eddy, an NIH researcher (figure 2.3). Eddy, who had a tissue culture lab at the NIH, took tumor extracts from Stewart and grew them first in monkey cells, then in mouse embryo tissue culture. After it incubated for one to two weeks, the fluid harvested from the cultures was found to have a high titer (amount) of a virus that induced parotid and a variety of other tumors in newborn mice. The oncogenic potential of the tissue culture extracts was much higher than extracts derived from serial mouse-to-mouse passages.

Papers flowed from the collaboration between Stewart and Eddy. In a famous article published in 1958, Stewart, Eddy, and their coauthor Ninette

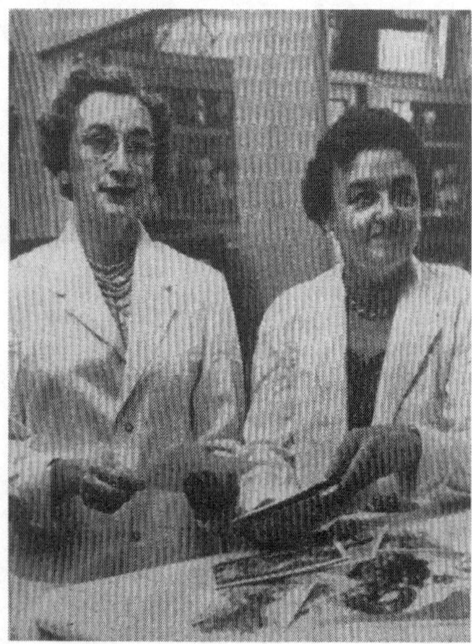

2.3. ***Time*** **magazine, July 1959. John Heller of the NCI on the cover for the lead story, "The New War on Cancer via Virus Research & Chemotherapy," and Sarah Stewart and Bernice Eddy on the inside. Original photo of Stewart and Eddy taken by Walter Bennett. From TIME. © 1959 TIME USA LLC. All rights reserved. Used under license.**

Borgese ended by asserting that the "tumor agent" was a virus: "Extracts and filtrates of certain mouse tumors and fluids from tissue-culture preparations of such tumors have consistently produced more than one anatomical variety of neoplasms when injected into mice within 24 hours of birth. Perhaps just one agent as yet unidentified is responsible. The most reasonable hypothesis is that it is a virus."[71] They did not cite Gross's earlier observations on the parotid virus.

By 1957–58, the tide was turning. The idea that viruses could cause tumors in mammals was becoming more acceptable. As Stewart wrote, "I can't get over how the Saturday morning [conference] session really went all out for Gross—I guess viruses are here to stay."[72] John Heller, summarizing the Third National Cancer Conference in Detroit, Michigan in June 1956, quoted Stanley, "I believe the time has come when we should assume that viruses are responsible for most, if not all kinds of cancer, including cancer in man."[73] By 1959, he stated that "right now the hottest thing in cancer is research on viruses as possible causes" (figure 2.3).[74] What had changed since 1953? Following the virologist George Klein, the historian of science Daniel Kevles has argued that the field took off because Jacob Furth successfully repeated Gross's experiments using the same sublines of mice and then notified the world.[75] But this is only part of the story. The discoveries of additional mouse leukemia viruses[76] by researchers inspired by Gross's reports of success were also important in changing the tide.[77] In 1956, Arnold Graffi (1910–2006) and coworkers in East Germany at the Deutsche Akademie der Wissenschaften zu Berlin used extracts from mammary carcinoma to induce generalized leukemia in mice.[78] Coming from the other side of the Iron Curtain, these results were all the more surprising. At the Sloan-Kettering Institute in New York City, Charlotte Friend discovered a leukemia virus in Swiss mice that was later named the Friend virus in her honor. Gross promoted these new discoveries in a guest editorial in the *Journal of the American Medical Association* in December 1956.[79] John Moloney (1924–2007) at the National Cancer Institute also discovered a virus using Sarcoma 37 extracts in 1959, as did Frank Rauscher (1932–1993) in 1962.[80]

Charlotte Friend's experience publishing her findings demonstrates the resistance to the idea that there were mouse tumor viruses and the importance of rigor in convincing critics. She sent her paper to the prestigious *Journal of Experimental Medicine*, edited by Peyton Rous. On the face of it, one might think that Rous would welcome papers on tumor viruses, given that his papers on chicken tumor viruses in the same journal had founded research in the field.[81] However, he had already advised Gross not to publish his research in the *Journal of Experimental Medicine* and instead to try a specialized cancer journal.[82] As he wrote to Friend, "Now prepare yourself for some hard words. I write them for the greater good, meaning thereby knowledge of cancer causation."[83] Wanting Friend to "settle all doubts" about whether a virus could cause leukemia in mice, Rous advised to carefully characterize the disease with the help of an expert pathologist. As Rous put it, doubts about a viral cause existed because of the complexity of Gross's experiments, the "large claims" drawn from them, and a "general feeling" that his experiments were not adequately controlled. "Many seasoned investigators write off the claims [of Gross] as worthless," he wrote. To help Friend protect herself from over-associating with Gross, Rous meticulously line-edited her paper to temper her discussions of Gross's findings. Where Friend had written, "Gross has shown . . . ," Rous rewrote, "Gross has reported . . ." It was not normal to give such detailed feedback, but for reasons that are unclear, Rous made an exception for Friend. Her paper was eventually published in the *Journal of Experimental Medicine*, providing additional evidence for the view that mammalian tumor viruses existed.[84]

The priority dispute between Gross and Stewart over the discovery of the mouse tumor virus simmered for a few years, but it came to a head in late 1958, when Jacob Furth wrote an article that attributed the discovery of the parotid tumor virus to both Gross and Stewart.[85] Gross objected with a long letter to Furth, stating that he did not accept "independent discovery." Furth suggested that a "neutral person" such as A. J. Goldfarb, secretary of the Society for Experimental Biology and Medicine should pinpoint credit. Stewart believed she had a claim to independent discovery because she had observed parotid tumors in the summer of 1952, before she met Gross.[86] But Gross argued that he had observed the virus first and had performed initial experiments to distinguish the parotid tumor virus from the leukemia virus. In contrast, he asserted that Stewart had continued to believe that there was one virus that produced different conditions in different mouse strains. Furth supported Stewart's contention that it was a case of independent discovery. Goldfarb declined to referee the dispute, suggesting instead that a committee of scholars from a national scientific society should decide the case of "Stewart vs. Gross."[87] While he did meet with both Gross and Stewart independently, examining documents from both of them, a national society was never involved in the dispute. The ultimate outcome appears to have been a standoff.[88]

What particularly annoyed Gross was that Stewart and Eddy gave the parotid tumor virus a new name: SE polyoma virus, with "SE" standing for "Stewart Eddy."[89] This renaming of the virus, he argued, violated established research etiquette allowing the discoverer of a new virus the right to name their finding. Gross believed it had the potential to diminish future researchers' view of his initial discovery. When Peyton Rous gave introductory remarks at a 1959 meeting of the American Cancer Society in which he attributed the discovery of polyoma virus to Stewart and Eddy, Gross wrote him a long letter summarizing work on the virus over the prior eight years. As he concluded, "The importance of the contributions made by Drs. Stewart, Eddy and their associates is considerable; it does not give them the right, however, to ignore my initial observations and to give their names to a virus which was isolated, recognized and first reported from my laboratory."[90] In his reply, Rous acknowledged Gross's priority. Gross also wrote similar letters to Charlotte Friend and Giampiero di Mayorca, who had published a paper in the *Proceedings of the National Academy of Sciences* showing that infectious polyoma DNA could be isolated from tissue cultures.[91] Di Mayorca argued that while Gross had priority, the work of Stewart and Eddy showing the variety of lesions caused by the virus and its whole host infectivity meant that the name parotid virus was "no longer fitting." He concluded that the name polyoma virus was a "logical" one.[92] The polyoma virus name stuck, but the SE designation was dropped; eventually everybody including Gross came to use the new name or, more recently, "polyomavirus" as one word.

Retrospectively, Eddy had regrets, and she later tried to set the record straight with a historian of science: "Sarah was very aggressive. We named it. We probably shouldn't have."[93] Stewart had wanted to name it omnioma virus, but Eddy suggested polyoma virus instead, since the virus did not cause tumors in all cases, but rather it caused many different types of tumors in many different animals. "He had that virus before we did. There was no question," Eddy conceded. "Sarah would never admit it. She was always sparring with him."

The Book Oncogenic Viruses

In May 1958, Gross was invited to England and France to give talks at the Institut du Cancer in Villejuif, the Institut Curie in Paris, and the Ciba Foundation in London. He spoke at the London School of Hygiene & Tropical Medicine on the "viral etiology of spontaneous mouse leukemia and its possible implications for the problem of cancer."[94] The next morning, he received a letter from Pergamon Press inviting him to write a book on "viruses and cancer." Gross happily accepted, appreciating the positive response to his project. Robert Maxwell of Pergamon Press, who had a personal interest in this topic, was enthusiastic to have the book as part of his development of the publishing house. (Maxwell and his family led controversial lives. He would later be elected as the Labour MP from Buckingham and build a large publishing empire that collapsed after his death in 1991.) The first title of the book was "Viruses in Leukemia," and Gross was given a deadline of June 15, 1960. This would be the first book on this topic, and Gross spent much time tracking down journal articles in several languages. He covered papilloma, benign tumors, frog kidney carcinoma, chicken sarcoma, and his own work on mouse leukemia. He ended the survey of various viruses with a chapter on polyoma virus, which he now called "parotid tumor (polyoma) virus," including a discussion of the work of Sarah Stewart in an extended footnote.[95] The chapter reflected Gross's view of the history, although he included pictures of Sarah Stewart and Bernice Eddy among the 22 biologists depicted. Gross considered the history of tumor virology important and spent a lot of time tracking down photographs of the major players.

Reviews of the book were generally favorable. It proved to be a useful reference guide, and no similar work existed.[96] Gross sent complimentary copies to Rous, Shope, Stanley, and several others. Rous read his copy on vacation. He liked it. "I have read the chapter on the parotid virus with special care," he wrote in a letter to Gross. "You make plain beyond question that you were the finder of the virus, and you do with dignity, and indeed generosity in showing pictures of S[tewart] and E[ddy]. They really deserve these because of their proof that the parotid virus causes growths in hamsters and rats."[97] Even Bernice Eddy remarked, albeit a few years later, "It is one of the most useful books that I have."[98]

In the last chapter, using an analogy with lysogenic bacteriophage, Gross argued that the virus theory of cancer could be approached by looking more at "vertical" transmission (from one generation to the next), rather than "horizontal" transmission (among members of the same generation), which biologists do with common communicable diseases.[99] He cautioned against confusing vertical transmission and inheritance, which often have similar effects but are fundamentally different in that "inheritance is conditioned by genetic factors, and is never acquired, nor is it related to extraneous parasitic agents."[100] In this appeal to vertical transmission, Gross came close to articulating the later influential "oncogene hypothesis" of Robert Huebner and George Todaro discussed in chapter 9.[101]

Because the field of tumor virology had expanded, Gross began working on a second edition, published in 1970. It grew to nearly 1,000 pages. Gross added new material on simian virus 40, Burkitt's lymphoma, bovine leukosis, and adenoviruses. Reviews were mixed this time around. Eddy gave it a positive review in the *Journal of the American Medical Association*.[102] However, biologist/science journalist John Tooze, who was also working on a book on tumor virology, thought that Gross had not included enough material from recent molecular biology. "Gross's preoccupation is with the history of discovery of tumor viruses to the exclusion of the more recent work."[103] Generational differences were apparent even in some of his praise: the book was "a work of history on a grand scale by one of the grand old men of oncogenic virology."[104] In the 1970s Gross thought about a third edition, but the editors at Pergamon were not as enthusiastic. Biology was increasingly becoming molecularized, and Gross was now part of the old guard. However, higher management disagreed with the editors and allowed Gross to proceed. The third edition would be published in 1983, having grown to over 1,200 pages. It was encyclopedic in its coverage of the various discoveries. The third edition also garnered a negative review from the newer generation of biologists, who were critical of Gross for not keeping up with the new developments in molecular biology.[105] Ironically, given his difficulties entering the community of cancer researchers three decades earlier, Gross now represented a conservative old guard to many younger researchers, who saw his approach to virology as old fashioned.[106] By the 1980s, it was clear that the cutting edge of tumor virology was now occupied by molecular biologists and the golden age of the solitary cancer virus hunter was over. From now on, further progress would come from teams of scientists.

Gross's struggle to demonstrate the existence of a mouse cancer virus clearly illustrates the power of persistence. Gross spent eight years at the Institut Pasteur and four years in the United States without a significant breakthrough, yet he did not give up. It also illustrates that, although risky, working in an unfashionable area can lead to a large payoff. Gross faced ridicule from physicians who thought that the idea of viruses causing cancer in mammals was a pipe dream. Many assumed that if viruses did cause cancer, one could observe medical personnel such as oncologists catching cancer from their patients, but this appeared not to happen. In thinking this way, they excluded the possibility of vertical transmission from mother to offspring. Gross also faced challenges obtaining the necessary strains of mice. That E. Carleton MacDowell would not give Gross his strain of mice illustrates the importance of intellectual property and the way leaders in the field of mouse genetics could control the field by controlling who had access to valuable sublines.

However, once researchers were able to replicate Gross's work on newborn mice, and additional leukemia and tumor viruses had been discovered by others inspired by his initial work, Gross finally received significant recognition. Hoping to extend and generalize Gross's initial findings, the NCI then invested significant financial resources into tumor virology, beginning the Special Virus Leukemia Program (SVLP).[107] They believed that further significant discoveries would require large teams of researchers. Thus, by the 1960s, the tide of acceptance had definitely turned. Consequently, Gross started to receive a number of prizes, including the UN Prize for Cancer Research in 1962, the Bertner Award in 1963, the Lasker Award in 1974, and the French Legion of Honor in 1977. In June 1972, he appeared in a long interview on WNBC-TV defending the view that cancer was caused by viruses. He was nominated for the Nobel Prize but did not win. The Nobel Committee did recognize tumor virology by awarding the 1966 prize to Peyton Rous for his early work on chickens and Rous sarcoma virus. It later showed a preference for human tumor viruses: Baruch Blumberg (1925–2011) won the prize in 1976 for work on hepatitis B virus and liver cancer in humans, and Harald zur Hausen (1936–) won in 2008 for work on human papilloma virus and cervical cancer. It is likely that the successes with murine leukemia and tumor viruses of the 1950s and the subsequent surge in research on tumor viruses made it easier for the committee to finally award the Nobel Prize to Peyton Rous in 1966, 55 years after his original research. Sarah Stewart also received some recognition for her work on polyoma and other viruses. She won the Lenghi Award in Italy in 1963 and the Medical Men [*sic*] of Georgetown award in 1964, among others.

Gross's tense relationship with Sarah Stewart illustrates how in addition to the resistance from the medical community at large, he also faced competition from the few like-minded scientists at the time. Her and Eddy's decision to rename the parotid tumor virus increased the probability that others would mistakenly overlook Gross's early work on the virus. Even Renato Dulbecco, who won the Nobel Prize for developing the molecular biology of animal viruses in 1976, credited Stewart and Eddy with the discovery of polyoma virus in 1958, rather than Gross in his Nobel Lecture.[108] The particular name of a virus was not merely a matter of semantics or classification; it also predisposed future scientists to attribute priority in a particular way. Depending on the name and the authors of the first paper coining it, future scientists gave different degrees of credit for the discovery.[109] The publication of *Oncogenic Viruses*, the first authoritative comprehensive book on cancer causing viruses, reinforced Gross's position as the pioneer in the new field of mammalian tumor viruses.

The discoveries of murine leukemia virus and polyoma virus and the subsequent rise of quantitative tumor virology[110] allowed for the large growth in NCI-funded tumor virology research in the early 1960s,[111] the growth in molecular biology in the late 1960s and 1970s,[112] and eventually for the discovery of oncogenes and tumor suppressor genes in the 1970s and 1980s.[113] With the development and refinement of cell culture, the cutting edge of molecular virology moved from the study of organisms to the study of immortalized cell lines.[114] Cutting-edge molecular virology was increasingly performed by well-funded teams of scientists at Cold Spring Harbor Laboratory and at the Imperial Cancer Research Fund, among other institutions. Both Gross and Stewart continued to work on viruses for the remainder of their respective careers. Sarah Stewart moved from the National Cancer Institute to Georgetown Medical School and died of cancer in 1976. Gross remained with the Veterans Administration Hospital for the remainder of his long career, which ended with his death in 1999.[115]

3

The Importance of Measurement

Renato Dulbecco, Marguerite Vogt, and the Rise of Quantitative Animal Virology

Renato Dulbecco (1914–2013) grew up in northern Italy near Turin. He attended the University of Turin and followed his mother's wish that he study medicine. In his second and third years, he studied anatomy with Professor Giuseppe Levi, who was well known at the university for his anti-fascist views. Dulbecco worked in Levi's laboratory, an environment that produced two other Nobel Prize winners: Salvador Luria and Rita Levi-Montalcini. Although much of Levi's work was qualitative, his interest in cell culture rubbed off on Dulbecco, who was also inspired by Alexis Carrel's book on the subject. Additionally, Levi's personality had an impact on the young Dulbecco, who later described his mentor as "highly encouraging" but also "critical to the maximum extent."[1]

Dulbecco became a physician in 1936 and was conscripted into the Italian army, where he spent two years meeting people from all walks of life, opening him to consider new life trajectories. He returned to Turin and spent time in the Department of Pathological Anatomy writing a thesis on how the liver is destroyed when the duct is obstructed. With World War II breaking out, he was redrafted into the army and sent to the Russian front, a place from where as few as 20% of soldiers returned alive. While traveling through Poland, he was horrified to learn that Jews, identified with yellow stars, were working as prisoners and would be shot once the work was completed. This was a turning point for Dulbecco—he knew he had a duty to stand against such atrocities. After being injured on the front, he returned to Italy but did not return to the army as he was supposed to. Instead, he joined a group of partisans encamped in the hills around Turin, which was occupied by the German army. He became the physician for a small village where the group was located, and even taught himself some dentistry when members of the group had dental problems. Incredibly, he continued doing research at the university, going so far as to sleep in the morgue when nothing else was available. Living this way, he managed to avoid the police until finally the Germans withdrew from Turin in 1945. The underground organization of which Dulbecco was a member, Movement of the Christian Left, took control of the city and the city council and worked with the Americans who moved in after the Germans left. After two months with the council, Dulbecco tired of the bureaucracy and left politics to return to science.

Following the end of World War II, Dulbecco attempted to establish a research career in biology in Italy, but success was elusive. Through discussions with Rita Levi-Montalcini, he realized that genetics was going to be important to the future of biology, but having no mentors from whom to learn, Dulbecco studied physics hoping that it would be helpful in building a career in future genetics-driven biology.

Salvador Luria had left Italy at the start of the war and established himself in the United States as one of the leaders of the "phage group," scientists who studied viruses that infect bacteria to advance biology and genetics.[2] On a trip back to Italy in 1946, Luria saw Dulbecco again. Dulbecco described his idea to study viruses using radiation and mentioned that he was studying physics to get a better handle on the problem. Luria said, "By gosh, that's what I'm doing!" and invited Dulbecco to join him at Indiana University in Bloomington.[3] Although Dulbecco had a lifetime appointment as a professor at the University of Turin, he decided to resign to join Luria in the United States. One factor in this decision was the postwar environment at the university, which was not great for a laboratory scientist—many buildings were destroyed by Allied bombing, which had targeted a nearby ball bearing factory but destroyed half of the building where Dulbecco worked. Eventually he procured a US visa and left Europe via ship. Coincidently, Rita Levi-Montalcini was on the same passenger ship as Dulbecco—the Polish MS *Sobieski*—crossing the Atlantic in September 1947.

Renato Dulbecco's Research in the United States

In Luria's laboratory, Dulbecco studied how UV radiation affected phage behavior. Luria had one large room on the top floor and, in addition to Dulbecco, a technician. James Watson joined the lab in early 1948 as a graduate student of Luria's, and he became good friends with Dulbecco. The two men talked a lot—sometimes about the work of Ole Maaløe, whose experiments involving radiolabeled DNA Watson thought were important.[4] Seeking to deepen his knowledge of genetics, Dulbecco took a class with Hermann Muller. Watson also took the class, and the two competed for the best grade. (Dulbecco claims he won.)[5]

Dulbecco had success investigating how visible light could reverse mutations induced by UV light—this process was called photoreactivation and was significant enough to be published in *Nature* in 1949.[6] He reached this surprising result after noticing that plates of infected bacteria left directly under a fluorescent light gave different results than plates of infected bacteria in the shade. When Luria saw the unexpected results, he was shocked: "He nearly died!" Dulbecco recounted.[7] Even before he learned of all the details of this work, Max Delbrück, another leader of the phage group, was impressed enough with Dulbecco to invite him to Caltech to work in his lab as a senior research fellow with a salary of $5,500 a year.[8] Delbrück was interested in the relationship between light and life more generally.[9] Watson advised Dulbecco to accept the invitation as he thought Caltech was possibly the best place in the United States for modern biological research.[10] And he did accept, writing to Delbrück in November of 1948, "Thank you for your kind offer—the position and kind of work are of the type I like and I therefore accept your offer."[11] He went on to describe his "almost accidental" discovery of photoreactivation as "a very strange phenomenon." They agreed he would start in the fall of 1949.

The uniqueness of Dulbecco's discovery was clouded somewhat by the earlier work of Albert Kelner, a biologist at Cold Spring Harbor Laboratory, who had discovered a general effect of photoreactivation on actinomycetes (a group of bacteria) and communicated it to Luria a few weeks before Dulbecco's discovery. Delbrück wrote to Dulbecco, "Kelner is very perturbed, and understandably so, by the fact that you discovered your photo-reactivation just a few weeks after he had communicated his discovery to Luria, and that nevertheless you or Luria claim complete independence on your part from the communication."[12] Once the priority of Kelner's work was clear to Dulbecco, he politely withdrew any claim to be the first to discover the phenomenon, even though he believe he had "found phage reactivation without any connection with Kelner's discovery."[13] This gentlemanly behavior was not uncommon in his dealings with fellow scientists. Nonetheless, Dulbecco continued to study the phenomenon, looking at different wavelengths and different viruses. The Dulbecco family moved to California in 1949, driving and camping all the way.

The Transition to Animal Virology

Dulbecco expected to continue working on phages with Delbrück, but in 1950 Colonel James G. Boswell, a wealthy cotton businessman in California and a Caltech trustee, donated $100,000 through his foundation to fund a new research effort on animal viruses at the school.[14] Colonel Boswell suffered from shingles and wanted a cure found. With the new money, Delbrück arranged a conference on virology, inviting the most significant figures from several areas in plant and animal virology. His desire was "to bring men who work on the three great groups of viruses, those that attack animals, plants, and bacteria respectively, into one room and to discuss whether and to what extent our respective charges can be brought under one hat."[15] The resulting talks were published in a small volume appropriately titled *Viruses 1950*. The first article in the book, Salvador Luria's discussion of bacteriophage reproduction, was circulated among the participants. Frederick "F. C." Bawden came from England to talk about plant viruses. Richard Shope presented his ideas about "masked" animal viruses. The last item in the book was a proposed syllabus on bacteriophage biology authored by nine members of the phage group, including Watson and Dulbecco. After the conference, many of the participants continued talking informally around campfires in Death Valley. Delbrück often took scientists camping in the desert for recreation and to build camaraderie. Despite the talent assembled at the conference, no clear path emerged on how best to propel animal virology forward.

Another tack was needed. Delbrück summoned Seymour Benzer and Dulbecco to his office. He wanted one of them to switch from phage to animal viruses. Dulbecco was excited by the opportunity: "To me [animal virology] sounded wonderful. I had been thinking perhaps with nostalgia, of my work with tissue culture, years before, in Giuseppe Levi's laboratory in Torino; so I immediately expressed my interest, before Benzer could say anything."[16] Given his medical background, Dulbecco made a fine choice to head the new effort. To get up to speed on the latest techniques in animal virology, Dulbecco planned to travel around the United States visiting virus laboratories in Denver, Ames, Chicago, Urbana, St. Louis, Ann Arbor, Boston, Philadelphia, and New York. One of his goals was to determine which virus-host system would be best suited to the Caltech initiative.

He spent a couple of weeks in Baltimore, visiting George O. Gey and F. B. Bang at Johns Hopkins. (Gey become famous for creating the HeLa cell line using a sample from Henrietta Lacks in 1951.)[17] He learned a lot from them but thought that "these people have not tried to study the elementary things, like how many cells in a culture are infected . . . but have jumped to the most complicated things."[18] A few days later in a letter he wrote again, "It seems to me quite obvious that important improvement in their knowledge would be [possible] if a more accurate system of assay [of the number of infected cells] were worked out."[19] What Dulbecco needed was more knowledge of tissue culture, which he got from Wilton R. Earle at the National Cancer Institute in Bethesda. In particular, Earle had developed techniques to grow a uniform layer of cells in a flask, much like a layer of bacteria in a petri dish. Dulbecco was also impressed with the work John Enders carried out on poliovirus in Boston.

Dulbecco returned from his trip and refined the new techniques he had picked up. He decided to use western equine encephalitis virus because it killed cells definitively, but Max Delbrück and division chair George Beadle worried about its pathogenicity:[20] the virus causes disease in horses but also occasionally in humans. Consequently, Dulbecco was relocated to the second basement of Kerckhoff Laboratories in a room at the end of the corridor. Once resituated, Dulbecco wanted to create uniform layers of host cells. He drew inspiration from methods to grow a lawn in the hot California climate—a common approach was to plant small patches of dichondra about one foot apart, which then grew and spread to cover all the ground. He tried to use stacks of razor blades to make small pieces of tissue with limited success. Following a suggestion from Earle, he took pieces of chick embryo and using the centrifuge, pushed them through a finer and finer mesh, dividing them into very small pieces that he could spread on a petri dish to grow together again in a layer. Another improvement was growing them in a CO_2-flushed incubator.[21]

Initially he could not create visible plaques—little "holes" about 2 millimeters wide in the uniform layer of cells caused by viral infections—but once he illuminated the cells from a different angle, they became visible. He invited Delbrück down to look at the plaques, and Delbrück told him to remember the exact date as it was a significant step forward for animal virology.[22] (Alas, it was forgotten.) As he continued to refine the assay, he talked with his colleagues. An old-timer at Caltech, A. G. R. Strickland, suggested he use a light stain that turns living cells red but not dead ones. The stain provided the contrast needed to see the plaques. By April 1952 he had made enough technical progress on his new technique to give a paper at the annual spring meeting of the National Academy of Sciences in Washington, DC. The talk was published in *Proceedings of the National Academy of Sciences* after being submitted by Delbrück.[23] In a press release from Caltech on the new assay, Dulbecco explained the strength of the assay: a single flask of cells could now give results that would have taken 100 chicken embryos previously. Using statistical reasoning, Dulbecco argued that each plaque was initiated by a single virus particle. This was important, as it would allow preparation of virus particles cloned from a single ancestor and so be genetically identical. He publicly argued for his approach in the summer of 1953 at the annual Cold Spring Harbor symposium, which that year was devoted to viruses (figure 3.1). This important meeting was attended by twelve future Nobel laureates and is well known to historians of science for the first public presentation of the DNA model by James Watson, but it also featured Dulbecco trying to convince virologists that his new assay could show that each plaque derived from a single virus particle.[24] Dulbecco even cited Albert Einstein, "the ultimate weapon," to make his point.[25] This new approach allowed animal virology to replicate earlier technical advances in plant virology (tobacco mosaic virus)[26] and bacteriology (bacteriophages) that also had quantitative assays to measure the infectivity of viruses.

Dulbecco saw the new assay as the beginning a new phase in his research:

> So that was the beginning. That put me in that field, and people started coming—postdocs and so on—from all over the world. I concentrated in the beginning, just on studying a little bit more about viruses—studying things like the role of antibodies in inactivating retroviruses, a variety of things. Then the people from the National Science Foundation[27] came and said that I should try to develop a system like that for polio, because they were in vaccine development at the time.[28]

In early 1952, Dulbecco started working on poliovirus after he was approached by the National Foundation for Infantile Paralysis (NFIP)—a large funder of poliovirus research in the 1950s—and given financial backing to develop a plaque assay for poliovirus. For safety reasons George Beadle did not want a poliovirus lab in the building, so he rented Dulbecco a small laboratory at the Huntington Hospital in Pasadena. At this time, Dulbecco gained his most important collaborator, Marguerite Vogt.

3.1. Renato Dulbecco and Harry Rubin chatting at the 1953 Cold Spring Harbor Symposium on Viruses. Image courtesy of Cold Spring Harbor Laboratory Library and Archives

Marguerite Vogt

Born and educated in Germany, Marguerite Vogt published her first scientific paper at the age of 14, and as she grew older she published many papers on fruit fly mutations, including one in *Nature*.[29] Her parents, Oskar Vogt and Cécile Vogt-Muggier were famous scientists in their own right, specializing in neurology. Oskar was eminent enough to be one of the scientists who examined Lenin's brain after his death in 1924. As he sympathized and socialized with Jewish people, the Nazi Party in Germany forced him from his position as the director of the Kaiser Wilhelm Institute in 1937, the same year Marguerite obtained her MD from the University of Berlin. After some time in Paris, she worked on fru it fly development at her father's institute in Neustadt in the Black Forest.[30]

Vogt came to the United States in 1950 (figure 3.2). She initially worked as a research fellow with Max Delbrück, but he encouraged her to start work-

3.2. Marguerite Vogt and Curt Stern at Cold Spring Harbor Laboratory in 1950. Image courtesy of Cold Spring Harbor Laboratory Library and Archives

ing on poliovirus with Dulbecco. Together Dulbecco and Vogt developed a plaque system for poliovirus using monkey kidney cells.[31] By June 1953, they perfected the new system and submitted a paper to the *Journal of Experimental Medicine*.[32] Monkey kidney cells worked best, but they also considered using human HeLa cells. Recovery of virus from a plaque provided a new method of purifying poliovirus. Dulbecco and Vogt determined that a plaque was started by a single poliovirus particle infecting a single cell, as was the case with western equine encephalitis virus. With this knowledge, they isolated genetically pure strains of the three types of poliovirus. The virus progeny of each strain had identical properties to the parent virus. This result would prove crucial for Sabin in developing his vaccine for polio. The assay itself transformed animal virology, as many virologists adopted the new quantitative approach. The 1954 paper by Dulbecco and Vogt has been cited more than 3,800 times in the biological literature, one measure of the influence of the assay,[33] which had been developed by replicating the plaque-based assays used by bacteriophage biologists like Luria and Delbrück.

Dulbecco and Vogt found many similarities in the growth characteristics of bacterial and animal viruses, but there were some differences. Infected bacterial cells would be burst by a batch of 200 or more bacterial viruses, or bacteriophages; infected animal cells would not be burst by the virus but would continue to release new virus particles for seven or eight hours, up to 1,000 particles per cell. Dulbecco and Vogt surmised that this difference was due to the different types of cell wall—bacterial cell walls are more rigid than animal cell walls.[34]

With the new system in place, Dulbecco and Vogt studied the growth characteristics of poliovirus. By 1954, they discovered a plaque type variant on type 1 poliovirus that had significantly less pathogenicity in the monkey nervous system. The variant formed plaques sooner and with a sharper rim, and so could be identified using the assay alone. This discovery allowed them to begin a genetic analysis of poliovirus using pure strains once various properties of the two strains were worked out. Vogt and Dulbecco collaborated with Albert Sabin, who sent them multiple strains of poliovirus that they tested under various conditions, including low pH, to find correlations between neuropathogenicity and properties with the potential to form plaques.[35] The quantitative assay and Dulbecco's approach were beginning to bear fruit.

In 1954, Dulbecco sought more funding from the NFIP, framing his work explicitly as bringing the bacteriophage approach to poliovirus. The purpose of the grant was "to study the biological properties of poliomyelitis viruses along lines similar to those followed in bacteriophage work."[36] Dulbecco applied for funds for a salary for Vogt, who was now a senior research fellow, for two technical assistants and for a centrifuge for his 500-square-foot laboratory. Monkeys cost $35 each, and he estimated he would need 500. NFIP awarded him $32,327 for a year of research. Vogt and Dulbecco began work on recombination between different strains of type 1 poliovirus, including new mutant strains that were heat resistant. They also investigated the neutralization of viruses by antibodies.

At an important meeting in 1956 at the Ciba Foundation in London, Dulbecco presented his new approach to a small group of the best virologists in the world. James Watson, Francis Crick, Rosalind Franklin, and Michael Stoker, among other luminaries, listened to Dulbecco present the virtues of the plaque system he and Vogt had created (see figure 4.1). At Berkeley and Tübingen, Germany, it had been established that RNA was the genetic material for tobacco mosaic virus. Dulbecco argued that having a system that used RNA as opposed to DNA viruses was useful. Citing his ability to obtain genetically pure virus particles using plaque isolation, Dulbecco concluded,

"Poliomyelitis viruses and Newcastle disease virus [an RNA virus that afflicts birds] seem therefore to constitute the best material for a genetico-chemical study of the system constituted by an RNA virus and its host cell."[37] The experiments performed by the Dulbecco lab suggested that the viral RNA behave as a "haploid genetic system," where the information in the RNA is independent from that of the cell. Nonetheless, considerations of ultraviolet radiation sensitivity suggested that "virus components and cellular components become integrated for virus multiplication. As soon as this integration has become established, the whole cell may be considered as a single virus producing unit." In a few short years, Dulbecco had emerged as one of the leaders of experimental animal virology. He began to attract scientists to in his lab. For the history of tumor virology, the appearance of a young, experimentally inclined veterinarian in Dulbecco's lab was important.

Harry Rubin Joins Dulbecco's Group

Harry Rubin went to veterinary school at Cornell, graduating in 1948. He spent a year in Mexico studying foot and mouth disease and then worked for the CDC Public Health Service in Montgomery, Alabama, where he was concerned with zoonoses, diseases that could be transmitted from animals to humans—mostly, rabies and eastern equine encephalitis. He discovered a new virus in the blood of wild snowy egrets that he thought was related to a human pneumonitis outbreak in 1943.[38] After a few years in Alabama, Rubin decided that he wanted to do basic research that was beyond epidemiology and applied for a fellowship to work with Wendell Stanley at University of California Berkeley.

Rubin became interested in tumor viruses at Berkeley after taking a virology class. He had played around with Shope papilloma virus but was considering a switch to Rous sarcoma virus. After a year he decided to go to Dulbecco's lab at Caltech to learn cell culture techniques, as no one knew them at Berkeley. In 1953, before he went to Caltech, he visited Peyton Rous at Rockefeller, as well as Ray Bryan and Francisco Duran-Reynals, who were also on the East Coast.[39] Rous gave Rubin two vials of RSV, but Rubin preferred to use a "hyped up" strain of RSV from Ray Bryan, called the "high tire" strain.[40] Rubin moved to Caltech to work with Dulbecco, but Dulbecco was working on poliovirus in his off-campus laboratory, the Huntington Lab, where he spent most of his time. Still, the two men spoke relatively often.

Rubin tried to develop a cell culture assay for tumor viruses by modifying Dulbecco and Vogt's assay. His attempts were unsuccessful, so he turned to chorioallantoic membrane of the chick embryo, which had been shown to produce small countable tumors. Unfortunately, there were a lot of fluctuations from chick embryo to chick embryo. In 1956, Rubin reviewed an article in *Virology* on culture by Robert A. Manaker and Vincent Groupé[41] and decided to return his research to Rous sarcoma virus after working on western equine encephalitis virus and Newcastle virus. His decision was driven by a new graduate student who frequently came down to get his embryos from Dulbecco's lab. In 1956, Rubin found Howard Temin "young and persistent," and Temin "essentially talked him into working on Rous sarcoma virus."[42] Temin worked on RSV under Rubin's direction. Among other properties of the virus, Temin was interested in the one strain of RSV from Rous that was thought to be able to transform chicken fibroblasts into bone cells. Rubin promised Temin could work on this strain after the assay was developed. This transformation was related to Temin's interests in developmental biology. (Unfortunately, they later found the 35-year-old virus preparations that Rous gave him were no longer viable.) One open question was why Rubin's earlier attempts at making an assay failed. Rubin had not used tryptose phosphate as Manaker and Groupé had in 1956. This proved to be a crucial difference. It is still not known why adding tryptose phosphate helped; one possibility is that small peptides are slightly toxic to cells and make them slightly sicker when infected, thus helping virologists to visualize the plaques of transformed cells.

Under Rubin's guidance, Temin lowered the concentration of cells so that they were still growing when the virus was added.[43] The cells were then left for a week to grow before being killed by fixing. The assay worked (figure 3.3).Temin put it this way in his Nobel Lecture:

> I soon found that addition of RSV to cultures of chicken embryo cells in a sparse layer, rather than in a crowded monolayer as then used for the assay of other animal viruses, led to the appearance of foci of transformed cells. The number of these foci was proportional to the concentration of virus, and the foci resulted from altered morphology and altered control of multiplication of the infected cells. The foci were cell culture analogs of tumors in chickens.

With the assay in place, Rubin and Temin worked to quantify the infectious properties of RSV. In the beginning, Rubin did not know that RSV was an RNA virus. In 1955, he had published an analogy with bacteriophage

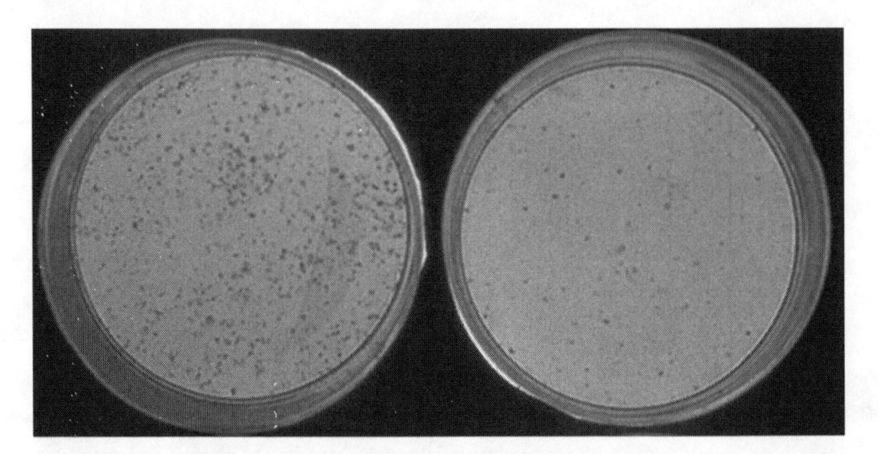

3.3. The focus assay of Temin and Rubin. Each dot reflects a transformed cell. The dish on the right is a 100-fold dilution of virus preparation. From Robin A. Weiss and Peter K. Vogt, "100 Years of Rous Sarcoma Virus," *Journal of Experimental Medicine* 208, no. 12 (2011): 2351–2355.

lysogeny that was his model for RSV, but at that time he mistakenly thought it was a DNA virus.[44] Others performed experiments on irradiating the virus that got results similar to bacteriophage experiments. But when evidence for RSV having an RNA genome emerged, Rubin backed of his lysogeny model.[45] Temin did not share Rubin's worries about the analogy between bacterial and avian viruses and doubled down on the lysogeny model for RSV, calling it the "provirus hypothesis," that is, that the nucleic acid of the virus integrated into the cellular genome of the host cell.[46] (See chapter 8 for more detail.)

Avian Leukosis Virus Discovered in the Virus Stock

One of the strengths of the assay for transformation by RSV is that it is normally reproducible. Occasionally, however, Rubin noticed that the number of foci was 100-fold fewer than expected. This reduction appeared to be caused by something in batches of the cells used to make the monolayer. The agent that caused this reduction was named resistance-inducing factor (RIF) and proved to be an avian leukosis virus (ALV). Using techniques available in 1961, RIF was morphologically indistinguishable from RSV but antigenically different; in other words, the viruses looked the same in the electron microscope, but antibodies could tell the difference. Interestingly, however, RIF was antigenically indistinguishable from another virus called RPL12, which afflicted chicken flocks. Testing for RIF, then, could be used to test for infections in chickens. Rubin also isolated a second virus from the ALV family that he called Rous associated virus (RAV)—this virus was antigenically indistinguishable from RSV but did not cause tumors. Instead it caused fatal erythroblastosis when injected into chickens. Rubin asked his new postdoctoral researcher, Hidesaburo Hanafusa, to try to make RSV isolates free of RAV, but the Japanese scientist wound up discovering that the RSV strain actually needed RAV to make new infectious RSV particles.[47]

Rubin studied these ALVs in his lab. Avian leukosis viruses can be transmitted through the egg, taking nine months to show up in the young chicken. Surprisingly, an infected rooster did not transmit it congenitally to the next generation. If the virus were in the DNA, then Rubin would expect transmission from fathers as well as mothers. The lack of such transmission deepened Rubin's doubts about Temin's lysogeny model of the RSV lifecycle.

In 1958, Rubin moved to UC Berkeley to work on leukosis viruses, moving his research focus to the regulation of cellular growth and the loss of regulation in the transformed cell. He found that magnesium was an important factor in regulating growth.

In the 1960s, Berkeley a hotbed of the anti-war protests, and Rubin played an active part in the movement. Eventually Rubin's scientific interests drifted away from tumor virology to cell biology, or perhaps the field moved away from him. The field was becoming part of molecular biology, and he thought of himself as a cell biologist, not a biochemist or molecular biologist.[48]

Dulbecco Moves into Tumor Virology

Like Rubin, who specialized in the biology of RSV to investigate tumor development, Dulbecco chose a tumor virus to study. This narrowing of focus from animal virology more generally to tumor virology is perhaps the most profound effect Rubin had on Dulbecco.[49] In his case, Dulbecco chose polyoma virus, first discovered by Gross and further characterized and renamed by Stewart and Eddy (see chapter 2). He procured samples of the virus from Bethesda and collaborated with John Smith in Cambridge to confirm polyoma virus was a DNA virus. Dulbecco and Vogt had to learn how to grow the virus. Vogt worked to find the best type of cells to use in culture, discovering that hamster cells best yielded identifiable plaques.

Dulbecco's first publication that involved adapting a plaque assay for polyoma virus came out in 1959. Inspired by James Watson's work on the two forms of DNA in Shope papilloma virus, Dulbecco and Vogt began to investigate the DNA from polyoma virus. Dulbecco was invited to present on viral DNA at a meeting in Houston, Texas, and he was keen to present something new to the audience. He enlisted his technician Maureen Muir to set up a Hershey column[50] used to fractionate nucleic acids to understand why there could be two distinct bands of polyoma DNA sedimenting differently.[51] It appeared that the two bands of viral DNA were different shapes—one circular and the other linear—and that the enzyme DNase (deoxyribonuclease) could convert the DNA from one form to the other. Dulbecco tried to enlist a Caltech microscopist to look at the DNA in the electron microscope, but the microscopist was having some personal problems and never got around to doing the work. Later, Jerome Vinograd looked at the DNA under an electron microscope and saw that the two forms were not linear and circular but rather circular and twisted. The cutting enzyme DNase cut only one strand of the double-stranded DNA—cutting both strands would turn a circle of DNA into a linear stand—but the single cut made it easier for the circle of DNA to twist up like a rubber band, in a process called supercoiling.

For their invention of quantitative virology made possible by the plaque assay and transformation assay, and in recognition of several of the discoveries that flowed from this new approach, Rubin and Dulbecco won the Lasker Award in 1964. Rubin was recognized for his discovery that RSV preparations actually consisted of two viruses, RSV and the helper virus RAV. Dulbecco was singled out for showing that polyoma DNA came in both a ring and linear form, both of which could cause cancer, with the ring form more active.[52]

The same year Rubin won the Lasker Award he became a professor of molecular biology at Berkeley. He would remain there for the remainder of his career, continuing his work on cellular transformation. He was elected to the National Academy of Sciences in 1978. In later years he was funded by the Council for Tobacco Research, publishing papers in *PNAS* on cellular transformation processes.[53] He passed away in 2020 at the age of 93.

By 1964, Dulbecco and Vogt moved to the Salk Institute, and Rubin had a position at UC Berkeley. Earlier, Jacques Monod had visited Dulbecco to tell him about Jonas Salk's plan to create an institute of biology with no teaching duties and plentiful research funds.[54] Salk wanted to recruit five or six of the best scientists in the world to start it, and Dulbecco was one. Dulbecco accepted this exciting proposition, in part because it offered him a change of scene and a chance to escape his foundering marriage, which eventually ended in divorce. He decided to go to the new Salk Institute in 1962, but his laboratory was not yet completed, so in the interim he spent a year with his friend Michael Stoker in Glasgow (see chapter 4). That Dulbecco had fallen in love with a Scottish woman, Maureen Muir, whom he met at Caltech, also played a factor in his decision to take a Salk-funded sabbatical in Scotland. She would become his second wife. While in Glasgow, Dulbecco discovered that polyoma virus infections cause the host cell to make more DNA by stimulating a particular enzyme (thymidine kinase).[55] After Glasgow, Dulbecco moved to the Salk and continued the research program on viral DNA that he had started at Caltech.

Vogt moved with Dulbecco to the Salk Institute, where she continued in her role as a senior research scientist. While Dulbecco was in Scotland, Vogt mentored a new postdoc named Lee Hartwell, who would later win the Nobel Prize for studies on the cell cycle. In a memorial after her death, Hartwell wrote fondly of Vogt and the tremendous intellectual impact that she had on him:

> I have had many mentors during my career. I learned different things from different people—to work on big problems, to have high standards, to not take oneself too seriously. From Marguerite, I learned the joy of being passionately lost in ideas. Marguerite made a habit of latching onto the latest postdoc to join the Dulbecco laboratory as a way of keeping herself abreast of technology developments. I was particularly lucky because I came to Renato [Dulbecco]'s laboratory at the time it moved to the Salk Institute but before the buildings were finished. Consequently, there was room for only two postdocs during most of my time there and I benefited by getting most of Marguerite's attention. We used to meet in La Jolla for tea and crumpets early in the morning. While Marguerite had many interests, her passion was ideas. It seemed like each new paper she read generated another theory about the origins of cancer. . . . Her inspiration was formative, and I have never ceased to find ideas the greatest reward in the life of science.[56]

Vogt did not share in Dulbecco's growing fame, partly because of the sexism in science in the 1960s and 1970s, although she never protested. She said that what was important was the science itself. She told the *New York Times* in 2001, "I'm happy not to have been bothered. . . . When you get too famous, you stop being able to work." And she did work, usually 12 hours a day, six days a week. Nonetheless, the different experiences she and Dulbecco had as European MDs making scientific careers in the United States illustrates how much more difficult it was for a woman to advance in the scientific community at mid-century.

The most important result that came from Dulbecco's laboratory in the late 1960s was Joseph Sambrook's demonstration that viral SV40 DNA can integrate into the host genome. (SV40 virus is closely related to polyoma virus.) Finding that the DNA of SV40 was circular was a "good inducement" to explore the hypothesis that it integrates since one possible mechanism of integration involved circular viral DNA joining with host DNA and linearizing. Sambrook was recruited by James Watson to run a new tumor virology lab at Cold Spring Harbor, which in many respects can be thought of as a continuation and extension of the research program established by Dulbecco and Vogt.

Dulbecco spent eight years at Salk before taking an extended five-year sabbatical at the Imperial Cancer Research Fund in London. Among other things, he was not happy with the political situation in the United States. Dulbecco's friend Michael Stoker was there and arranged a position for him. In 1975 Dulbecco was awarded the Nobel Prize with Howard Temin and David Baltimore. "Then a fit of enthusiasm caught me, and I decided that I should work more directly for mankind: namely, to work on cancer—not cancer virus, but cancer." He chose breast cancer partly because of personal experience: his friend and fellow biologist Seymour Benzer had lost his wife to breast cancer. Feeling better about the general political climate in the United States, Dulbecco returned to the Salk Institute in 1977, bringing his family with him. He served as the institute's president from 1988 to 1992. In 1986, Dulbecco was among the first to call for the sequencing of the human genome. In his later years, he used his fame as a Nobel Prize winner to promote science in Italy, publishing books in Italian. He died in 2012 at the age of 99.

Vogt worked at the Salk for decades, becoming the institute's oldest working scientist and publishing her last paper in 1998. She taught many visitors to the Dulbecco lab tissue culture procedures and was known as a generous person who helped many students financially. She was politically active and, like Harry Rubin, protested against the Vietnam War. In 1973, with Dulbecco in the UK, Vogt became a research fellow and established her own laboratory. The new title allowed her to apply for research grants. The 1964 paper on the poliovirus assay she coauthored with Dulbecco is her (and his) most cited paper, with more than 3,800 citations. She died in 2007 at the age of 94.

4

Cell Lines and Cat Leukemia

Michael Stoker, Bill Jarrett, and the Early Fruit of the Glasgow Institute of Virology

Michael George Parke Stoker (1918–2013) was born into an Irish Protestant medical family. His father was a doctor, and his great-uncle was the surgeon on the famous explorer Henry Morton Stanley's second expedition to Africa in 1886–89.[1] Stanley himself was godfather to Michael Stoker's father, who joined the Royal Army Medical Corps and served in the trenches of World War I, earning the Military Cross. When Michael Stoker was 15 or 16, his often-uncommunicative father asked him what he was going to do after school by giving him three options: "Well, there is the church, the army, and medicine." Ruling out the church and the army, Stoker chose medicine. He studied at Sidney Sussex College at the University of Cambridge, beginning in 1936. After leaving Cambridge, he began clinical training at St Thomas' Hospital in London, which was interrupted by the blitz of World War II but completed by 1942.

In November 1943, Stoker was drafted overseas with the Royal Army Medical Corps. He left behind a new bride, now pregnant, whom he would not see again for three years. Although his ship was bombed—unsuccessfully—on its voyage through the Mediterranean Sea and the Suez Canal, he finally reached India, where he was posted to a movable field hospital in Lahore. During the year, the hospital was moved by camel and train to Hyderabad in central India. While in this clinical setting, Stoker came to see the need for proper laboratory investigation of tropical diseases. His military responsibilities, however, prevented him from pursuing that type of research. He had orders to join the Chindits, a special force devoted to guerrilla warfare behind enemy lines, to help with a secret operation that involved being dropped in gliders into southern Burma (now Myanmar). To his relief, the operation was canceled because of the rapid retreat of the Japanese, and Stoker was instead sent to Poona (now Pune) for a laboratory course, a much more attractive option. There he met experimentally trained medical doctors including William (Bill) Hayes, who would become a leader in bacterial genetics and "sex" in bacteria, publishing *The Genetics of Bacteria and Their Viruses* in 1964. Stoker performed excellently in the course and was consequently offered a position on the staff. His first publication was an article in *Nature*, coauthored with Douglas Black, on the level of plasma iron levels in babies born to anemic mothers.[2] Before leaving the army, he moved to the Typhus Research Unit, an assignment that led his career into virology and cell biology.

Stoker returned to England in 1946 to reunite with his wife and meet his son for the first time. After demobilization in 1947, he joined the Department of Pathology at the University of Cambridge and began to work on viruses with John Miles, the incumbent virologist. Late 1940s virology was largely driven by advances made on two fronts; first with bacteriophages by "the phage group" led by Max Delbrück and Salvador Luria; and second with plant viruses by Wendell Stanley and his colleagues in the United States and by F. C. Bawden, Norman Pirie, Kenneth Smith, and Roy Markham in Cambridge. Stoker was also attracted to the emerging new approach to biology exemplified by the new electron microscope installed by Vernon Cosslett, and the work of the Cavendish Laboratory group—John Kendrew, Max Perutz, Francis Crick, and James Watson—who were pursuing the structure of hemoglobin, myoglobin, and later DNA using x-ray crystallography.[3] These people, he would later note, "had a great influence on me and we have remained friends for most of my life."[4] Not surprisingly, he was one of the first to be shown Watson and Crick's famous model of DNA based on Rosalind Franklin's data in 1953.

In addition to taking a role in the emerging discipline of molecular biology, Stoker worked on Q fever, a bacterial disease that afflicts humans, cows, goats, and sheep, then he switched to herpes virus in 1954. He showed that herpes virus has an "eclipse phase" much like bacteriophage and thus does not divide by binary fusion like bacteria, as some still believed. The eclipse phase is the temporal gap between the virus infecting the host cell and the creation of new viruses by the cell. Collaborating with Robert Horne, who had expertise with the electron microscope, Stoker was able to accurately visualize the substructure of a variety of viruses for the first time. They saw

4.1 Michael Stoker, Francis Crick, Milton Salton, and James Watson (*left to right*) at the 1956 CIBA symposium on viruses in London. Photo in the author's collection

regular bumps on the viruses' surfaces, now called capsomeres. Other viruses were enclosed in a lipid envelope, which was also now visible. Virology was edging into its molecular phase.

Stoker was elected a fellow of Clare College and later become a tutor for medical students. When James Watson returned to Cambridge to work with Crick in 1955–56, Stoker taught him how to grow cancer cells, and the two men cemented their friendship in the laboratory and at scientific meetings (figure 4.1). Peter Wildy, an expert with herpes virus who had trained in Macfarlane Burnet's laboratory in Melbourne, joined Stoker in Cambridge. Wildy would leave Cambridge and move north with Stoker to establish an important center for virus research in Scotland.

The Institute of Virology at Glasgow

During this period, the first chair in virology in the UK was being established at the University of Glasgow. The holder would lead the new Institute of Virology funded by the Medical Research Council (MRC) and housed in a building designed by the eminent architect Sir Basil Spence. (Among other structures, Spence designed the New Zealand government building known as the "Beehive.") Stoker's scientific and administrative talent did not go unnoticed, and he was offered the position without having to apply. Complicating matters, the University of Edinburgh also offered him a position at the same time. In March 1957, Stoker wrote to Watson for advice: "strictly on the scientific issue," he asked his friend, what were the relative merits of each institution? "You ought to be highly flattered," Stoker wrote, "that I should approach you—an American—about two British Universities."[5] In reply, Watson pointed out that the geneticist Guido Pontecorvo (known as Ponte), who was at Glasgow, was a fun and stimulating person to be around.[6] In the end, Stoker decided to go there, bringing along Peter Wildy and recruiting other talented scientists as well: Ian Macpherson, Kenny Fraser, and Lionel Crawford, among others. The following nine years at Glasgow would be by his own assessment "the most productive period" of his life. The pages of the new specialist journal *Virology*, founded by George Hirst, Salvador Luria, and Lindsay Black in 1955, often featured the research of the Glasgow group.[7]

Peter Wildy and others continued to work on herpes virus while Stoker and Ian Macpherson turned to the cancer-causing polyoma virus. (See chapter 2 for the discovery of polyoma virus.) Stoker wanted to use advancements in the molecular biology to transform animal virology, as the phage group had transformed biologists' understanding of bacterial viruses.[8] Virology was on the leading edge of the growing field of molecular biology: for example, virologists had shown that DNA was the genetic material and that phage genomes could fuse with host genomes in a process called lysogeny, among other exciting discoveries.

Stoker's group was the first to see the substructure of polyoma virus using electron microscopy and newly invented "negative staining" (figure 4.2).[9] Earlier electron microscopists could see the rough shape of viruses but not any internal structure. Stoker wrote to Marguerite Vogt, the longtime collaborator of Renato Dulbecco, "The polyoma virus structure is proving interesting. It is one of the series with hollow polygonal subunits of two types (5-sided and 6 sided). It has 12 pentagonal and 20 hexagonal prisms. There must be some sort of coding for these two types presumably—though each is probably made of several protein molecules (? 5 & 6 respectively). Even if the proteins are identical there must be two types of condensation system or such like for assembly."[10] Here Stoker drew on the theoretical speculation of Francis Crick and James Watson, who had hypothesized that "spherical" viruses actually possessed icosahedral symmetry in virtue of being built though a regular assembly of identical protein subunits. The x-ray crystallographers were also investigating virus structure and would build upon these theoretical insights.[11]

In addition to the structural studies, Stoker aimed to emulate Temin and

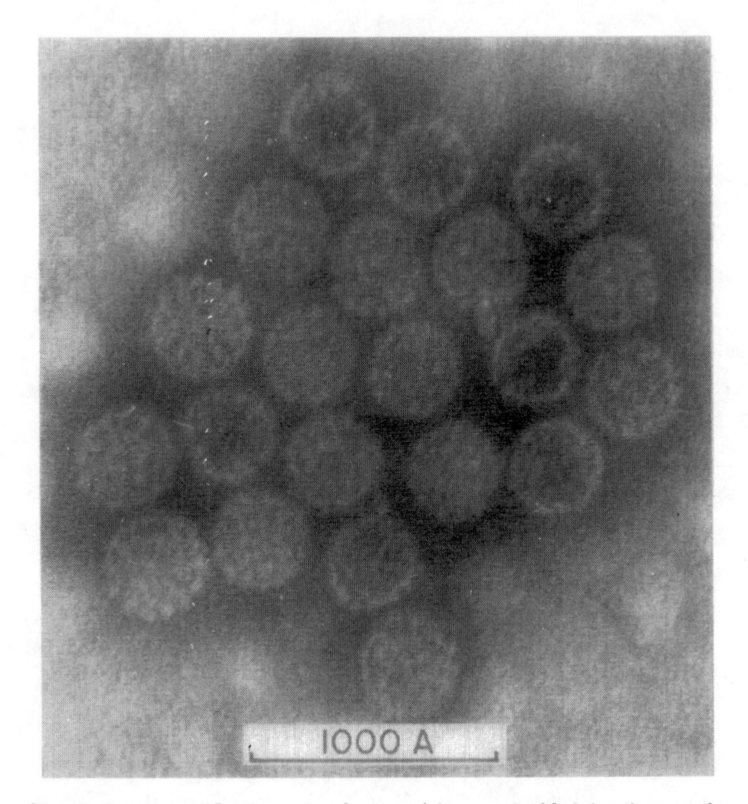

FIG. 1. Low magnification of polyoma virions embedded in electron-dense phosphotungstate; note the central region of some of the particles, filled with phosphotungstate. The particles appear to be spherical with surface projections. Magnification: × 360,000.

4.2. Electron micrograph of polyoma virus in 1960 taken by Stoker and his colleagues. From P. Wildy et al., "The Fine Structure of Polyoma Virus," *Virology* 11, no. 2 (1960): 444–457.

Dulbecco's success by creating a quantitative assay to measure transformation of normal cells into cancerous cells by polyoma virus. By September 1960, he could say, "Our transformed line of hamster cells has an easily recognized colony different from normal, and we are trying to use this to get the thing onto a quantitative basis. No luck so far, but hope prevails."[12]

One of the challenges of creating a new animal virology fashioned in the same tradition as the phage group was the production of standardized host cells. A relative advantage of working with bacterial viruses—phages—is that the host cells are bacteria, such as the widely used *E. coli*, that are easy to grow and clone. To replicate, animal viruses needed animal cells, which are more difficult to grow and standardize outside of animals. Stoker chose to work on polyoma virus because it had a short incubation period and was highly efficient at inducing tumors in hamsters. Its cytopathogenic effect on mouse cells could be used to make plaque assays, as Dulbecco showed, just as bacteriophage workers had done with phage.

Stoker had spent six months in Dulbecco's Caltech lab in 1958 and was clearly aware of the trend toward working with cell cultures instead of whole animals. By early 1960, he would write to the Dulbecco lab, "The plaque assay is fairly reliable so we can go ahead with genetic studies."[13] Plaques could be used to measure accurately the number of infectious viruses. Nonetheless, some of Stoker's early polyoma work was still with hamsters themselves. He showed that the Toronto strain of polyoma virus, which was highly pathogenic, would cause fatal kidney and liver cancer in two weeks. By injecting genetically pure samples of polyoma virus at decreasing concentrations into hamsters, Stoker inferred that the virus could cause cancer at multiple sites including the heart, peritoneum, lungs, and testes. Relatedly, he showed that the potential to cause cancer in many tissues was inherent in a single particle.[14]

Further developments would depend on more sophisticated in vitro work and the development of cell tissue lines that could be manipulated in vitro. Developing new in vitro assays with new immortal cell lines was the goal. In working toward this goal, Stoker saw himself as part of a long tradition in tissue culture. Such research is classically thought to begin in 1912 when Rockefeller Institute biologist Alexis Carrel, who won the Nobel Prize the same year, was able to grow chick heart tissue by adding embryo extract to the tissue.[15] The cultures could be repeatedly split in two and maintained much longer than the lifespan of a chicken.[16] In fact, he reported that he kept the culture growing for 20 years, although some biologists suspected it was not the same culture. Over the next three decades, tissue culture techniques were developed in what became Strangeways Laboratory in Cambridge, but they were used in only a limited number of laboratories around the world and usually only with short-term cultures. (Long-term cultures were more difficult in part because they were harder to keep sterile.) In an important result published in 1948, K. K. Sanford, W. R. Earle, and G. D. Likely, a team from the National Cancer Institute, demonstrated that it was possible to grow a population of cells from a single ancestral cell and thus grow a clonal population.[17] Such a population would be genetically pure, rather than a mixture of different cell types. The key to growing a population from a single cell was the surrounding fluid and nutritive compounds, or media. If the media were conditioned by prior growth of cells and were contained in very small volumes, then single cells could proliferate. In the mid-1950s, this conditioning was done by using irradiated "feeder cells" that could not reproduce but could release the necessary metabolites into the medium to allow for the growth of nonirradiated cells. Wildy and Stoker used microdrops of media under oil, but these techniques were cumbersome and varied in their efficiency.

In the 1950s, investigators at Johns Hopkins University in Baltimore, notably Gey (mentioned earlier), cultured the first and perhaps most famous immortal human cell line, called HeLa, after Henrietta Lacks, the woman from whom the cervical carcinoma cells had been extracted without her consent.[18] HeLa cells, however, were less useful to Stoker, as he needed noncancerous cells to study how tumor viruses transform healthy, normal cells into cancer cells. Other developments in the 1950s were more useful. In the mid-1950s, Harry Eagle at the National Institutes of Health methodically analyzed the nutritional needs of HeLa cells and mouse fibroblasts to determine which chemical compounds were necessary to keep these cells alive in a minimal synthetic media.[19] He established which vitamins, amino acids, sugars, and salts growing animal cells needed: 13 amino acids, in addition to other nutrients. Despite cells still needing something extra from mammalian blood serum to grow, the resulting mixture, called Eagle's medium, was widely used by animal virologists and cell biologists.[20]

The Creation of the BHK21 Cell Line

Perhaps the most important development of Stoker's team at Glasgow was the development of the BHK21 clone 13 cell line. Derived from baby hamster kidney cells, these cells were not cancerous but could be transformed to malignancy by polyoma virus. The importance of good cell lines is perhaps underappreciated—they are necessary for an effective virus-host cell experimental system. Stoker and Macpherson built on the work of Vogt and Dulbecco in 1960, and Leo Sachs and Dan Medina in 1961, who found that polyoma virus could transform mouse and hamster cells in vitro.[21] One difficulty was that fresh preparations of hamster and mouse cells contained a mixture of different types of cells. After two months of continuous cultivation of baby hamster kidney cells, Macpherson and Stoker noticed that cells from litter 21 were growing abnormally fast.[22] By repeatedly splitting the culture, these cells could be given fresh medium and grown indefinitely—essentially, they were "immortalized." They were composed of cells of one type, fibroblasts that grew in oriented parallel bundles, and had a high plating efficiency: up to 70% would form colonies. In the cells from other litters only 2% would. Although this cell line was "immortal," it was closer to "normal" cells than the HeLa cell line. BHK21 cells also had the benefit of mostly not being killed by the polyoma virus. With the BHK21 cell line, Stoker and his group showed that the transformation of an individual hamster cell is induced by a single polyoma virus particle. He was generous with this new tool, sending the cell line to others, including Dulbecco,[23] before he published his discovery. He proclaimed the BHK21 cell line "God's gift to virologists."[24] His chance to present these findings to the world's leading virologists en masse occurred in 1962 when Cold Spring Harbor Laboratory devoted the annual symposium to virology. Dulbecco provided a paper on the properties of cells infected by polyoma virus.

In addition to being a respected scientist, Stoker was an extremely capable administrator. Under his direction, the Institute of Virology at Glasgow became a world-class center. He recruited scientific talent. He organized meetings including a small informal conference at the Burn, ("burn" means stream or river in Scottish English) a study center in the countryside in the northeast of Scotland, where many leading biologists attended including André Lwoff, Salvador Luria, Francis Crick, and Leo Sachs. For younger scientists, it was a chance to talk to Nobel laureates, past and future, and get expert feedback on their speculative ideas.[25] By design, there was plenty of time to walk around the surrounding countryside and fish in the river. Lwoff and Stoker took time to indulge in their passion for landscape painting. Dulbecco went fly fishing. In addition to meetings, prominent researchers visited the institute, many spending extended time there including Renato Dulbecco, Luc Montagnier, and George Hirst. Several of the staff of the Institute of Virology became leading scientists in their own right. Eventually, most would leave Glasgow to spread knowledge of molecular virology across the UK and beyond. Throughout his time in Glasgow, the Scots saw Stoker as an "English gentleman." He was generous and built bridges with fellow scientists.

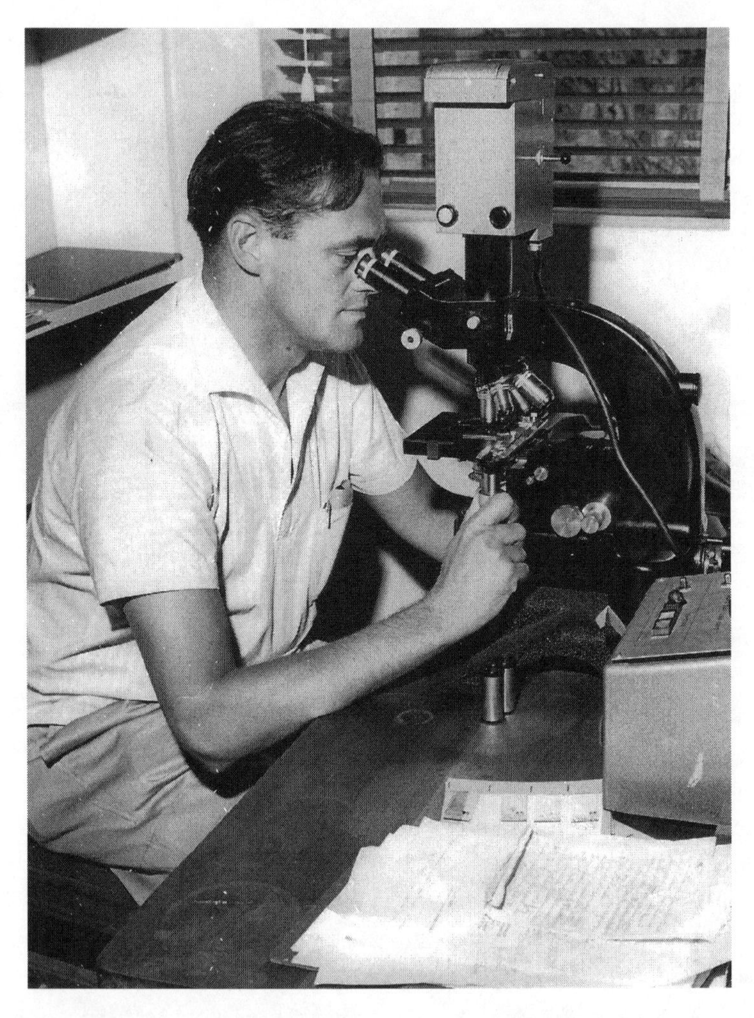

4.3. Bill Jarrett and microscope. Image courtesy of Os Jarrett

The Work of Veterinarian Bill Jarrett

One of the most important bridges Stoker built was to the University of Glasgow Veterinary School, where one of the leading veterinary pathologists of his generation, Bill Jarrett (1928–2011) was also interested in virology and particularly tumor virology (figure 4.3). In 1961 Jarrett and his collaborator Bill Martin obtained funding from the British Empire Cancer Campaign to investigate viruses as the cause of leukemia in animals. Stoker provided laboratory space in the Institute of Virology. Jarrett and Martin searched for a leukemia-causing virus in cattle, culturing cells from lymphomas in the hope of releasing and identifying a virus from the cells. They were unsuccessful because, as they found in subsequent epidemiological studies, the type of lymphomas caused by a virus (enzootic bovine leukosis) did not occur in the United Kingdom at that time. A bovine leukemia virus was discovered in the United States in 1969 and was introduced later to the UK via imported cattle from Canada where it was endemic.

In contrast with the cattle work, Jarrett and his colleagues had spectacular success with cats. Noticing that the incidence of leukemia was much higher in cats than it was in humans, Jarrett asked local veterinarians to contact him if they found cats with lymphoma he could study. A local vet, Harry Pfaff, alerted Jarrett to a cluster of cases of lymphoma in the residence of a local "cat lady" who had "rescued" more than 100 cats. Jarrett took material from the lymphomas, removed all the cellular material, and injected the cell-free suspension into four newborn kittens in a transmission experiment. He waited on tenterhooks to see whether the kittens developed lymphomas. All four developed lymphomas in 9–18 months. Jarrett asked Elizabeth Crawford, the electron microscopist at the Institute of Virology to examine sections of the tumor. Crawford and her husband Lionel had spent time in Caltech in Renato Dulbecco's lab, and she was acquainted with viruses, having worked with Rous sarcoma virus. The tumor cells from a kitten were examined, and a large number of virus particles were identified in the electron micrographs. The similarity between these virus particles and those already known to occur in leukemia in domestic poultry and some strains of laboratory mice was "striking."[26] The virus was subsequently named feline leukemia virus (FeLV).

A Fast Test for Feline Leukemia Virus

Jarrett's group subsequently published a paper showing that FeLV was detectable by electron microscopy in the white blood cells and platelets of infected cats. William (Bill) Hardy Jr., a young veterinarian training with Lloyd Old (1933–2011) at Memorial Sloan Kettering Cancer Center (MSKCC) in New York City, read this paper and wondered whether he could develop a rapid blood test to detect infection, using immunofluorescence to demonstrate FeLV antigen in the white blood cells in blood smears from the cats (figure 4.4). He proceeded to learn the technique of immunofluorescence from his colleague Yashar Hirshaut, who was using the method to detect antigens

4.4. Bill Hardy at the ASPCA in the 1970s. Image courtesy of Bill Hardy

in cancer cells. Hardy then isolated the virus, grew it up, and immunized a rabbit to obtain antibodies for his test. The immunofluorescent antibody test was vivid, staining FeLV infected blood cells bright green. Hardy quickly presented the test at the Fifth International Symposium on Comparative Leukemia Research meeting at Padua, Italy, in 1971. As most biologists at this time did not think about patenting their discoveries, there was no thought to patent the test, although it would have brought a significant amount of money to him and the MSKCC.

With the ASPCA hospital, Hardy and his colleagues explored the epidemiology of the virus as cats progressed in their disease. He examined samples taken from cats living in high-rise buildings in Manhattan and free-roaming cats from the New Jersey suburbs. About 3%–4% of cats were infected with the virus, which was easily transmitted through their saliva. Like Bill Jarrett, Hardy studied the catteries of "cat ladies," one of whom was living with more than 100 in her fifth-floor walkup. She would not let the scientists into her apartment but would bring every one of her cats to be tested over a five- to seven-year period. Hardy tested every one of her cats, many multiple times. He saw how the virus was transmitted from cat to cat in her apartment. Although she was initially skeptical of this theory, the cat owner eventually separated cats that tested positive from the others. His work was reported in *Time* magazine as well as academic journals.[27]

It was not just cat owners who were skeptical of a viral cause of leukemia. Some veterinarians did not accept that leukemic cats could be infectious and would not believe that a virus could transmit the disease from cat to cat. But Hardy was not deterred. He first recommended that cats that tested positive for FeLV should be euthanized, as it was unclear whether FeLV could potentially cause leukemia in humans. Subsequent studies indicated that the virus could not affect humans, so infected cats were then quarantined, not killed. Separating infected cats from susceptible cats made it possible to break the cycle of infection, and Hardy's method became widely used. In 1972, Hardy started a private testing facility, the National Veterinary Laboratory in Franklin Lakes, New Jersey, to process FeLV tests in addition to his lab at MSKCC. Similar diagnostic facilities were established throughout the world. From the results of the Jarrett laboratory in Glasgow, it was estimated that control measures in the UK in the period 1980–2000 saved more than 1.5 million cats from death due to FeLV. It is now standard practice to vaccinate pet kittens against FeLV with what is arguably the first cancer vaccine.[28]

Michael Stoker's Research at Glasgow and the ICRF

Throughout the 1960s, Michael Stoker continued to work with BHK21 cells. For example, he investigated anchorage dependence—the need for some cells to be anchored to or in contact with another layer of cells to survive—and how contact with normal cells could suppress the transformed phenotype. Perhaps the single most important feature of BHK21 cells is their capacity to support the growth of a wide variety of viruses.[29] Stoker generously donated the cells to virologists at the Animal Virus Research Institute, now the Pirbright Institute, in England, where scientists found that foot-and-mouth disease virus (FMDV) grew well in them. Subsequently, this system was used by the Burroughs Wellcome company to manufacture many millions of doses of a vaccine to protect cattle from FMDV. (Previous vaccines had limited production because they were grown on cattle tongue cells.) A British patent for the vaccine production was applied for to "block unscrupulous Americans."[30] The incidence of foot-and-mouth disease fell drastically in many European countries in the late 1960s and early 1970s because of this vaccine.[31]

The talent that Stoker marshaled to turn Glasgow into a world-class center for virus research did not go unnoticed. In 1968, Stoker was invited to become director of research at the Imperial Cancer Research Fund laboratories in London. He had already turned down earlier offers from Cambridge and Oxford, but the opportunity to bring molecular biology to a large cancer center—ICRF was five times bigger than Glasgow Institute of Virology—proved irresistible, and Stoker returned to England. Senior scientists Lionel Crawford and Ian Macpherson decided to move with him, as did Bill House, Stoker's highly competent laboratory manager. The eminent geneticist Guido Pontecorvo, who was one of the attractions for Stoker to come to Glasgow in the first place, also decided to move to the ICRF. Before taking the reins there, Stoker spent the summer of 1968 in California working again in Renato Dulbecco's laboratory. The friendship between the two men was instrumental in getting Dulbecco to visit the ICRF for four years, including 1975, when Dulbecco won the Nobel Prize.

Stoker would spend 12 years at the ICRF. His time there "heralded dramatic alterations in the overall research strategy."[32] He focused research on single cells with the hope that the emerging power of molecular biology could be harnessed to shed light on the general mechanisms of cancerous growth. He also made research at ICRF more flexible by removing hierarchy and allowing the individual laboratories more autonomy in choosing their research topics. Ian Macpherson continued the emphasis on tumor virology by heading a department of the same name.

Bill Jarrett was made professor of veterinary pathology in 1968 and remained a researcher at Glasgow until 1990. His younger brother Oswald (Os)

Jarrett earned his PhD with Stoker and Macpherson in Glasgow and became a virologist.

Michael Stoker was knighted in 1980. After his time at the ICRF, he returned to Cambridge, where he continued his work on cell growth and also served as master of Clare Hall for seven years. He died in 2013 at the age of 95. The ICRF was eventually incorporated into Cancer Research UK and is now part of the vast Francis Crick Institute in London. The Institute of Virology was renamed the MRC Centre for Virus Research and because of its presence Glasgow remains a world center for the study of viruses. In 2015, a new building was opened to house the center, fittingly named the Sir Michael Stoker Building.

5

Insights from the Field

Anthony Epstein, Denis Burkitt, Werner and Gertrude Henle, and the First Human Tumor Virus

In March 1961, a young pathologist named Anthony Epstein attended a talk on "the commonest children's cancer in tropical Africa." The speaker was Denis Burkitt, "an unknown bush surgeon" who had spent the previous eighteen years in Africa.[1] The talk would radically change the trajectory of Epstein's career.

In 1943, in the middle of World War II, Burkitt, a member of the Royal Army Medical Corps, was posted to East Africa with the British colonial troops (figure 5.1). He saw Kenya, Somaliland, and Uganda, the latter of which, being peaceful and predominantly Christian, appealed to the religious Burkitt. Seeking to stay after the war ended, he successfully applied the Colonial Medical Service and was appointed to the Largo District, where he administered a 100-bed hospital for a population of around 300,000 people. There he made his home. "I didn't go out to Africa for science but because what I believed to be God's call, and I've had no complaints"[2]

In 1957, a colleague at the Mulago Hospital brought him an unusual patient: a child with swelling in all four angles of the jaw. The swelling did not appear to be caused by a typical cancer or a typical infection. A little while later, Burkitt noticed another case of swelling of the jaw in a child. This second case piqued his interest because, as he later said, "a curiosity can occur once, but two cases indicated something more than a curiosity."[3] Burkitt searched for similar cases in the patient records at Mulago Hospital, discovering that such a condition was relatively common in young children; indeed, dozens of kids had been identified with sarcoma of the jaw in the previous seven years. It was normally fatal. He wrote up a description of the 37 cases and had it published in the *British Journal of Surgery* in 1958.[4] Unfortu-

5.1. Denis Burkitt and two young patients. Image from the Wellcome Collection, https://wellcomecollection.org/works/s7vv6xvv/images?id=mr4x5b8k

nately, this first publication on what would be called Burkitt's lymphoma did not attract much attention, at least in the beginning.

From the outset, Burkitt took an interest in the geographical distribution of the children exhibiting the jaw tumor, and it was the geographical distribution of the tumor that made it special.[5] Summarizing his findings in a 1962 article in *Nature*, Burkitt noted that the tumor was a malignant lymphoma that affected both boys and girls between the ages of 2 and 14. It accounted for over 50% of the cancer found in children in Uganda, and it could be found in different tribes. Burkitt sent out questionnaires across Africa to map the occurrences of the disease. He also undertook at 10,000-mile safari to investigate in person. The cancer was found in a band across central Africa. After considering altitude, which is correlated with climate, Burkitt proposed that it was a minimum temperature of about 60°F in the coldest time of the year (July) that was relevant. Places where it was colder than 60°F did not appear to produce the cancer. The fact that the tumor distribution is dependent on climatic factors strongly suggests that some vector is responsible for its transmission. Perhaps mosquitoes carry a cancer-causing virus.[6]

Denis Burkitt Meets Antony Epstein

In March 1961, Burkitt presented his results outside of Africa for the first time. Speaking at the Middlesex Hospital Medical School in London, he titled his talk "The Commonest Children's Cancer in Tropical Africa." The flier, which was posted on the notice board, fortunately caught the eye of Anthony (Tony) Epstein, who attended out of curiosity (figure 5.2). Epstein was "electrified" by Burkitt's description of the tumor's rapid progression and unusual epidemiology. Even during the presentation, Epstein was thinking about the possibility of an arthropod vector spreading an oncogenic virus. Within the first 20 minutes of the talk, Epstein decided he would stop his work on Rous sarcoma virus and look for this virus.[7] After the talk he took the flyer down from the board for safekeeping. He invited Burkitt for tea at his lab. The two met to talk a few days later, and it was decided that Burkitt would send Epstein biopsy samples from Africa. Perhaps surprisingly, given the lack of even preliminary data, Epstein convinced the British Empire Cancer Campaign to fund a trip to Africa, where he arranged for Burkitt's lymphoma biopsies to be flown overnight from Kampala, Uganda, to London, where his laboratory was located. His experience with bureaucracy in post–World War II India helped him navigate the Ugandan colonial administration. Over the next two years, Epstein labored over these samples, trying to detect a virus. He injected the tumor tissue into chicken eggs, newborn mice, and other kinds of cell cultures. But nothing worked and he had no evidence of a virus. He also looked for a virus in the samples directly using

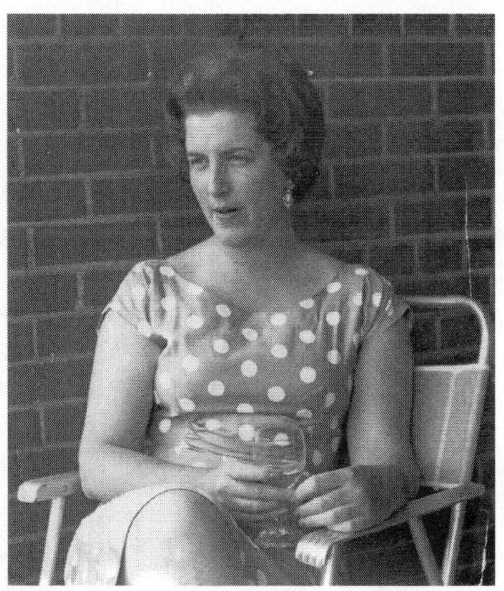

5.2. *Left,* Anthony Epstein. Image courtesy of the Master and Fellows of Trinity College Cambridge. *Right,* Yvonne Barr in 1962. Image courtesy of her daughter, Kirsten Balding

an electron microscope. Again nothing. Epstein had high hopes for microscopy, having been trained by the Nobel Prize–winning cell biologist George Palade in 1956 at the Rockefeller Institute. Palade was at the forefront of the use of electron microscopy to describe subcellular structure and was later awarded the Nobel Prize. However, the electron microscope also yielded negative results and was "especially disappointing."[8]

Despite the lack of results, Epstein secured a $45,000 grant from the US National Cancer Institute. He hired two assistants, Yvonne Barr and Bert Achong, in late 1963. Achong was to help with the electron microscopy and Barr with the tissue culture. Epstein decided to try to grow Burkitt's lymphoma cells in vitro away from the host's immune defenses with hopes of detecting a latent oncogenic virus. Such a strategy had been successful with chicken cell cultures. The major obstacle was that no one had ever been able to culture human lymphoid cells. Numerous attempts to create a culture by many methods including from floating lymphoma fragments failed. It was disheartening. The team needed a change in luck, and they got it in December 1963.

A Lucky Day in 1963

On the first Friday of December 1963 there was fog at the airport in London. The flight from Kampala with the most recent batch of biopsy samples from Africa was diverted to Manchester. It took until the afternoon before the small green thermoses containing the biopsies were delivered to the Epstein lab. At first this delay looked like a disaster. The tissues floating in the guinea pig serum were unusually cloudy, which Epstein initially thought indicated a bacterial contaminant. However, before leaving for the weekend he put a sample of the fluid under the light microscope and was amazed to discover that the turbidity was due to massive numbers of free-floating tumor cells. The free-floating cells immediately reminded Epstein what he had learned at Yale Medical School: mouse lymphomas could be cultured only from single-cell suspensions grown, in their case, in mouse abdomens; the dispersed cells of the biopsy delayed by the weather resembled such cells closely, and Epstein gave them to Barr to culture in suspension for the first time (figure 5.2). A continuous cell line grew, which was designated EB1 (Epstein Barr 1). Using the same method of suspension culture yielded additional cell lines EB2 and EB3, and others, including Sarah Stewart (see chapter 2),[9] were able to grow Burkitt's lymphoma–derived lymphocytic cells lines using the technique. Independently, researchers in Africa grew a cell line that they called "Raji" after the boy whose tumor was used.

Eventually Epstein grew enough cells that they could be put in a centrifuge and pelleted into a solid, fixed, and prepared for electron microscopy. He was convinced that microscopy could be used in discovering and classifying viruses. Practically all virologists at the time disagreed with him. They wanted a different kind of evidence: to see injecting the purported virus into an animal or tissue culture cause disease or an immune response. In any case, Epstein marched to his own drum, as he later recalled:

> It was on February 19th, 1964, and it was in the very first sample that I looked at of cultured tumor cells in the electron microscope. It was in the first grid square—a cell stuffed full of particles which I immediately recognized as being of the herpes family because I had worked on herpes virus as well and I knew about electron microscopy and I knew perfectly well that you could identify what family a virus was morphologically in the electron microscope, just as had been done with the light microscope and bacteria for 100 years, but it was not be-

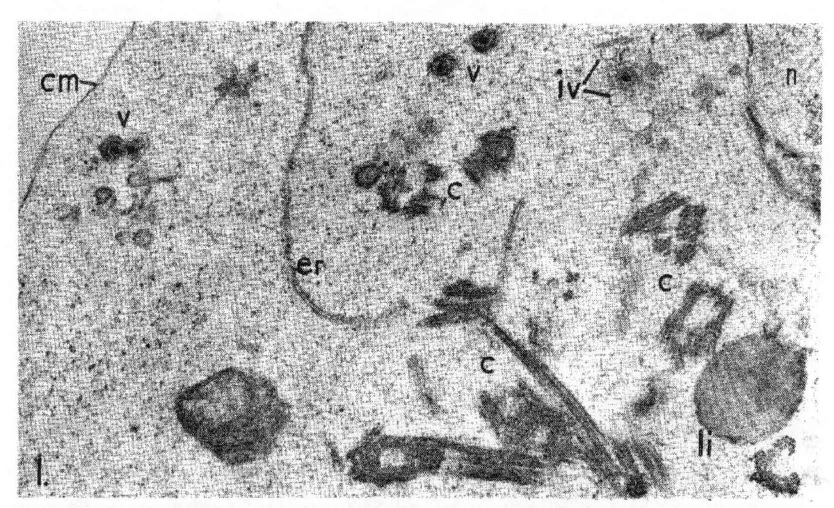

Fig. 1—Part of a cultured lymphoblast derived from a Burkitt lymphoma. The cell membrane (cm) crosses the top left corner and the nucleus (n), bounded by its double membrane, lies in the upper right portion of the field. The intervening cytoplasm contains several mature virus particles (v) within spaces enclosed by fine membranes, some immature particles (iv), and crystals (c) cut in various planes; a large lipid body (li) and endoplasmic reticulum (er) can also be seen. In addition profuse free ribosomes lie scattered throughout the cytoplasmic matrix. Electronmicrograph × 42,500.

5.3. Early electron micrograph of Epstein-Barr virus. From Epstein, Achong, and Barr, "Virus Particles in Cultured Lymphoblasts from Burkitt's Lymphoma," *Lancet* 283 (1964): 702–703, with permission from Elsevier.

> lieved at the time that that was how you identified viruses. It was snowing. I was so anxious that the specimen would burn up in the electron beam so I switched off and went out into the snow without a coat or anything, just a lab coat, and walked around the block to calm down. When I got back I switched it on and took the electron micrograph images.[10]

The discovery was written up and published in the *Lancet* (figure 5.3). In two crisp pages, Epstein, Achong, and Barr described a virus that "resembles Herpes simplex" but is smaller by about 20%. They concluded that "it seems that a 'passenger' role can be assumed since the agent has persisted in vitro in the dividing cells for may weeks. Any more significant connection [between Burkitt's lymphoma and the virus] remains to be established."[11]

The team worked to establish the connection, but again when they injected the virus into cell culture, chicken eggs, and newborn mice, no transmission resulted. Thinking that perhaps they needed help from other virologists, Epstein approached two different British experts in herpes biology, but neither was interested in collaboration. Their refusal may have been a symptom of how traditional virologists undervalued electron microscopy in identifying new viruses.

Epstein's Collaboration with Werner and Gertrude Henle

Seeking help further afield, Epstein developed partnerships across the Atlantic. Werner and Gertrude (Brigette) Henle ran the Virus Diagnostic Laboratory at the Children's Hospital of Philadelphia (figure 5.4). Epstein had met them in 1956 when he traveled the United States meeting prominent virologists, including Dulbecco, Bittner, and Salk. The husband-and-wife team had immigrated to the United States in the 1930s to escape anti-Semitism in Germany. Werner's grandfather, Jakob Henle, was a famous pathologist and anatomist who counted among his achievements the discovery of a structure in the kidney now known as the loop of Henle. By the time Epstein approached them about a collaboration, in 1965, Werner and Gertrude had built their virus laboratory into one of the biggest in the United States. Unlike the British virologists Epstein had earlier approached, the Henles welcomed the opportunity to work on a potential cancer virus.

After receiving samples from Epstein, the Henles ran a battery of tests that showed that the virus was unlike any known herpesvirus. It was immunologically unique. They called it Epstein-Barr virus after the EB cell line containing the virus that Epstein sent them. This uniqueness was good news, but it did not show that the virus caused cancer. It would take much more evidence to convince skeptics of a cancer-causing human herpes virus.

The Henles decided to study people infected by EBV, dropping other projects concerning mumps and influenza viruses.[12] They used an antibody that attached itself to EBV-containing cells that was found in sera from patients with Burkitt's lymphoma. They modified an immunological technique invented in the 1950s by coupling fluorescent labels to the antibody, which allowed the EBV-containing cells to show up bright green under the microscope. All 17 of the Burkitt's lymphoma samples were positive for EBV-containing cells. This was good news for Epstein and the Henles and was to be expected. What were unexpected, though, were the results of tests on samples taken from American individuals: 90% of American adults tested were also positive for the virus. The figure was lower for children, around 30%, but nonetheless it appeared that EBV was a very common virus that infected the majority of Americans. Some virologists thought that perhaps the team's immunofluorescence test was generating false positives, but the

5.4. Werner and Gertrude Henle examining a flask. Image courtesy of College of Physicians of Philadelphia / Children's Hospital of Philadelphia.

Henles were confident the test worked properly. Nonetheless, these results complicated the story, since Burkitt's lymphoma was extremely uncommon in the developed world. How EBV was connected to Burkitt's lymphoma was more complicated than first expected. More research was needed.

During trips back to Germany in the 1960s, the Henles visited Heidelberg and interviewed two young medical doctors who were interested in their research. Harald zur Hausen (see chapter 15) and Volker Diehl impressed the Henles and were offered research positions on EBV in Philadelphia. Zur Hausen worked on virus-induced chromosomal changes at Dusseldorf and was open to the idea that human cancer could be caused by a virus in a time when most medical professionals were skeptical. He had learned about lysogeny in bacteriophage during his medical training and wondered whether tumor viruses were likewise integrated into the genomes of transformed cells. He proposed this to the Henles as a possible avenue of research sometime after he arrived in the United States in late 1965. Werner was open to it, but Brigitte thought it better if he followed her protocols; after all, she was in the habit of managing junior researchers in her lab relatively closely. This focus on EBV DNA would have to wait.

Diehl arrived at the Henles' laboratory in June 1966. He started a project to see whether EBV could transform normal lymphocytes in the lab. In other words, could EPV added to lymphocytes make an immortal cell line? He used EPV purified from the Jijoye human cell line and then ingeniously mixed them with lymphocyte cells from a different sex than the Jijoye cells, allowing him to distinguish between the lymphocyte cells and the Jijoye cells based on sex. This way, he could check any cells that grew to see whether they were normal or cancerous. For the next few months, Diehl looked after his cells, but nothing grew. Care of the cells often involved late nights at the lab. Then one night at 2 a.m., he noticed cells growing. Too excited to wait until morning, he called Werner and woke him up to tell him about the new growth.[13] To be sure that this was the progress he had been waiting for, he still had to check that the growing cells were the correct sex to be Jijoye cells, which they were. The Henle laboratory now had a significant piece of evidence that the EBV could transform cells and was a cancer virus.

The Connection with Infectious Mononucleosis

A major open question was what diseases, if any, were caused by EBV in the wealthy Western world. On this question the Henle laboratory got a major break. The Henles had analyzed blood samples from their laboratory personnel for EBV. The results from one member of the lab, Elaine Hutkin, were especially useful. Unlike 90% of adults, she had not been infected with EBV—she had leukocytes that did not grow in culture, and her plasma was devoid of antibodies to EBV. The lab was therefore able to use her samples as a negative control in some of their experiments. On August 10, 1967,[14] she fell ill and did not come to work. After a few days, she came into the lab and was examined by Diehl. She still felt unwell and had enlarged lymph nodes, a fever, and inflamed tonsils. On the basis of a positive heterophile test she was diagnosed with infectious mononucleosis (IM), otherwise known as glandular fever. Blood was taken from Hutkin, and this time her leukocytes started to grow after four weeks. At this point, 1%–3% of the cells revealed immunofluorescence, showing they harbored EBV. The presence of EBV in her cells appeared to be correlated with her newly diagnosed disease.

Diehl and the Henles took this piece of scientific good fortune and looked for corroborating evidence. The next step was to obtain sera from patients with IM to see whether there was a general correlation. The team obtained samples from three laboratories: one at Yale, one at the University of Pennsylvania, and one at Children's Hospital in Los Angeles. All of the sera from patients in acute stages of IM were found to contain antibodies for EBV, a result that "suggested a definite link between the virus and the disease."[15] The data from Yale were most informative, as researchers there had collected baseline samples that showed that all patients were devoid of EBV antibodies before they were afflicted with the disease. A control group showed 24% of people have some reaction to EBV but at lower titers than those infected with IM, suggesting that EBV was relatively common in the population. In the discussion section of the resulting paper, Diehl and the Henles considered indirect evidence for EBV being the cause of IM. For example, some of the titers increased as the patients progressed through the stages of IM, suggesting an active infection. They also noted that antibodies to herpesvirus have no correlation with IM.[16]

An obvious question arose about the nature of the relationship between Burkitt's lymphoma (BL) and IM: Could a virus cause two different diseases? Does learning that EBV causes IM weaken the evidence for EBV as a cause of BL? The Henles concluded that it does not—the same virus could be responsible for two different diseases in two different populations and environments. But more evidence for the oncogenic potential of EBV was needed.

Harald zur Hausen also contributed to EBV research. Although he worked first on adenovirus to gain skills in molecular biology, he published on EBV in 1970.[17] By the end of the 1960s it had become possible to isolate DNA from partially purified EBV particles. Zur Hausen saw the opportunity to test Koch's first postulate: that the purported disease agent be present in all cases of the disease. His approach was to label EBV DNA with radioactive tritium, an isotope of hydrogen. He could then measure how strongly it hybridized to tumor cell DNA. The more it hybridized, the better the evidence of EBV DNA in the genome of the host cells. His results from the Raji cell line were the most interesting. Unlike the EB cell line created by Epstein that created EBV viruses, the Raji cell line created from a Burkitt's lymphoma patient in Africa did not create new virus. This line was also negative in the EBV immunofluorescence tests pioneered by the Henles. Nonetheless, zur Hausen found that based on the strength of hybridization between Raji host cell DNA and EBV DNA, there was the equivalent of six EBV genomes per Raji cell. Furthermore, zur Hausen was able to show that the EBV DNA in the Raji cells transcribed RNA, suggesting that the viruses that had taken up residence in the Raji cell genome were not completely inert. It was still unclear what the function of EBV RNAs was within the cell, however.

A natural place to continue investigating EBV and Burkitt's lymphoma was in Africa. It was one thing to find that EBV could transform cells in the lab, it was another to show that EBV causes Burkitt's lymphoma in the field. Cancer virus skeptics would find evidence from the field harder to dismiss. At the end of 1968, 40 scientists and physicians met in Nairobi to plan how to study EBV in the field. The Henles attended, as did George Klein of the Karolinska Institutet in Sweden. One approach was to collect blood samples of young children and then wait several years and look for differences in the blood samples of children who developed tumors. Since only roughly 1 in 1,000 children developed Burkitt's lymphoma, many several thousand samples would have to be collected for the study to have any statistical significance.

Testing the Viral Cause of Burkitt's Lymphoma in Africa

One of the attendees was Guy de Thé, an aristocratic Frenchman who had gone into virology to the horror of his family.[18] De Thé was well connected as a new head of a human tumor virology department at the World Health Organization (WHO)–funded International Agency for Research on Cancer. After the 1968 Nairobi meeting, de Thé proposed an enormous study to investigate whether there was an epidemiological link between EBV and Burkitt's lymphoma, and if so, how long the gap was between exposure to the virus and cancer. After consulting with a WHO statistician, de Thé realized that the study would have to enroll at least 40,000 children and follow them for nearly a decade. It would cost roughly $1 million per year, a huge sum at the time. Luckily, funds from de Thé's agency and money from the NCI Special Virus Leukemia Program allowed the ambitious project to proceed.

As there were minimal laboratory resources in Uganda, much of the program had to be built from scratch. Resources and medical supplies had to be imported and driven 300 miles by truck to Kuluva Hospital in the Northern West Nile District of Uganda. Sixty personnel were trained as lab technicians, phlebotomists, and various hospital workers. Getting the vehicles, medical supplies, and laboratory equipment there and set up required much planning, but by 1971 the project began.

Coinciding with the launch of the new initiative was the rise to power of

Idi Amin, the brutal Ugandan dictator. Luckily for de Thé and the project, Amin grew up in the West Nile District and considered the project "his baby,"[19] although it still had supplies like gasoline stolen by the secret police. Nevertheless, and despite the crumbling institutions in Uganda, the project kept going. Project leaders recruited children from remote villages, sampled their blood, and shipped the samples to Lyon for storage and analysis. By the fall of 1972, 44,000 children had taken part. After six years, the funding was pulled in November 1977. A total of 14 children enrolled in the study developed Burkitt's lymphoma, fewer than expected but nonetheless informative. All 14 had antibodies for EBV long before developing Burkitt's lymphoma. They did not have antibodies to other well-known viruses, and the controls, similar children without Burkitt's lymphoma, showed less exposure to EBV. Exposure to malaria did not seem to be an obvious factor either, and the role of malaria in Burkitt's lymphoma remained enigmatic.[20] What made these 14 children different from others who were infected with EBV but did not develop Burkitt's lymphoma was still unknown.

Work done by Anthony Epstein and colleagues reinforced the causative role of EBV in Burkitt's lymphoma. Epstein was able to cause tonsillitis in one of three owl monkeys that he injected with EBV. This monkey later died with tumors in the lymph glands, from which EBV-carrying cell lines were developed. The sample size was small but suggestive.

Further evidence came from the Klein laboratory in Stockholm, Sweden. George Klein barely escaped the train station where Nazis were forcing Jews to board trains for concentration camps. He lived underground until Hungary was liberated by the Russians and then fled the Communists for Sweden in 1947 with his wife, Eva.[21] Klein put the chance that they would be captured before getting to Sweden at 80%, but they beat the odds.[22] They began careers in medical research in Sweden, specializing in tumor immunology, which they viewed as a second chance at life. In particular, they began investigating tumors in mice but switched to human tumors in the 1960s. They teamed up with Peter Clifford, a surgeon in Nairobi, who sent them samples of Burkitt's lymphoma tumors almost every week. The Kleins found an antibody in the blood of patients suffering from Burkitt's lymphoma that reacted with 10% of cells in Burkitt's lymphoma cell lines, including those cell lines that did not produce virus. This was a different antibody than the one the Henles had found earlier. It was called "membrane antigen," as it was found on the membranes of cells.[23]

Additional evidence from the Karolinska Institutet showed that the EBV DNA formed small circles of DNA that replicated in the nucleus but did not fuse with the host cell DNA as could be the case for viruses like SV40. A natural question was what protein products were encoded and transcribed, if any, from the EBV DNA. Viral proteins presumably had something to do with the oncogenic process. Beverley Reedman, an Australian visiting scientist at the Klein laboratory, used enhanced immunofluorescence staining to begin to answer this question. She was able to get staining in the nucleus of all Burkitt's lymphoma cells, even those cells that had recently been flown from Nairobi and were not immortalized. The virus was not inert but rather was creating something that was not found in uninfected cells. Reedman and Klein named this antigen EBNA, for EBV-encoded nuclear antigen.[24] It was the first of several nuclear antigens discovered and clear evidence that the virus had a measurable effect on all infected cells.

Additional pieces of the puzzle were discovered in the Klein laboratory. Following a suggestion from George Klein, a visiting husband-and-wife team of Bulgarian scientists, George Manolov and Yanka Manolova, used new staining techniques to examine banding on the chromosomes of tumor cells, comparing them to normal cells. Through meticulous examination of more than 200 bands that could be distinguished on the chromosomes, in 1972 the Manolovs discovered an extra band on chromosome 14 in cells in 10 out of 12 tumors investigated.[25] Additional work by Lore Zech showed that a piece of chromosome 8 had been inserted onto the end of chromosome 14. Geneticists call such a change a translocation. The majority of Burkitt's lymphoma had the 14:8 translocation. A minority of them had a 2:8 or 8:22 translocation. The meaning of these translocations would not become clear until the discovery of oncogenes in other tumor viruses.

There was progress on other fronts too. Reporting in *PNAS* in 1973, biologists from Yale University including Robert Shope, son of Richard Shope (chapter 2), reported that they had caused malignant lymphomas in cotton-top marmosets by injecting them with materials containing Epstein-Barr virus.[26] Their research constituted solid evidence that EBV could cause cancer in primates.

In the fall of 1982, two different American groups submitted very similar papers to *PNAS*.[27] They were both accepted and published back to back. They showed that the significant mutation in Burkitt's lymphoma cell lines was a translocation of the oncogene to a new position in the genome. Importantly, the translocation changed the regulation of the gene, presumably increasing its transcription and causing the consequent transformation of the cell into a permanent growth phase. If anything should be called the proximate cause of Burkitt's lymphoma, it would be the relocation of the *c-myc* gene. So, if EBV is not a tumor virus in the sense of having oncogenes, how does it cause Burkitt's lymphoma and how does it immortalize cell lines? One piece fell into place. Malaria infections cause a vast increase in the number of B lymphocytes as the human body tries to fight the parasite. B lymphocyte levels in African children could be 100- to 1,000-fold greater than in European children. In virtue of the higher number of cells, the chances of a mutational translocation that changes the expression of the *myc* gene is greatly enhanced in African children. In the 1980s, Japanese researchers found evidence that EBV decreases the chance that an infected B lymphocyte undergoes apoptosis, or controlled cell death. Perhaps EBV does not cause cancer directly but rather increases the chance of cancer-causing mutations occurring in B lymphocytes, as the infected B cells live longer.[28]

Two years later, in 1984, the EBV genome was sequenced by a team at the MRC laboratory in Cambridge, England. At the MRC, Fred Sanger and Sydney Brenner had pushed to sequence important viruses with larger and larger genomes.[29] The EBV genome was challenging because of its size—172,282 base pairs long and as many as 84 genes. The approach, led by Bart Barrell, was to clone and sequence relatively large restriction fragments of the EBV genome and then put the pieces together to get the final sequence. The MRC attempt also drove the sequencing technology forward with longer reads of sequence per clone and new software to assemble the many different clone sequences into longer sequences. To date, there have been more than 30,000 peer-reviewed publications about EBV.

The Henles continued their work as virologists in Philadelphia until they retired in 1982, and even then Werner continued to work. They received a number of awards including the Robert Koch Prize (1971), William B. Coley Award (1975), Bristol-Myers Award (1979), and Gold Medal of the Children's Hospital of Philadelphia (1983). He died in 1987 at the age of 76, and she died in 2006 at the age of 94.

Yvonne Barr got her PhD in 1966 at the University of London under Epstein's supervision. She moved to Australia to raise a family. After performing some contract work at Monash University and finding it "a bit of a boys club," she got her diploma in education and became a schoolteacher.

Later she would humbly say that the virus discovery was a "minor thing" of her life.[30] She died in 2016.

Anthony Epstein continued his career in virology as a professor at the University of Bristol from the late 1960s to the early 1980s. In 1976 he proposed the creation of an EBV vaccine.[31] With funds of the MRC and the Cancer Research Campaign, Epstein and his collaborators developed a prototype vaccine that was effective on cotton-top tamarins in the 1980s.[32] The human vaccine project came to a halt when it was realized the glycoprotein target of the vaccine, gp340, was patented by Elliott Kieff of Harvard Medical School, who had sequenced the gene.[33] SmithKline Beecham resumed studies in the late 1990s but despite some successes in phase 1 and phase 2 clinical trials, no vaccine for EBV has been approved. Epstein joined Wolfson College, Oxford, in 1986 and remained there as a fellow until 2001. He was elected a fellow of the Royal Society in 1979 and knighted by the Queen Elizabeth II in 1991, becoming Sir Anthony.

6

Persistence despite Political Challenges

Jan Svoboda and Tumor Virology behind the Iron Curtain

Not all the advances in tumor virology were made in the West. There were pockets of activity in the Communist East as well. This chapter focuses on the career of Jan Svoboda, a Czech virologist who specialized in working with Rous sarcoma virus (figure 6.1).[1]

Jan Svoboda (1934–2017) felt lucky as a young boy in Czechoslovakia immediately after World War II. He had been largely insulated from the Nazi occupation, and he loved natural objects, which he collected. Frogs and snakes were a special interest. Childhood life was generally good for the curious Svoboda, and he explored ideas from the East and the West.

His world, and the world of the Czech people more generally, changed drastically in February 1948 with a Communist coup d'état. Svoboda's teachers felt inhibited from discussing sensitive topics. One of his classmates suspected Svoboda was an enemy of the state and turned him in to the authorities, which led to Svoboda being denied entry into the humanities and medicine at university.

Luckily, he was able to enroll at the Faculty of Science at Charles University in Prague in 1951. Biology was taught there, but it was driven by anti-Mendelian Lysenkoism.[2] Influential and well connected in the Communist state, the Soviet biologist Trofim Lysenko denied Mendelian inheritance, natural selection, and even the concept of the gene, for mostly political reasons. He imposed his views on a generation of Soviet scientists, who risked imprisonment if they did not renounce Mendelian genetics. He convinced political leaders that inheritance of acquired characteristics was consistent with Marxist thought and Mendelian genetics was not. Svoboda had to buy old used books to learn about classical genetics. Discussions with visiting bi-

6.1. Jan Svoboda. Image courtesy of Jan Svoboda Jr.

ologist Conrad Waddington armed Svoboda with anti-Lysenkoism arguments.[3] At Charles University, he "fell in love with tissue culturing." PhD student Mojmír Brada introduced him to Rous sarcoma virus and the way RSV rapidly transformed cells into cancer cells. Unfortunately, Brada suffered personally and took his own life, but not before he hooked Svoboda on RSV research.[4]

Svoboda investigated immunological tolerance by observing how exposing turkey embryos to chicken blood made them more susceptible to RSV infection.[5] He also extended this to ducklings, an interesting result that indicates how a virus might jump from one species to another. For reasons that are still not fully understood, the jump to ducks works only for ducklings two weeks old or younger.

In 1956, anti-Soviet revolts broke out in Eastern Europe.[6] Svoboda played a role in demanding more democratic institutions for Czechoslovakia. As a consequence of this political activity, he was accused of "anti-socialist activity" and brought before a special committee. He was given only a slap on his wrist: a one-year delay in admission to a PhD program. He was lucky that one of his judges was the biologist Milan Hasek, a more liberal scholar and defender of Czech science.

Svoboda's work on immunological tolerance to avian viruses was noticed, and he traveled to Moscow for a meeting on transplantation in 1959. The following year, he published on how RSV causes hemorrhagic disease in rats. Rats also developed tumors, presumably caused by RSV.[7] The strains of RSV studied in the United States generally could not infect mammals, but the "Prague strain" of RSV was different somehow. Svoboda thought the process was "enigmatic" because no trace of the virus could be detected in the tumor. These data seemed to support the "virogenetic" theory of Lev Zilber that a virus could initiate the formation of a tumor, but the tumor need not produce more viruses.[8]

The Creation of XC cells

The study of how RSV caused tumorigenesis became Svoboda's most important research focus. He injected minced tumors into chicks and rats. One day, in cage number 90, Svoboda discovered an important tumor in a rat that he called XC, using the Roman numeral for 90. The XC tumor could be propagated for many generations in rats. XC cells when injected into chicks caused sarcomas. Initially, Svoboda thought the sarcoma produced new virus, but more study found that infectious viruses were not produced by XC cells.[9] However, intact RSV particles could be produced (or "rescued") by associating and fusing XC cells with chicken fibroblast cells. He later confirmed that XC cells were rat cells. Svoboda speculated that perhaps the RSV genome had integrated into the genome of the XC cells, just as bacterial viruses do during lysogeny.[10]

Just as his research was getting interesting, Svoboda was sent against his will on a trip to Communist China by the Czechoslovak Academy of Sciences. Svoboda found that Chinese virology was significantly behind Europe, and his time in China did not help his own research. His Chinese hosts did give him some Chinese hamsters to bring back home, and he would use this species for his experiments when he needed a mammalian system in addition to rats. On the return trip, in Moscow he met with Lev Zilber, a prominent Russian biologist who promoted the idea that viruses caused cancer. They talked about XC cells, and Zilber encouraged Svoboda to continue to investigate them, agreeing that the system would be an effective model to investigate the transforming activity of an oncogenic virus.

The XC tumor work was noticed outside of Czechoslovakia. Howard Temin, the rising University of Wisconsin virologist, wrote to Svoboda in September of 1962, "I have read with interest your recent paper [on] Neoplasma on the XC tumor. I was especially interested in your suggestion of the analogy with the lysogenic state."[11] Temin had similar results (see chapter 8). He suggested that Svoboda try to "rescue" the virus using different strains of RSV to infect the XC cells, the idea being that although XC cells produce no infectious virus, they might be induced to produce it after being infected with a different virus. Svoboda had tried to do this already and it had not worked, suggesting a significant difference between the RSV that Temin used in Wisconsin and the "Prague strain" of RSV.

Svoboda devoted more time to XC cells. In a paper submitted in December 1962, he published his "final picture" of XC cells containing the RSV integrated as a provirus.[12] As he put it, "The XC cell is virogenic and contains Rous virus in the form of a provirus."[13] Like Temin and others interested in endogenous retroviruses, Svoboda drew some inspiration from André Lwoff, who had promoted the idea that bacteriophage can integrate into the host genome as a provirus. Endogenous viruses are embedded in the host's genome.

Svoboda Visits the United States

In March 1964, Svoboda finally was able to meet Western virologists. It was also the year that the National Institutes of Health began promoting this type of research through the Special Virus Leukemia Program.[14] The National Cancer Institute sponsored a conference on avian tumor viruses at Duke University in Durham, North Carolina, and invited many of the leading names in the field. There was much material presented, and the proceedings would stretch to 800 pages. Originally the conference was going to be in New Orleans, but a hotel could not be found that treated white and black people equally.[15] That the virology community cared about such an issue suggests that leading virologists were politically aware and did not conduct their research in a vacuum. Svoboda spoke with Harry Rubin, Howard Temin, Ludwik Gross, Hidesaburo Hanafusa, and Peter Vogt, among others. Like Svoboda, Temin, Rubin, and others used their stature to promote various progressive political causes.

Svoboda was struck by how well equipped American laboratories were with centrifuges and other machines. His laboratory would struggle to keep up, and he decided that he must keep his research tightly focused on XC cells to remain competitive. While he was there, he took the opportunity to visit a plumbing store to buy a pump for a CO_2 incubator used to maintain a stable pH and started a friendship with Howard Temin.

Before returning to Prague, Svoboda visited New York City to meet with Peyton Rous, who had said that Svoboda's work "has my intense interest."[16] Among other things Rous was interested in Svoboda's views about RSV's ability to infect mammalian cells.[17] Rous coordinated with Frank Horsfall at the Sloan-Kettering Institute for Cancer Research to have Svodoba give a talk there. There were no longer many at Rockefeller interested in cancer, Rous pointed out.[18] Svoboda enjoyed his time talking with Rous, and they exchanged their papers on RSV. In addition to the usual letter of thanks, he sent him a biologically inspired Christmas card, with a chromosome for the *X* in Xmas, wishing him, "Good luck with your experiments in 1965."

Work on "Virus Rescue"

Returning to Prague, Svoboda turned to understanding the mechanism of virus rescue—isolating infectious virus particles from cells. Through his research he learned that, when enough virus-infected cells fuse, the virus could be rescued. Svoboda and his colleagues added Sendai virus could promote the rescue by stimulating fusion, showed. This viral approach to cell fusion was inspired by the work of Yoshio Okada, a Japanese virologist.[19] Svoboda's lab also studied tumor-specific transplantation antigen (TSTA), an antigen that, as the name suggests, is present only in tumors. Cross-reactivity between rat tumors and mouse tumors suggested that TSTA was related to the provirus.

Svoboda discussed these ideas at a 1965 international symposium in Sukhumi, USSR, on the Black Sea to celebrate the 70th birthday of Lev Zilber. This conference proved to be a meeting of East and West, with European, American, Japanese, and Israeli virologists present. Svoboda mediated an informal discussion about virus rescue of RSV and SV40 with Albert Sabin and Hilary Koprowski, two Polish American virologists. Sabin was skeptical of the possibility of both viruses exhibiting the phenomena, because RSV was an RNA virus and SV40 a DNA virus, so, prima facie, different mechanisms would be expected to be in play. Svoboda suggested that it was possible that the transforming part of RSV was in fact DNA. Their discussion presaged a new age of international scientific collaboration that was evidenced elsewhere at the conference, as well. For example, the American virologist Robert Huebner, whom Svoboda later described as the meeting's most impressive participant, laid out a lofty plan for international cooperation in cancer research. In additional to being a capable scientist, Huebner was skilled at the administration needed to make such scientific cooperation happen. He would go on to foster a relationship with Svoboda in the years that followed. Others noticed Svoboda's unique approach and reached out to him, fostering East-West cooperation. For example, Bob Harris from the London ICRF invited Svoboda for an extended visit.

While in London, Svoboda continued to work on virus rescue. He created a quantitative assay that involved using Sendai virus to fuse virogenic (provirus-containing) Chinese hamster cells and chicken cells. It was tedious work, in part because there was no way to biochemically test for the provirus—instead Svoboda needed to use cruder methods[20]—but nonetheless his results led him to think that "the virus genome in a nonpermissive mammalian cell was not expressed fully until complemented by chicken cell machinery." Most virologists were skeptical, including Howard Temin, who was not afraid to endorse unpopular views. A few years later, however, Howard Temin's PhD student John Coffin confirmed Svoboda's work.[21] Some of these fusion techniques were then used in the 1980s by fellow Czech biologist Mika Popovic when trying to isolate the virus that causes AIDS (see chapter 14).

During this period, Svoboda met Warren Levinson, who was at the University of London in Michael Abercrombie's lab and also worked on RSV. Levinson would later bring RSV to University of California, San Francisco (UCSF), where he would pass the system to Michael Bishop, Harold Varmus, and Dominique Stehelin (chapter 9). Svoboda also visited Michael Stoker and Ian Macpherson at the Institute for Virology in Glasgow (see chapter 4). The provirus hypothesis got more support when Macpherson discovered normal cells in a population of transformed cells. There was no virus to rescue in these cells, so it was natural to assume that the provirus had been lost in them.

Svoboda was also able to attend virology conferences in the West. He organized a meeting on virus induction by cell association at the Wistar Institute in Philadelphia in late 1967.[22] This meeting was planned at the 1965 meeting in Sukhumi, USSR, and brought together virologists interested in RSV from nine countries. Svoboda then attended an April 1968 meeting organized by Michael Stoker and Lionel Crawford on the molecular biology of viruses at Imperial College London as part of the Society for General Microbiology.[23] Svoboda presented on defective viruses, that is, those that need a helper virus to replicate. If an oncogenic virus is defective, it will be difficult to detect without the helper virus and the right type of helper cell. Svoboda saw the meeting as, among other things, a confrontation between French and British virologists over who was making the most progress. From his point of view, Luc Montagnier's presentation that raised the possibility that RSV replication might involve double-stranded RNA was the most important talk. Although his participation at these conferences allowed Svoboda to develop international connections and to enjoy the openness of the West, he could not escape the difficult realities of the era's geopolitics.

The Soviet Occupation of Czechoslovakia

In July 1968, Svoboda attended the First World Congress of Virology, in Helsinki. While at the Helsinki conference, Albert Sabin warned him that the Soviet army was gathering at the Czech border. This time was known as the Prague Spring, when the reformer Alexander Dubček liberalized many aspects of Czechoslovak society, moving it closer to democracy.[24] In response, on August 21, 1968, the Soviet army and other Warsaw Pact armies invaded Czechoslovakia. Svoboda, like many, was not happy with the Soviet occupation and took to the streets of Prague to demonstrate. Some American students took part in the demonstration, and Svoboda was surprised to learn that they knew of Harry Rubin's involvement in the civil rights movement in the United States.

The invasion was a disaster for Svoboda's scientific work. Necessary supplies could not be procured, and a planned joint study of oncogenic viruses with Warren Levinson had to be abandoned. The English virologist Robin

Weiss, who had been invited to begin a postdoc in Prague, received a telegram from Hasek with characteristic humor: "Please postpone visit. Stop. Uninvited guests arrived first. Stop."[25] Svoboda considered emigrating using his US visa to escape the chaos. He also had an official invitation from Robert Huebner to a post as an independent scientist at the NIH. However, duty to Czech science called more urgently than these other opportunities. Milan Hasek, the current director of the institute in Prague, wrote to him from Vienna: Hasek was not returning to Prague, and Svoboda should take over as director. It was a what Svoboda called a "Danaan gift"—a gift that imposed a large cost on the receiver.[26] Svoboda decided to stay in Prague for the good of the institute; a year later, Hasek returned and Svoboda handed the reins back to him.

Svoboda kept working on virus rescue by cell association for the next couple of years. The Prague group showed that virus rescue was not caused by endogenous viruses but rather governed by the machinery of the chicken cell. He was even able to leave to present work at a French conference.

However, in 1972 the situation deteriorated. The puppet government closed the border. Purges began. Svoboda was questioned by a committee whether he supported the Soviet occupation. "I was asked what I think about the entry of our friends and I told them that I don't agree with it and will never agree. So, I was not cleared. . . . Then I was stripped of everything, even of the laboratory. So, I remained as surviving person and started the whole thing from scratch again."[27] He later would be banned from publication. His earlier grant money from the United States was used against him, and he was accused by the secret police of engaging in espionage. They interrogated him for hours. Luckily, Svoboda could show that he used the money to buy books for his research, and he was released.

The government's crackdown not only crushed Svoboda's productivity at home; it also stymied the international reach of his work in debilitating ways. In 1971, President Richard Nixon, hoping to emulate John F. Kennedy's success with the Apollo space program, launched the War on Cancer. Leading scientists including Svoboda were invited to attend the launch of new effort, which included generous additional funding for the Special Cancer Virus Program. Svoboda felt honored by the invitation and thought it would be possible for him to attend, but the police at the airport would not let him board his plane. Then, in 1974, James Watson invited Svoboda to the CSHL Symposium on Tumor Viruses, even writing a special letter to the president of the Czechoslovak Academy of Sciences arguing for the importance of his colleague's presence. Again, Svoboda was not allowed to leave. He had his text smuggled to the symposium, and his paper on centrifugation of XC fragments appeared in the distinctive red volumes published by CSHL, but his absence from the meeting cut him off from important collaborations.[28]

The cumulative effect of his lack of access to international research and new techniques meant that Svoboda was unable to stay at the forefront of RNA tumor virus research. He was unable to secure chemicals that were becoming increasingly important, such as radiolabeled nucleotides and restriction enzymes, and was forced to design experiments that did not use them. Unfortunately, we will never know what he might have achieved had he remained connected to the larger biological research community and not been denied a decade of potentially productive research.

Svoboda Reconnects with the West in the Late 1970s

It was not until the end of the 1970s when restrictions on travel started to loosen up somewhat that Svoboda was finally able to leave his home country. He traveled to Dominique Stehelin's laboratory in Lille, France, where they performed hybridization experiments using radioactive probes.[29] In a month Svoboda tried to get up to speed in the latest techniques of molecular biology. He found that his non-virogenic tumor line contained pieces of a provirus that could transform cells but could not make new virus.[30] He also strengthened his connection with Stehelin, who helped him to modernize his lab once he returned to Prague.

Svoboda also began a productive collaboration with Ramareddy Guntaka, formerly of the Bishop-Varmus lab. Impressed with Guntaka's work on the methylation of viral DNA, Svoboda visited Guntaka's lab four times in the 1980s and early 1990s. They investigated proviruses, including one that contains only the oncogene *v-src*.[31] After the first visit to the Guntaka lab at Columbia University in October 1982, federal agents turned up to investigate what Svoboda was doing in the United States.[32] Guntaka reassured them that his visit was purely scientific.

Robert Gallo and Mikulas Popovic, from US government labs, gifted Svoboda a Sequenase kit to help with DNA sequencing. He had to carry it in two large boxes filled with dry ice back to Prague. At the stopover in Frankfurt airport, while on the lookout for police officers, he switched out the dry ice, causing a large puff of CO_2 gas. He was nervous, but luckily no one noticed, and he brought his "treasures" safely back to Prague. The 1980s then proved to be less repressive for Svoboda. Mikhail Gorbachev's liberalization policy, Perestroika, also freed up international science. The Fourteenth Triennial International Congress of Biochemistry was held in Prague in 1988 and drew leading virologists like Howard Temin, Hidesaburo Hanafusa, and Ramareddy Guntaka, who joined Svoboda in a session on retroviruses. The Prague group's work on proviruses and their combinations was as good as any Western group's work.

Svoboda had political fortitude, and he was active in eventually overthrowing the Communists. It was fitting that he became the director of the Institute of Molecular Genetics in 1991, a post he would hold for eight years. He continued to work on virus rescue with the new techniques, investigating mRNA expression and splicing, the folding of envelope proteins, and virus entry into the cell. He received many scientific honors, but perhaps the most rewarding was being elected as a foreign member of the US National Academy of Sciences in 2015. He had been barred from publishing in *Proceedings of the National Academy of Sciences* during the occupation. He was still at the bench doing experiments when he was 80. He died in 2017 at the age of 82.

7

A Surprising Discovery in the Blood

Baruch Blumberg, Harvey Alter, and Hepatitis B Virus

Baruch (Barry) Blumberg was born in Brooklyn, New York, in 1925, and came of age during the Great Depression. He attended Far Rockaway High School, famous for graduating Richard Feynman, the great American physicist, and Burton Richter, the Nobel laureate who codiscovered the J/ψ meson in 1974. He had a childhood fascination with explorers like Lewis and Clark.[1] With the disruption of the Second World War, Blumberg joined the navy at 17 but was still able to study physics. He served as a commissioned officer on several small ships. Following advice from his college advisor and his father, Blumberg entered medical school at Columbia University in 1947. After four years of hospital training in New York City, Blumberg left for Oxford University in the UK to pursue research that led to a doctorate, a DPhil, in biochemistry, which he completed in 1957. He then returned to the United States and became an investigator at the National Institutes of Health.

Blumberg was drawn to studying heritable and environmental variation in disease susceptibility, an interest that he traced to experiences working with heterogeneous populations and environments in New York City.[2] During his time at Oxford, he read the great population geneticists—E. B. Ford, J. B. S. Haldane, R. A. Fisher, and Sewall Wright. He also became familiar with the new method of gel electrophoresis to separate proteins developed by Oliver Smithies in Toronto. Blumberg realized that examining human genetic polymorphisms with an understanding of population genetics and gel electrophoresis would be a powerful framework for studying disease susceptibility.

His approach was the opposite of a typical medical approach to disease. Instead of starting with people afflicted by a disease and then looking for a cause, Blumberg sought to characterize polymorphisms with the view that they would help explain the different reactions people have to disease. The travel needed to collect blood samples from around the world was fun, too. Some of his samples from Inuit people helped to show that Native Americans and Inuit originated from two different remote spots in Asia.

In 1960, Blumberg was joined by his Oxford colleague Tony Allison at the new Geographic Medicine and Genetics section at the NIH[3] and in the field in Spain and Alaska. Allison, born in 1925 like Blumberg, stimulated Blumberg's interest in polymorphism and would come to be known for his work on sickle cell anemia and malaria. They modified their approach to look for new blood polymorphisms by taking advantage of the reaction of the immune system to blood transfusion. If a person is given a blood transfusion that contains a new protein variant not inherited by the recipient, the immune response will often make antibodies against the new protein variant. A rare polymorphism would create an immune response in almost all the recipients of blood transfusions. The reaction between an antigen and antibodies could be visualized in agar gel plates called Ouchterlony plates after the Swedish bacteriologist Örjan Ouchterlony, who invented the technique in the 1940s. Serum from the transfused patient is placed in a well in the middle of agar, and the test sera are placed in wells surrounding the transfused test sera. Transfused patients have encountered "foreign" blood proteins from the transfused blood and will typically have more antibodies than non-transfused patients. From the center well, antibodies diffuse into the agar, and from the test wells antigens also diffuse into the agar. If the antibodies encounter their specific antigen, they react to form a solid precipitate that can be seen as a line in the gel between the two wells. The diffusion and precipitation took some time, so Blumberg would often set up wells one day and wait until the next morning to see whether any "precipitation arcs" would form.

Blumberg and his colleagues were guided by the belief that if heritable blood protein variation existed and persisted over time, it must play some biological role—otherwise, it would be selected against and disappear. The first significant new polymorphism was found using the blood of "Mr. C. deB," a Hungarian immigrant who had moved to the American Midwest. It turned out that he had antibodies for a low-density lipoprotein[4] that later was shown to be related to an increased risk of heart disease. Blumberg and his colleagues called the new protein "Ag protein." It was inherited in a classical

Mendelian fashion known as autosomal recessive. While the clinical implications of the polymorphism would take decades to understand, the discovery of Ag protein showed that Blumberg's immunological approach could yield new biological knowledge.

Blumberg was not the only one using Ouchterlony plates to investigate otherwise invisible variations in blood. Harvey Alter was working at the NIH blood bank using Ouchterlony plates to search for noncellular causes of febrile (fever causing) transfusion reactions. A New Yorker, Alter got his bachelor's and medical degree from the University of Rochester. While doing his residency he applied for the Public Health Service, the "Yellow Berets," as they were called, but before he was commissioned, he received his Vietnam draft papers.[5] Luckily for the history of medicine, he made it to Fort Dix to be commissioned in the Public Health Service before his draft date of November 1961. Once he joined the USPHS, he worked at a laboratory at the blood bank that was part of the NIH. One day in 1962 a young investigator at the blood bank, Richard Aster, told him that he heard an interesting talk by Baruch Blumberg on an approach similar to Alter's. He met with Blumberg and started a collaboration.

Australia Antigen

The polymorphism that made Blumberg's career and started Alter's was discovered in the first year of their collaboration.[6] In 1964, Blumberg was expanding his search for protein diversity by using from patients who had received multiple blood transfusions that contained antibodies for new polymorphisms to screen test samples from around the world. People who received transfusions are likely have antibodies to foreign proteins they encountered in the transfused blood. Harvey Alter, working in the Blumberg laboratory, serendipitously noticed a precipitin reaction that looked different from the rest (figure 7.1). The arc of visible protein was broader than usual and the staining was lighter. Lipoproteins usually stained light blue, but this arc did not take the blue stain and in fact took a red azocarmine counterstain, indicating a high protein content. At first Alter called it "red antigen." Interestingly, the unusual precipitin reaction was rare in American patients but more common in Taiwanese, Vietnamese, Koreans, and Australian Aboriginal peoples. The reaction with Australian Aboriginal blood was the first to be noticed, and for that the reason Blumberg renamed it Australia antigen (Au), following conventions in hematology that used the origin

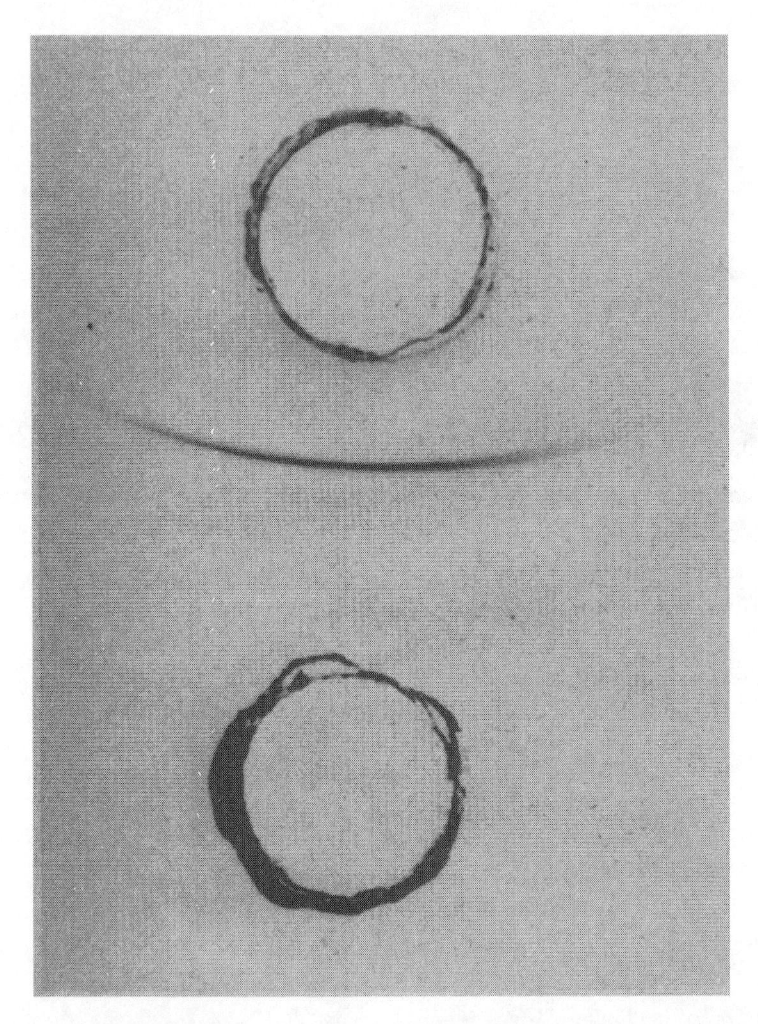

7.1. Ouchterlony plate. Image courtesy of American Philosophical Society

to name new hemoglobin discoveries. The antibody used in testing for Au came from hemophilia patients.

Blumberg and his team did not immediately realize the importance of this discovery. They were generating a lot of data, and this piece of information was just one among many. Blumberg liked to speculate, and he listed a large number of hypotheses on the wall of his office at the NIH.[7] As he put it retrospectively, "We started our new search on the assumption that the

Australia antigen (Au) was also an inherited protein. But it soon became clear that it had additional very interesting characteristics; we experienced a 'Eureka' event, so often described in the annals of scientific discovery. In our case, it wasn't exactly a single leap but a series of unfolding events."[8]

Throughout 1964, Blumberg investigated the geographic distribution of Au. Luckily, the antigen was stable in frozen serum, so samples from around the world could be tested. The antigen was common in Asia, the Pacific, and eastern and southern Europe, and uncommon in the United States, where it was found in only 1 in 1,000 samples tested.[9] It appeared to be more common in places with poorer public health. It also appeared to be clustered in certain families, which was consistent with it being an inherited protein. The group published their results in the prestigious *Journal of the American Medical Association*, or *JAMA*. They noted that the Australia antigen was correlated with leukemia, a disease that it could be used to help diagnose, at least in the United States. Much of the theoretical discussion was speculative: "The available data are too few to support a genetic hypothesis, but none of the family studies are inconsistent with a simple recessive inheritance of the specificity."[10] Subsequent work by other investigators shed doubt on the simple Mendelian inheritance of Au.

A New Blood-Borne Virus

A number of pieces of evidence suggested a blood-borne virus. First, many of those testing positive for Au were patients who had been transfused. Second, in 1966, only one person out of hundreds tested positive in Evans County, Georgia. This county had better records because a medical practitioner there collected a large number of blood samples. Curious whether there was anything special about that patient, Blumberg called Curtis Hames, the practitioner who had collected the samples. That patient had hepatitis, sometimes called "yellow jaundice," which suggested a link between Au and hepatitis.[11] This clue being a single data point was not conclusive, however. Third, another patient, this time from Rongelap Atoll in the Marshall Islands, also was suggestive. This patient was being medically monitored after the atomic testing on Bikini Atoll. After a single blood transfusion in 1965, he was found to have the Australia antigen. Relatedly in 1967, one of Blumberg's technologists, Barbara Werner, who had been the negative control for Au in the lab, developed hepatitis and then tested positive for Au. Relatedly, the correlation with leukemia was also suggestive—perhaps Au causes leukemia, or leukemia generates Au, or both have a common cause. But later work suggested that the connection with leukemia was not simple causation.

Down syndrome children also provided some key data. They were more prone to leukemia, and Blumberg hypothesized they would have more Au positives, which he did show. However, this was true only of Down syndrome children in large institutions, not of such children in smaller institutions. Presumably, infectious diseases spread more easily in larger institutions. One patient had results that were negative and then a later test that was positive. Liver function tests showed that the child had developed a mild case of hepatitis. Blumberg wrote about the case and submitted his account to *Annals of Internal Medicine*, but to his surprise, the paper was rejected. There had been other unconvincing reports of a hepatitis virus, and the paper's reviewer wanted more evidence. Nonetheless, in 1967, Blumberg and his associates published a paper that explicitly stated that Au could be associated with a hepatitis virus.[12]

One of Blumberg's rivals, Alfred "Fred" Prince of the New York Blood Center, likewise suggested that the Au virus was in fact the same virus that had been identified in earlier hepatitis research,[13] later called hepatitis B virus to distinguish it from hepatitis A virus, which is transmitted when fecal matter is ingested. Prince's work was the result of an abortive collaboration with Blumberg.[14] Prince thought they had enough evidence to call Australia antigen a virus; Blumberg did not. When Prince published his paper, he did not cite Blumberg's group's papers, and it caused some bad blood between the two. Blumberg's colleagues thought Prince was overemphasizing the novelty of his work; Prince argued that, without Blumberg's reagent, he couldn't make the required tests to be more definitive.[15]

There was much resistance to Blumberg's ideas in the community of hepatitis researchers, a phenomenon that he interpreted through a Kuhnian lens. Kuhn had famously argued that when new ideas emerge to challenge an old paradigm, established scientists are reluctant to endorse them.[16] In this case the established hepatologists were reluctant to believe that a geneticist, Blumberg, and a hematologist, Alter, could overturn entrenched views about the causes of hepatitis. After reading Kuhn, Blumberg was not surprised at the resistance.

Electron microscopy provided additional evidence for a viral link to hepatitis. In 1964 Blumberg moved to the Institute for Cancer Research, which became the Fox Chase Cancer Center. He also became affiliated with the University of Pennsylvania in 1967. Philadelphia was home to an impressive electron microscopy group led by one of the pioneers in the field, Thomas Anderson. With Manfred Bayer and Barbara Werner, who worked with Anderson, Blumberg observed particles 21 nanometers in diameter.[17] However, further investigation showed that they did not contain DNA. Blumberg and his colleagues speculated that the protein particles might be infectious and able to replicate making protein from protein. If this was correct, then it would violate the central dogma of molecular biology—roughly that there is transfer of information from nucleic acid to protein, but there cannot be transfer of information from protein to protein or from protein to nucleic acid. (Later prions would be discovered that operated in a similar manner—from protein to protein.) But these wild speculations were terminated by further work in microscopy. In 1970, a group of British scientists took electron micrographs of a 42-nanometer particle that they inferred was the virus. Smaller particles of 21 nanometers were probably excess coat protein—that is, viral coat protein without any viral DNA.[18]

Blumberg tried to grow the virus in tissue and cell culture, drawing on the talent of Scott Mazzur, who had trained with the Henles in Philadelphia. They were unsuccessful. Instead, the Blumberg lab tried to demonstrate transmission of the virus in animal studies, an approach suggested by one of Koch's postulates. They injected Australia antigen injected into Rhesus monkeys, which developed more Australia antigen, which then could infect a further monkey.[19] Australia antigen was increasingly looking like an infectious agent.

During this period, two members of Blumberg's group, Tom London and the immunologist Irv Millman, made an important and surprising discovery.[20] They found that highly purified Australia antigen injected into animals did not lead to infection but less purified samples did. The smaller protein particles that did not contain DNA did not cause infection. As would become clearer later, this would be an important fact in constructing a vaccine.

Hepatitis and the Blood Supply

Blumberg's work had practical effects relatively quickly. In the 1960s, with the rise of new types of surgery along with the rise of health insurance,[21] the number of surgeries in the United States requiring blood transfusions grew significantly. Up to 50% of people who had received multiple transfusions during surgery developed posttransfusion hepatitis. Many blood bank scientists were unaware of Blumberg's test. Furthermore, there was some skepticism whether Au was a virus and further evidence was needed to force any changes to blood bank procedures. Blumberg and his colleagues designed a study to test whether screening for Australia antigen would decrease the number of postoperative hepatitis cases. That he, Tom London, and Alton Sutnick were attending physicians at Philadelphia General Hospital made it easier. They would test blood for Australia antigen and then track the patients who received Au+ blood, comparing them to patients who received Au– blood, beginning in late 1968. A parallel study was started in Japan the next year. Results came in from the Japanese study first, partly because a higher percentage of the blood tested Au+ and because the donation volume was smaller and subsequently recipients of donor blood got more donors per operation. Nineteen of the 58 patients who received Au+ blood became Au+ after 110 days; only 10 of 99 patients who received Au– blood did so. This was a statistically significant result. The Philadelphia study produced a similar result, and by July 1969 the clinical trial was stopped so that patients no longer received Au+ blood, and it became routine at the Philadelphia hospital to test blood for Au. Following the new rules, the rate of posttransfusion hepatitis fell from roughly 30% to 9%. Not all the decline was due to testing for Au. At the same time, rules for blood donation changed from paid donors to volunteers. Only roughly 25% of cases of posttransfusion hepatitis case were related to Au.

The National Research Council considered whether to adopt the rule nationally and certify the test. Its members decided that because the test detected only about 25% of all hepatitis, more research was needed. The blood bank at the NIH disagreed, noting that the 25% of cases the test captured represented 40,000 cases of hepatitis in the United States that could be prevented.[22] As Blumberg put it, "There was no reason to allow the perfect to drive out the good."[23] It might have helped that a former member of Blumberg's lab, Harvey Alter, was working at the blood bank. After this decision by the NIH was covered by the *New York Times*,[24] the NRC changed its recommendation to for laboratories with the capacity to test blood for hepatitis to do so. In 1971, New York passed a state law to require testing, and in 1972, the Federal Register required blood to be tested if it was used in interstate commerce. In 1973, the test was required of all blood banks, and by the mid-1970s, posttransfusion hepatitis cases caused by the virus were negligible.[25]

In its early days, the new test for hepatitis B virus seemed to raise a number of ethical issues. Should an HBV-positive person be allowed to practice various forms of medicine or be treated at a dialysis center, for example? Should someone be denied a job, for safety reasons, because of their HBV status? Should Vietnamese children fleeing their war-torn region be allowed into the United States if they were HBV positive? But as research continued, some of these issues became less pertinent. It became clear that HBV was not as infectious as many other viruses, and medical practitioners did not seem to pass the virus to their patients. The US Public Health Service decided not to test Vietnamese children.

The Hepatitis B Vaccine

With the term of his major grant ending and prompted by the possible loss of significant NCI funding for their research, Blumberg decided that he and his associates should file a patent for a hepatitis B vaccine. After thousands of tests, rarely had someone both been a carrier of HBV and also had the antibody against the surface antigen (anti-HBs). This result suggests that people with the antibody to surface antigen are able to ward off an infection from HBV. Work with vervet monkeys showed that the pure surface antigen could be separated from the larger infectious particles using a centrifuge. The pure surface antigen was composed of the smaller particles seen with the electron microscope. Adding additional procedures to destroy any residual live virus gave a method that could be patented. This was a novel way of making a vaccine. It was not made with an attenuated virus like the Sabin poliovirus vaccine or a killed virus like the Salk poliovirus vaccine. Rather, it took advantage of the strange property that HBV makes excess viral coat protein, which was then harvested from infected individuals. The application to the patent office for an HBV vaccine was submitted in October 1969, and the patent was issued in January 1972.[26]

Commercializing a vaccine was outside the scope of the Fox Chase Cancer Center, so Blumberg looked for a pharmaceutical company to develop and test it. After considering an offer from a French company, Blumberg and Fox Chase decided to go with the New Jersey–based Merck, in part because, given the support he had been given by the US government and the Commonwealth of Pennsylvania, it seemed right to go with a US company. Maurice Hilleman, who was involved in the discovery of SV40 (see chapter 10), headed up the program and had corresponded with Blumberg since the early 1970s.[27]

He perfected the procedures to guarantee that no infectious virus or other contaminating viruses ended up in the vaccine, no small feat. Hilleman first tested his purified hepatitis B virus surface protein vaccine on Merck executives, who he monitored for six months to make sure that they did not develop hepatitis. His contribution was significant, and he later complained that Blumberg did not give him enough credit in his history of the vaccine.[28]

Once the vaccine was manufactured, clinical trials began. It made sense to focus these trials on high-risk populations like gay men. The Polish American physician Wolf Szmuness and his colleagues ran a double-blind trial in 1978 and 1979.[29] During these two years, they enrolled more than 1,000 people, a little over 500 each in the placebo and vaccine wings of the trial. The results were remarkable: 52 subjects developed HBV, 45 of whom were in the placebo wing. Of the seven in the vaccine wing who developed HBV, only one had completed the three doses of the vaccine. In 1983, a study in Taiwan showed that the transmission of HBV from mothers to babies fell from over 90% to less than 10% when the infants were vaccinated.[30] In only two years, the FDA approved the vaccine, which became part of the standard set of vaccines given to children in the United States. Over a billion doses have been administered, and it is on the WHO list of essential medicines. In 1986, a recombinant version was produced, in which the viral coat protein was cloned into yeast; this product is considered safer than the original blood-derived version.[31]

Throughout the 1970s, Blumberg investigated the causal connection between HBV and hepatocellular carcinoma (HCC), a common type of liver cancer. Starting at the beginning of the decade, researchers noticed a correlation between antibodies to HBV and liver cancer: a significant percentage of patients who had chronic HBV infections would develop liver cancer.[32] Blumberg looked at cases of HCC in Africa and other underdeveloped places in the world. Like Denis Burkitt, Blumberg used international geographical analysis to uncover a virus that caused cancer. Tumor virology was becoming increasingly international.[33]

Blumberg won the Nobel Prize in 1976. He remained at the Fox Chase Cancer Center for most of his career but traveled often, promoting the HBV vaccine around the world. In 1989, he became the master of Balliol College at Oxford University, where he had received his doctorate. In 1999, Blumberg became the founding director of the NASA Astrobiology Institute (NAI). He was tasked with investigating three large questions: How does life begin

7.2. Baruch Blumberg and Harvey Alter. Image courtesy of Harvey Alter

and evolve? Does life exist elsewhere in the universe? And what is life's future on Earth and beyond?[34] In March 2003 he joined the Board of Trustees of the SETI (Search for Extraterrestrial Intelligence) Institute, expanding his involvement in the realm of astrobiology. In 2005 he was elected president of the American Philosophical Society, and he remained in that position until his death in 2011 at the age of 85.

Harvey Alter continued to work on viral hepatitis (figure 7.2). In the 1970s he showed that most cases of posttransfusion hepatitis were not due to hepatitis A or B virus. Researchers looked for an elusive third hepatitis virus for 10 years, but the traditional approaches to virus hunting failed. Michael Houghton at Chiron Corporation found DNA fragments from an unknown virus in an infected chimpanzee. These fragments made proteins that reacted to antibodies made by patients with chronic hepatitis. This work provided evidence for third virus, an RNA flavivirus that was subsequently called hepatitis C virus (HCV) in 1989.[35] Charles Rice at the University of Washington and later the Rockefeller University analyzed the genome of HCV and showed that RNA transcripts from the genome could cause an

7.3. *Left,* Winner of the 2020 Nobel Prize: Dr. Michael Houghton in his laboratory at Li Ka Shing Institute of Virology at the University of Alberta. Photo Michael Holly. © University of Alberta. Image courtesy of Michael Houghton and used with permission. *Right,* Charles Rice, with his postdoc Arash Grakoui examining a plate for mutants. Image courtesy of Rockefeller University

HCV infection after being injected into chimpanzee livers. He and his collaborators also created a genetically modified HCV that could serve as a robust cell-based model system in the laboratory.[36] Houghton, Alter, and others created a diagnostic test for HCV. After testing for HCV in blood, rates of posttransfusion hepatitis fell close to zero in the late 1990s.[37] The Lasker Foundation gave its award to Alter and Houghton in 2000 and to Rice in 2016. In 2020 Alter, Houghton, and Rice won the Nobel Prize in Medicine or Physiology for their work on HCV (figure 7.3). Work on an HCV vaccine is ongoing.

8

A Breakthrough and a New Tool

Howard Temin, David Baltimore, and Reverse Transcriptase

If anyone is the scientific hero of these cancer virus hunter stories, it is probably Howard Temin.[1] He had the courage to resist other scientists' derision for much of the 1960s when the emerging field of tumor virology was in its adolescent phase. He remained resolute that his ideas were correct, and eventually he was vindicated when he, along with David Baltimore, modified biologists' understanding of the central dogma of molecular biology.

Temin was born in 1934 in Philadelphia, the middle child in an academically talented family. His younger brother would go on to become an economics professor at MIT. As a child, Temin was a voracious reader who excelled at school and gave the valedictory address at Central High School in Philadelphia in 1951.[2] Science was cemented as Temin's calling by his participation in a summer school for high school kids at the Jackson Laboratory in Bar Harbor, Maine, the same laboratory that played a key role in Bittner's career, described in chapter 1. Interviewed about his experience by the *Philadelphia Inquirer*, he observed, "Scientists aren't an eccentric bunch at all, the way they're pictured in some books and movies. They're perfectly normal, hard-working men and women. This is the first time I've come into close contact with working scientists, and I've found them democratic, friendly, and helpful. Now more than ever I plan to go into science myself."[3] He would repeat his hands-on experience at Jackson Laboratory for the next three summers, in the last as an instructor. Frederick Avis, Temin's instructor, wrote, "I can't help but feel this boy is destined to become a really great man in the field of science."

Unsurprisingly, when he got to university, Temin majored in biology.

However, he was frustrated that the biology curriculum at Swarthmore College in the early to mid-1950s had not incorporated the new molecular biology exemplified by Watson and Crick's structure of DNA in 1953. By selecting Caltech for graduate school, Temin remedied this omission: it was one of the centers of the new biology. While there, Temin lived with Matt Meselson, Frank Stahl, Jan Drake, and John Cairns, who were also destined for successful careers in molecular biology. Although he started in embryology, Temin shifted to virology and socialized with the phage group led by Max Delbrück. He joined the Dulbecco laboratory and worked with Harry Rubin to develop a quantitative assay to measure neoplastic transformation of animal cells by tumor viruses (see chapter 3). The 1958 publication with Rubin on the transformation assay would be his first widely cited paper. By 1959 he reported to his parents, "I have established that there are 'genetic factors' in the cell and in the virus for [resistance to infection] and that the variation arises in a mutation like way."[4] He wrote up his results for *Engineering and Science*, a Caltech publication. After summarizing what was known about Rous sarcoma virus, Temin put his views this way:

> The existence of such a small number of genetic units of the virus in the cell, and the regular inheritance of these units, shows that the virus, in some structural sense, as well as the functional sense . . . becomes a part of the genome of the cell. Probably it does not attach to a chromosome, and may not even be in the nucleus, but becomes part of the general apparatus of the cell which controls what a cell is.[5]

The idea that the viral genetic information somehow fused with the host's genome would motivate Temin's research program for the next decade. An offer from the University of Wisconsin came with the laboratory he needed to develop and test his ideas. Van Potter, a professor of oncology, encouraged Temin to join Wisconsin to form a new research focus in virology.[6]

Temin Begins a Laboratory at the University of Wisconsin

In the fall of 1960, Temin launched his research program in Wisconsin as part of the McArdle Laboratory, one of the first basic cancer research centers housed at a university. He stayed in contact with Marguerite Vogt, Renato Dulbecco, and his French collaborator Boris Ephrussi, with whom Temin had worked on RSV with chick iris epithelium instead of fibroblasts. But the precocious young researcher sometimes faced obstacles. For example, a grant application he wrote in 1960 was nearly rejected on account of Temin being only 25 years old. The founding director of McArdle Laboratory, Harold Rusch, a particularly supportive scientist and administrator, had to step in and vouch for him to secure the funding.[7]

In July 1961, Dulbecco wrote to Temin, "I have been told by many people that you have a great discovery at your hands, but nobody knows what it is about. Why don't you tell us something more precisely rather than being so mysterious?"[8] Temin replied that he was not being mysterious and detailed his results: "After many wrong ideas I finally have a consistent experimental picture. . . . [W]hat I have found is: infected chick cells plated on mouse cells give rise to foci of altered cells which differ greatly in FFU [focus forming units] / viable cell from 0 to 2."[9] In other words, he had discovered two different RSV infected cells, distinguished from one another by their ability to form foci on plates (and also by looking at the shape of the infected cells). Because of these differences, and some additional results, Temin concluded that there were two different types of virus-cell complexes and an inducing factor that might be viral.

His results attracted praise from fellow virologists. W. Ray Bryan of the National Institutes of Health wrote, "There is no question that you have found an approach to a problem of very great importance at the present time; I refer to the induction of one tumor virus (in occult state) by another, or by some factor in supernatants or extracts from viral tumors."[10] Sarah Stewart echoed Bryan's praise.[11] When Bryan invited him to talk at the NIH, Temin proposed a title that captured how he was thinking about RSV in early 1962, "Rous Sarcoma Virus: A Paraepisome." By "paraepisome," presumably he meant an independent genetic element that could replicate inside the host cell but was not part of the host genome.

During this period, other aspects of Temin's life were also going well. In 1962 he married Rayla Greenberg, who studied population genetics under James F. Crow. It was a good match, both romantically and scientifically, and the newly married couple spent part of their honeymoon at Cold Spring Harbor Laboratory at the 1962 meeting on basic mechanisms in animal viruses.

Presenting the Provirus Hypothesis

The 1962 CSHL meeting was an important conference in the history of animal virology. Dulbecco, Harry Rubin, Michael Stoker, and David Baltimore all gave talks in addition to Temin. In a session chaired by W. R. Bryan,

Temin summarized his research on RSV. He had discovered some converted non-virus-producing cells—what he called CNVP cells—suggesting that viral replication and host conversion were two distinct functions of RSV. Additionally, he had found two different strains of RSV. One strain, which he called morph^{r}, caused host cells to turn into round cells. Another strain, morph^{f}, turned host cells into elongated fusiform (spindle-shaped) cells. Interestingly, if Temin took some fibroblasts that were infected with morph^{r} and superinfected them with morph^{f}, the second strain turned the cells from round to elongated. As Temin noted in his talk, "These results suggest that the change in cell morphology is a process like lysogenic conversion."[12] Additional experiments and an analogy with polyoma virus led Temin to infer that the integration of the viral genome at a specific site in the cells causes carcinogenesis.

The obvious difficulty with Temin's conclusions was that the RSV genome was made of RNA, not DNA. He was well aware that Lionel and Elizabeth Crawford had published this exact result in the journal *Virology* in 1961, and they had likely discussed it, as they were friends. In fact, they wrote him a long letter congratulating him on his marriage the exact day he gave his talk at CSHL.[13] If RSV genetic material integrated into the host cell DNA, then there would have to be a mechanism to first convert RSV RNA to RSV DNA, a process for which no known mechanism existed. In fact, Francis Crick had codified a basic assumption in molecular biology that he called the central dogma of molecular biology. It was often understood by molecular biologists schematically as the claim that DNA makes RNA, which in turn makes protein, but understood cybernetically, in terms of the flow of information from one type of molecule to the next, not the conversion of one molecule into another.[14]

Following the CSHL conference, Temin began to search for more biochemical evidence that RSV replication requires DNA. He wrote to Marguerite Vogt in August 1963, "I am trying to become a biochemist now. I have decided to look for new enzymes. I have postulated that RSV-RNA makes a new DNA when it becomes provirus and am going to hunt for a polymerase. At every step I go across to the biochemists and find out what to do. So far it is fun."[15] Like many communications among colleagues, his letter gives a sense of Temin's many interests and concerns beyond his specialized corner of science. "We applauded the test ban treaty and wired our senators," Temin wrote, referring to the Limited Nuclear Test Ban Treaty that President Kennedy had signed that week. "I hope it is ratified."[16]

Temin's publishable results, written up in a variety of journal articles, did not involve a direct search for the polymerase that made DNA from an RNA template; he was constrained by the tools that were available in 1963. One of his first stabs at the problem was to inhibit the functions of DNA with chemicals and see whether doing so also inhibited RSV production. Temin used actinomycin D, later used in chemotherapy treatments, to show that virus production was reversibly inhibited. He ended his paper with the following conclusion: "Since the effect of actinomycin depends on interaction with DNA, the effect on RSV production is probably mediated through DNA. It appears . . . that the treatment prevents formation of viral RNA. Therefore, it is suggested that the template responsible for the synthesis of virus either is DNA or is located on DNA."[17] This hypothesis that the RNA RSV goes through a DNA template as part of its life cycle came to be known as the "provirus hypothesis," and it garnered some positive feedback when this paper was published. Richard Barry, working in the Department of Pathology at the University of Cambridge wrote to say that he had gotten similar results from influenza virus, which also appeared to be dependent on DNA for its replication.[18]

An international conference at Duke University on avian tumor viruses in April 1964 offered an opportune time for Temin to summarize his data for his DNA provirus hypothesis. In his presentation, he pointed out two ways of determining the nature of a provirus in Rous-infected cells: first, consider the morphology of infected cells, and second, look at the genotypes of released viruses. Temin estimated the mutation rate of a provirus—about 1 mutation in 10,000 cell divisions, a number that he considered high. In addition to actinomycin D experiments, Temin described his experiments with amethopterin and 5-fluorodeoxyuridine and amethopterin. These chemicals inhibited DNA synthesis while also inhibiting virus infection of cells, thus suggesting the virus needed thymadine to infect the host cell. (One of the bases of DNA is thymine, often abbreviated T, and it is not found in RNA).

Temin encountered strong resistance to his idea of a DNA provirus. As Geoffrey Cooper and Bill Sugden write, "Throughout the 1960s the DNA provirus hypothesis was given little credence in the scientific community, instead being met with scorn and even anger by many of Howard's colleagues."[19] As Michael Bishop put it, the hypothesis earned Temin "years of derision."[20] Detractors thought that the experimental evidence Temin marshaled was too weak for his bold claim that an unknown mechanism for creating DNA sequences from RNA sequences. But Temin stuck to his guns.

The Early Career of David Baltimore

In the 1960s, Temin corresponded with a number of scientists, including many featured in this book: Dulbecco, Watson, Stoker, Macpherson, Huebner, Fenner, Crawford, Vogt, and Svoboda, among others. Among the many letters that passed between them, one written at the beginning of the decade would stand out for its significance by the end. David Baltimore, a biology student at MIT, wrote to Temin for advice in 1961. After explaining that he was becoming interested in animal virology, he asked "for any advice-to-a-novice about pathways of development in biology which appear to you most fertile."[21] In reply, Temin told him not to worry about the direction biology was headed but rather just learn how to perform good experiments. Little did Temin know that in the 1970s Baltimore's and Temin's careers would converge.

Baltimore was raised in Great Neck, Long Island, New York. In 1955, his junior year of high school, Baltimore successfully applied to the same summer research program at Roscoe B. Jackson Memorial Laboratory in Bar Harbor, Maine, that Temin had attended five years earlier. Temin had returned to the program in 1955 as a mentor to the younger students, and Baltimore thought of him as a biology "guru."[22] During that summer, Baltimore performed experiments in mouse genetics and blood cell formation. He spent many hours at the microscope identifying different blood cells. He found a role model in Temin—someone to be venerated for his deep knowledge and commitment to science. And the overall Jackson Laboratory experience cemented Baltimore's desire to build a career in science.

The first step he took was to attend Swarthmore College near Philadelphia, again following in Temin's footsteps. He developed a circle of friends including Peter Temin, Howard Temin's brother, and Gil Harman, later a Princeton philosopher. Not surprisingly, Baltimore majored in biology, albeit mostly focusing on classical biology in the classroom. To supplement his formal education, he spent time reading journal articles in the new molecular biology. Baltimore was smart and competitive. Although they did not overlap—Temin graduated a year before Baltimore arrived—the legend of Howard Temin as a mythically good student at Swarthmore was still alive. As his biographer put it, Baltimore "saw Howard's fame as a challenge. Because Howard wasn't at Swarthmore any longer, this was a competition with a shadow, a myth, a sort of academic ideal, and it pushed David to try to be the greatest biology student the campus had ever seen."[23]

David Baltimore knew that the way to master biology was to get more hands-on laboratory experience. He asked his microbiology professor where he could get bacteria and bacteriophage to play around with, and the professor suggested Cold Spring Harbor Laboratory, not far from Baltimore's childhood home in Long Island. At Easter break in 1959, he drove to CSHL and spoke there with Hungarian biologist George Streisinger, a former student of Salvador Luria's. Streisinger showed Baltimore how to grow bacteria in glass flasks and how to plate infected bacteria on petri dishes. He also invited Baltimore to return to CSHL in the summer for an NSF-funded program designed to give undergraduates access to laboratory research.

Returning to Swarthmore, Baltimore switched his major to chemistry so he could get more laboratory experience. Biology did not have the same opportunities. He learned how to purify protein with Philip George, who worked on the chemistry of ATP, an important chemical in cellular energy production. This training in biochemistry would serve him well.

In addition to introducing Baltimore to experimental virology, Streisinger introduced him to Cy Levinthal and Luria, two heavyweight molecular biologists at MIT. On the basis of Streisinger's recommendation, Baltimore was accepted into MIT for graduate school starting in the spring of 1960.[24] There he took classes and conducted phage research in Levinthal's lab. He was interested in animal viruses, but nobody at MIT was working in this area. Luria gave Baltimore guidance: "I don't know whether there are things to learn about animal cells that are so different than bacteria. But I think it's a good question and animal viruses are a fascinating group of viruses. I tell you what: You are going to have to have to decide this for yourself. I'll help."[25] Luria arranged for Baltimore to work on animal virology in the New York laboratory of Philip Marcus and also to take the new animal virology course at Cold Spring Harbor Laboratory. At the course, Baltimore met Richard Franklin, an assistant professor at Rockefeller University. Franklin was investigating questions about the biochemistry of viruses that were exactly the ones that interested Baltimore, so, with Luria's help, Baltimore moved to Franklin's lab. Franklin had shown how mengovirus, a small RNA virus, shut off nuclear synthesis of the host cell and then induced new viral RNA synthesis in the cytoplasm. Baltimore worked at the lab for two full years, completing a thesis titled "The Diversion of Macromolecular Synthesis in L-Cells towards Ends Dictated by Mengovirus." He gained experience working with RNA viruses, including poliovirus, and became interested in RNA itself.

After two years at Rockefeller, Baltimore returned to MIT and spent a year working with James Darnell, later a coauthor with Luria of the second edition of the textbook *General Virology*. At each new laboratory, Baltimore focused on acquiring new skills. From Darnell, he learned how to use sucrose density gradients and electrophoretic gels to probe RNA synthesis. Following Darnell's move to Albert Einstein College of Medicine at the end of 1964, Baltimore joined Jerald (Jerry) Hurwitz's laboratory there. Hurwitz was one of the discoverers of RNA polymerase, having studied with Arthur Kornberg. Hurwitz was interested in nucleic acid biosynthesis, and while working with him, Baltimore learned how to do in vitro biosynthetic work.

Baltimore had chutzpah—he thought nothing of bouncing around from lab to lab, picking up the skills he needed. As he put it, "I was on a fast track in life . . . building myself up so I could become the person I wanted to be."[26] In the same amount of time, a typical young biologist would have less than half the experience. Scientists of the older generation took note of his determination and drive. When Renato Dulbecco was lecturing in New York City in late 1964, he arranged to meet with Baltimore to offer him a job. He said, "Look, I would love to have you join me. I am going to move to the Salk Institute from Caltech. And I can offer you a position .. . . I'm going to have a huge amount of space. . . . I'll help you develop [as a scientist]. I won't get in your way. You can do anything you want!"[27] An offer like that was too good to refuse, so Baltimore moved west to the Salk, staying there from April 1965 to the end of 1967, at which time he returned to MIT and the relative ease of an established institution. Returning to the East Coast with Baltimore were his first student, Michael Jacobson, later a founder of the Center for Science in the Public Interest,[28] and Alice Huang, his second postdoc and later his wife.

Baltimore, now an associate professor, began to think about trying a different virus-host system. Up until this point Baltimore had worked with poliovirus, but he shifted his focus to vesicular stomatitis virus (VSV) on Huang's advice; she thought poliovirus research was "getting a little monomaniacal." Huang had devoted her thesis to VSV, which was a good virus for use in experiments because it grew fast to a high titer. Baltimore's strategy was to apply the approach and tools worked out on poliovirus to VSV.

A new graduate student mentored by Baltimore and Huang began to investigate the nucleic acid synthesis of VSV. The Baltimore lab had "walked into a new world."[29] Poliovirus was a positive-stranded RNA virus, which meant that its genome could serve as messenger RNA once inside the host cell. The Baltimore lab had good evidence that VSV, on the contrary, was a negative-stranded virus.[30] In other words, the genome of VSV was complementary to the messenger RNA. The difference between the two types of ssRNA genomes was significant, and Baltimore would later invent a classification system of viruses based on the seven types of genome.[31] VSV is an example of a group 5 virus; poliovirus, a group 4. (SARS-CoV-2 is also a group 4.) Clearly, VSV has a more complicated replication cycle than poliovirus—it uses complementary RNA to start and has many more RNA products than polio. Presumably there had to be an enzyme, a polymerase, to copy the negative-strand information into a positive strand that then could be translated into protein, what molecular biologists would eventually call an RNA-dependent RNA polymerase. Baltimore's background in biochemistry, which extended back to his graduate school days, helped as he and his colleagues looked for the polymerase. It did not appear to be in the cell, but another possibility was that the polymerase was packaged along with the RNA in the virus particle itself; in other words, the virus brought its own polymerase. There was some evidence that a virus could do this: in 1968, a polymerase that used double-stranded RNA as a template was found in a reovirus.[32] Once Baltimore and his team thought to look in the virus itself for polymerase, it was a relatively small task to run the biochemical experiments. And "sure enough, it was there."[33]

Early in 1970, Baltimore began to wonder what other viruses might have this same mechanism. Some viruses, like polio, did not. Reovirus and vaccinia did.[34] Did merely "oddball" viruses carry the polymerase, or was it more widespread? At the time, it was unknown whether there were other negative-stranded viruses, so the Baltimore lab looked at Newcastle disease virus, which had a polymerase, and influenza virus, which appeared not to. (Later work showed that it did, but it required an experimental trick to see it.) The last group of viruses Baltimore's team looked at was the RNA tumor viruses.

Baltimore was aware that Temin had been working on RNA tumor viruses for a decade, and that he claimed that RSV made a DNA provirus. While he respected Temin as a scientist, Baltimore, like almost everyone else, did not find the evidence convincing. Nonetheless, he sought the virus to work with. Peter Vogt sent him RSV, which he assayed for RNA polymerase, but he came up empty handed. Then he called another friend, fellow Swarthmore graduate George Todaro at the NIH. Luckily, the Special Virus Leukemia Program of the NCI had been stockpiling tumor virus for years and would be only too happy to send Baltimore what he needed.

The NIH sent Baltimore a vast amount of Rauscher murine leukemia virus in a vial. Baltimore added detergent to the vial to remove lipid membranes and then assayed for polymerase. The results were close to zero activity, but there was one nonzero result that kept him going. He then centrifuged the virus sample to make it more concentrated and tried again. Then Baltimore had a clear positive result for DNA polymerase, which he then repeated for RSV. The key experiments took less than a week, but that was largely because years of biochemistry training provided Baltimore with the skills to design and run the experiments quickly. He was delayed by several days, when on April 30, 1970, the US military invaded Cambodia, and Baltimore, like many at MIT, temporarily closed his laboratory in protest.

Baltimore started to write up his results. He called Temin with the good news, knowing that he would be interested in the results. The conversation went something like this:

> "Howard, there's DNA polymerase in the virion of RNA tumor viruses!"
>
> "I know, but where did you hear it?"
>
> "I didn't hear it. I did it."
>
> "You did it? We did it!"[35]

By "we" Temin meant him and his Japanese postdoc Satoshi Mizutani. Temin announced the discovery on May 28 at the Tenth International Cancer Congress in a short talk in Houston, Texas, but Baltimore was not at the conference. Temin and Baltimore agreed that they should publish their independent discoveries together, noting that "the findings are remarkably similar."[36] In June 1970, their respective papers were published back to back in *Nature*, just 12 days after they were submitted. Temin and Mizutani concluded their article, "If the present results and Baltimore's results with Rauscher leukemia virus are upheld, they will constitute strong evidence that the DNA provirus hypothesis is correct . . . [and] have strong implications for theories of information transfer in other biological systems." Both approaches showed that the polymerase activity was dependent on deoxyribonucleotides, the building blocks of DNA. An enzyme that destroys RNA, ribonuclease, also destroyed the formation of DNA, presumably by destroying the RNA template. They were right that the discovery of what would become called reverse transcriptase would have theoretical implications. It would also have drastic practical implications because it allowed for the study of RNA by first reverse-transcribing it into a single strand of what has come to be called complementary DNA, or cDNA. After degrading the RNA, biologists then can use a regular DNA polymerase enzyme to synthesize the second DNA strand, creating a DNA double helix, which is more stable and easier to study than RNA.

Temin was correct that the biggest theoretical consequence concerned what is called the central dogma of molecular biology. Francis Crick in 1958 had hypothesized that once biological information got into protein, it could not get out.[37] Although he called this principle a dogma, Crick later acknowledged that he did not know what the word meant when he coined the phrase—he thought it meant an idea for which there was no reasonable evidence.[38] Nevertheless, the central dogma did function as a foundational assumption of the new biology and was accepted by practically every molecular biologist. Now, more than a decade later, Crick took notice of Temin's work, writing in July 1970, "Congratulations on both your farsightedness and on your excellent and important work." He also informed Temin that he had an open mind and had never said that an RNA to DNA information transfer was impossible.[39] Temin knew this, as Crick had spoken in 1965 with the virologist Dick Matthews, who was working in Temin's lab at the time. In response to the discovery of reverse transcriptase, Crick wrote an article for *Nature*.[40] Building on his 1958 hypothesis about information transfer in biological systems, he now argued that there were general transfers of information from DNA to RNA (transcription), from RNA to protein (translation) and from DNA to DNA (replication) that could be found in every cell. A second class of special transfer were rare, RNA to DNA, RNA to RNA, and DNA to protein. The third class were biologically impossible: protein to RNA, protein to protein, and protein to DNA. Since reverse transcriptase did not overturn any of these postulated impossible transfers, it did not overturn the central dogma as he now formulated it.[41]

Upon receiving Crick's *Nature* article from him, Temin wrote a careful reply. After all, Crick was one of the seminal thinkers in molecular biology. In fact, Temin's letter went through three drafts before he sent it.[42] Temin suggested that the RNA-to-DNA transfers might not be confined to RNA tumor virus life cycles. He also was skeptical that the third class of transfers were impossible: "After reading your article (and rereading it), I still wonder why finding the unknown transfers would be more disturbing than finding the special transfers. In fact, with the presently unresolved questions of embryogenesis, long-term memory, and the immune response, I expect the unknown transfers do exist."[43] Crick replied in his typical English politeness that he "quite disagrees" with Temin about the unknown transfers. It was his opinion that the mechanisms needed to make the transfers would have to be "so elaborate" that they wouldn't exist.[44] They continued to correspond, and Crick's last letter grew to four pages of dense argumentation.

Over the course of this correspondence, the discussion drifted from the central dogma to the related "sequence hypothesis": the idea that the sequence of bases of DNA determines the sequence of amino acids in a protein.[45] Temin thought this was more important than the central dogma and mentioned Sydney Brenner's rendition of the idea, notably that given the sequence of the genome of an organism, one could describe (or compute) the organism. Crick thought that the important question was how much extra information is required, in addition to the DNA and code, to make a cell work. Some information was in the environment, but Crick wondered how much.[46] He ended his letter with an invitation to Temin to visit him next time he was in Cambridge so they could hash out these issues in person.

The discovery of reverse transcriptase also had important practical implications. Some thought that an assay for that activity could be an assay for cancer cells. Robert Gallo (see chapter 14) thought that he found reverse transcriptase activity in three leukemia patients, hinting at the existence of a human leukemia RNA tumor virus. Sol Spiegelman of Columbia University reported finding reverse transcriptase in 120 people suffering from leukemia.[47] Research money for this sort of work came easily as President Nixon began the War on Cancer in 1971, and the Special Virus Cancer Program[48] of the NCI viewed the discovery of reverse transcriptase as opening a new fruitful approach to search for a human cancer virus.[49] Baltimore thought the ben-

Newsweek
February 22, 1971 / 50 cents
The War Against CANCER
A PROGRESS REPORT
Researcher Howard M. Temin

8.1. Howard Temin on the cover of *Newsweek* in February 1971. Used under license from EnVeritas Group, Inc.

efits of reverse transcriptase were more intellectual, but the field as a whole, and Baltimore's and Temin's labs, benefited from free-flowing federal grant money. The importance of the discovery was seen rapidly, and Baltimore and Temin won the Warren Triennial Prize in 1971, the same award Watson and Crick had won in 1959.

MIT was a recipient of a great deal of money devoted to finding a cure for cancer, from both private and governmental entities. A private donor gave the university $1.75 million to build a center for cancer research, and Baltimore himself was awarded an American Cancer Society professorship as long as he pursued cancer research. The Center for Cancer Research at MIT was completed in late 1973, and Baltimore moved his lab there.

Meanwhile, Temin continued to work on viruses. The discovery of reverse transcriptase vindicated his views about the DNA provirus and convinced the skeptics, making him widely seen among the biology community as a deeply insightful scientist. Temin also became a public figure. *Newsweek* magazine put him on the cover of their February 22, 1971, issue (figure 8.1). Quoted in the article, Baltimore pointed out that, for the virologist Temin, the discovery of reverse transcriptase capped off a decade-long search, whereas Baltimore himself was a relative newcomer to tumor virology. Baltimore's advantage was his background in biochemistry, whereas the more genetically inclined Temin had taken longer to work out the necessary biochemical assays.

Only five years after their blockbuster discovery, Temin and Baltimore were awarded the 1975 Nobel Prize (figure 8.2). This followed the Lasker Award of 1974, which Temin shared with Sol Spiegelman, Ludwik Gross (see chapter 1), and two others.[50] Fittingly, Temin and Baltimore's Nobel Prize was shared with Renato Dulbecco, under whom Temin had worked as a graduate student at Caltech and whose approach had begun what might be called quantitative animal virology (see chapter 3). The official reason for the award was "for their discoveries concerning the interaction between tumor viruses and the genetic material of the cell." Baltimore was 37, Temin was 41, and Dulbecco was 61. Baltimore was informed by his wife, Alice Huang, who heard the news while visiting in Denmark, several hours ahead of the United States.

Despite being selected for the Nobel Prize based on tumor virology's investigative potential, all three men also emphasized the role of environment in cancer in their respective talks. Temin took the opportunity to chastise the crowd for smoking tobacco, saying that he was "outraged that the one measure available to prevent much cancer, namely the cessation of smoking, has not been more widely adopted."[51] Temin was apparently not intimidated by the aristocratic crowd; news reports noted that the queen of Denmark was one of those smoking. He also asked and was granted a private audience with the king of Sweden for his two young daughters.

Following the Nobel Prize, Temin continued to work on RNA tumor vi-

8.2. Howard Temin, David Baltimore, and Renato Dulbecco smiling as they receive their checks for the Nobel Prize in 1975. Leif T. Jansson / TT News Agency / Sipa USA

ruses, or retroviruses, as they came to be called in the late 1970s and early 1980s. He also was politically active—trying to improve conditions for Jewish scientists in the USSR oppressed by the KGB. He stayed in Wisconsin at the McArdle Laboratory until his untimely death from lung cancer at the age of 59 in 1994. He did not smoke. Temin was awarded the National Medal of Science by President George H. W. Bush in 1992. He was also active in science policy, participating in the National Cancer Advisory Board (NCAB), and speculated on how to design an AIDS vaccine. HIV is a retrovirus, so Temin's background was well suited to the project.

Baltimore also continued to study the basic biology of cancer. His increasingly large research group studied Abelson murine leukemia virus (AMLuV) and issues in immunology, among other topics. In the early 1980s, he helped establish the Whitehead Institute at MIT, a self-governed institute focused on basic biomedical research. While there, Baltimore's lab discovered NF-kappaB, an important transcription factor that controls which DNA is transcribed into RNA. This transcription factor work led to more citations from other biologists than the reverse transcriptase work. Eventually, Baltimore became an executive-level administrator in addition to a scientist, though his career path in that direction was bumpy at first. He led Rockefeller University as president in 1990–91 but resigned following negative publicity surrounding his support of MIT scientist Thereza Imanishi-Kari, who was accused of scientific misconduct. This complex fracas, further entangled with the NIH and congressional politics, was the subject of a book by the historian Daniel Kevles titled *The Baltimore Affair*, published after he was exonerated. In 1997, Baltimore was appointed president of Caltech, a position he held for eight years. President Bill Clinton awarded him the National Medal of Science in 1999. David Baltimore remains at Caltech today as an emeritus professor and is still active in science.

9

The Molecular-Genetic Basis of Cancer

Michael Bishop, Harold Varmus, Dominique Stehelin, and Hunting of the Oncogene *src*

The humble, likable J. Michael (Mike) Bishop (1936–) was the son of a Lutheran pastor and raised in rural Pennsylvania. His early education was in a two-room school, with many grades of students taught in the same room. He excelled as a student and went on to Gettysburg College, a small, private liberal arts school where there was little opportunity to conduct scientific research. Although drawn to music, he made the pragmatic choice to develop his scholastic talent and begin a career in medicine, gaining acceptance into the medical schools of Harvard University and University of Pennsylvania. The decision between the two schools was easy. Even the associate dean of Penn who interviewed Bishop advised country boy to go to Harvard.

In Boston, Bishop saw the importance of research for an academic career and secured an opportunity to work in the laboratory of Edgar Taft, who had earlier worked on photoreactivation with Albert Kelner, mentioned in chapter 3. In his spare time, he taught himself molecular biology and read voraciously. In his third year, he took an elective course with Elmer Pfefferkorn, later the chair of microbiology at Dartmouth Medical School, who took Bishop into his laboratory and showed him the exhilaration of research at the laboratory bench. After graduating from Harvard, Bishop trained clinically at Massachusetts General Hospital, but this experience only reinforced his desire to go into research. Luckily for him, the National Institutes of Health had a special postdoctoral program for physicians like him who wanted to make the transition from medicine to basic science.[1] He increasingly disliked the clinical work at the hospital and could not wait to become a researcher. In fact, to mark the transition out of clinical medicine, on his last day at

Massachusetts General, Bishop took off his bulky pager and threw it against the wall, destroying it.[2]

At the NIH in Bethesda, Maryland, Bishop joined the Public Health Service as a Yellow Beret, although he never wore his uniform as a lieutenant commander. An added benefit of working for the Public Health Service was that it allowed him to serve his country without being sent to the Vietnam War by the US Army. Leon Levintow (1921–2014), who also did his residency at Massachusetts General Hospital, was Bishop's mentor, and he advocated for the young scientist by expressing confidence in his future to others. In Levintow's lab, Bishop worked on the replication of poliovirus. Although Salk's and Sabin's vaccines had muzzled the danger of the virus, little was known about how exactly poliovirus replicated in animal cells. Halfway through Bishop's postdoctoral research fellowship, Levintow left the NIH for a faculty position at University of California San Francisco. A couple of years later, he would woo Bishop to move west as well, luring the younger scientist away from a permanent position at the NIH and competing offers from East Coast universities. Although at the time UCSF was less prestigious than some of his other options, Bishop wanted to be where he felt needed—and Levintow made him feel needed in San Francisco.

After arriving at UCSF in February 1968, one of Bishop's first orders of business was to write an NIH grant application to fund his new laboratory. His mentors advised him to improve his chances of winning the grant by asking for less money, but he ignored their advice and ultimately prevailed, receiving five years of generous funding for poliovirus research. In the adjoining laboratory, Warren Levinson was working on Rous sarcoma virus. His research intrigued Bishop, who was interested in virus replication and was aware of Howard Temin's controversial claim that RSV could make DNA from its RNA genome (see chapter 8). Temin's reasoning was appealing, and Bishop began some experiments into whether the necessary enzymatic machinery to reverse the flow of biological information could be found in the virus-host system. In what he would later describe as one of his biggest regrets, Bishop terminated his experiments prematurely, thinking they represented too much of a departure from his NIH grant objectives and taking advice from his older colleagues. One year later, Temin and David Baltimore independently discovered the enzyme, later called reverse transcriptase, that catalyzed the creation of DNA from an RNA template. They would be awarded the Nobel Prize only five years later. "The discovery of reverse transcriptase was a devastating blow for me," Bishop later recalled. "A momentous secret of nature, mine for the taking, had eluded me. I grieved for months."[3] From this event, Bishop learned to trust his instincts and not to fear going against received wisdom.

But Bishop knew that his work had to continue and he had to regain his focus. His laboratory grew and became home to a number of postdoctoral researchers. Perhaps the most significant postdoctoral fellow was Harold Varmus, who arrived in late 1970.

The Early Career of Harold Varmus

Varmus had taken a trajectory into basic science similar to Bishop's (figure 9.1). He had studied liberal arts at Amherst College, majoring in English and writing his senior thesis on Charles Dickens. He then wavered between a career in the humanities and one in the sciences, starting a PhD in English at Harvard University but then switching to medicine at Columbia University. Strongly opposed to the Vietnam War, Varmus was attracted to the Yellow Berets, as Bishop was. He arrived at the NIH in July 1968 and was matched with Ira Pastan, whose lab studied hormones produced by the thyroid gland. Before Varmus arrived, however, Pastan switched the direction of his lab to study cyclic adenosine monophosphate (cAMP) and gene regulation in bacteria after making the significant discovery that lactose metabolism in bacteria can be regulated by cAMP.[4] Varmus knew even less about bacterial gene regulation than he did about thyroid hormones, but he went to the library to read Jacob and Monod's seminal work on the subject. Pastan assigned Varmus the task of determining whether cAMP affected production of RNA copied from the regulated set of genes called "the lac operon." Practically, this meant that Varmus had to measure the amount of RNA transcribed from a single gene despite a background of RNA produced from the remainder of the bacterial genome. To complete the task, Varmus had to learn about nucleic acid hybridization and virology.

The NIH ran classes, and Varmus took John Bader's course, which introduced him to RNA tumor viruses. Bader had published some supporting evidence for Howard Temin's hypothesis that RNA tumor viruses make viral DNA that can be inserted into the DNA genome of the host.[5] About the same time, Varmus's mother was diagnosed with breast cancer, further deepening his interest in tumor virology, and he began looking for laboratories where research in this area was being conducted. An overture to Renato Dulbecco

9.1. Harold Varmus and Michael Bishop at Cold Spring Harbor Laboratory in 1978. Image courtesy of Cold Spring Harbor Laboratory Library and Archives

at the Salk Institute in La Jolla was rebuffed, but Varmus had better luck with Harry Rubin at UC Berkeley, who told him about a new group working in this field at the medical school at UCSF: Michael Bishop, Leon Levintow, and Warren Levinson. They were looking for new research fellows, and when Varmus sought them out, he thought the meeting went really well: "Since all four of us were medically trained [only Warren also had a PhD] since three of the four of us had worked at the NIH, and since we all spoke about the power of viruses to reveal biological truths, the warm welcome may have also come from the recognition that we were already part of the same professional club, with a shared point of view."[6] It was quickly decided that Varmus would join the group in the summer of 1970, after he finished his Public Health Service requirements at the NIH.

Steve Martin's Temperature Sensitive RSV Mutant

The year 1970 was a good time for research in tumor virology on two fronts. First, as chapter 8 discussed, reverse transcriptase was discovered,

9.2. Steve Martin. Image courtesy of Steve Martin

solving a puzzle about how RNA tumor viruses replicated—they used a DNA intermediate. Second, an important mutant of RSV was discovered by Steve Martin, who had started a postdoc with Harry Rubin in September 1968 (figure 9.2). Martin was interested in the mechanism of malignant transformation. He knew about other avian viruses that seemed similar to RSV but did not transform cells or cause solid tumors in birds, which suggested that there might be a genetic difference between the two types of viruses. This was the reason he chose to work with RSV. By this time, Rubin himself was primarily investigating the properties of transformed cells, and he allowed Martin the freedom to do what he wanted with the virus.

Martin tried to identify temperature sensitive (ts) mutants of RSV, having learned about isolation of ts mutants more generally in his time as a graduate student working with Sydney Brenner on *E. coli*. The advantage of using ts mutants in the laboratory was that researchers could change the phenotype of the mutant simply by changing the temperature. Martin induced mutations in the virus using a chemical mutagen and then screened for mutant viruses that transformed cells at 36°C but not at 41°C. It took him about a year of playing around with different approaches before he "hit the jackpot," finding RSV mutants that affected transformation and the maintenance of transformation but not replication of the virus.[7] Peter Vogt and Peter Duesberg had discovered non-conditional mutants for transformation that were missing part of the RNA genome.[8] Subsequent work using genetic crosses showed that Martin's ts transforming mutations mapped to the region deleted in Vogt and Duesberg's non-conditional mutants.[9]

Martin's ts mutants were important given the background work done on the West Coast. In Harry Rubin's lab in 1962, it was discovered that the RSV strain he was working with was defective in replication and needed a helper virus to replicate. Hidesaburo Hanafusa, Rubin's postdoctoral fellow, "nailed down" this aspect of RSV biology. Theoretically, this insight suggested that it was not necessary to make infectious virus to transform the cell; in other words, transformation and replication might be independent processes. Most people working on RSV in the 1960s wanted to understand infection. From work in Temin's and John Bader's labs, it was known that inhibiting DNA production inhibited virus production. Temin proposed the provirus hypothesis—that RNA tumor viruses go through a DNA stage in their life cycle. Rubin, meanwhile, told his students not to believe Temin's hypothesis because the experimental evidence was weak (see chapter 8). The field was preoccupied with this riddle.

In the 1960s, Peter Vogt was pulled into this problem because he was investigating the helper virus (figure 9.3).[10] He designed experiments to inhibit DNA synthesis in the cell to see whether the need for DNA synthesis was virus specific or cell specific. If Temin was correct, the DNA synthesis needed to replicate an RNA virus is specific to the virus. Using a double-infection experiment, Duesberg and Vogt infected cells with an RNA tumor

9.3. Peter Vogt. Courtesy of Peter Vogt

virus and then stopped DNA synthesis before infecting them again with a different virus. They concluded that whatever DNA is induced by the first virus does not remove the requirement for further DNA synthesis by the second virus, which implies that what is needed for RSV replication is not cellular DNA but specific viral DNA.[11] Duesberg and Vogt did not take the extra step and look for reverse transcriptase; as Vogt later recalled, "We considered the possibility of a viral polymerase, but this did not sound plausible."[12] The field was not focused on oncogenesis until replication was solved by finding reverse transcriptase.

"Then," said Vogt, "we thought, why don't we mutagenize and look for mutants?" With Japanese virologist Kumao Toyoshima he found a ts mutant.[13] They blasted the virus with so much mutagen that it was temperature sensitive in both replication and transformation. (This was one year before Steve Martin's work showing it was possible to have a ts mutant in transformation only.) About the same time, Vogt examined RSV strains from all around the world, finding that most had a complete genome that could reproduce and transform. Using these "non-defective" viruses made genetic experiments easier by eliminating the need for a helper virus.

Vogt teamed up with Duesberg (figure 9.4) to look at the RNA, finding that the RNA of the non-defective virus was larger than RNA from a defective virus. In fact, it looked like the defective virus had a gene deleted from its genome. Vogt asked Duesberg to look at the virus that could not transform anymore; its genome was found to be shorter, as measured by gel electrophoresis. They speculated that the extra section of RNA was the gene for oncogenesis. "It turned out to be true," Vogt later remarked, "but it was a wild guess at the time."[14] Now there was physical evidence of oncogenic transformation. Later, Vogt and Duesberg used RNA fingerprinting to arrange the viral genes and oncogenes in a physical map.

The West Coast Tumor Virus Cooperative

A unique cooperation among tumor virologists began in 1971, when Bishop and Varmus requested some virus from Peter Vogt's lab, launching a robust network of collaborative and competitive research. What became called the West Coast Tumor Virus Cooperative included researchers from the Bishop-Vogt group and the Duesberg lab at first and then later also others from the Salk Institute, Caltech, the UC schools, and the Fred Hutchinson Cancer Research Center in Seattle. For nearly a decade, these researchers met regularly in different West Coast cities to share their unpublished results, their shared interests and sense of purpose overriding the risks of being so open with data and ideas. The West Coast Tumor Virus Cooperative lasted until the interests of the various group members diverged and the development of recombinant DNA technology advanced enough to make viruses less important tools of research in molecular biology. During its lifespan, many important collaborations were formed.

9.4. Peter Duesberg in the 1970s. Used with permission of Peter and Sigrid Duesberg

Dominique Stehelin Joins the UCSF Group

In retrospect, the most important research project of the West Coast virologists in the early 1970s was Bishop and Varmus's plan to make a radioactive probe to detect the oncogenic potential of RSV. The scientist who would master the experiments to look for the RSV oncogene using a specially designed probe came from France. Dominique Stehelin studied chemistry and biochemistry at the Université Louis Pasteur in Strasbourg, progressively developing an interest in tumor virology while there. Competing with scientists from all over France, Stehelin won a permanent job as *attaché de recherche* (research associate) in January 1969, which allowed him to be paid by the French government but retain the flexibility to choose laboratories in which to work. He completed a postdoc under André Lwoff at the Institut de recherche sur le cancer (IRSC) in Paris-Villejuif. Lwoff's lab at that time was directed by Marc Girard. Stehelin worked on SV40 proteins and replication.

Stehelin used the flexibility of his French funding to gain experience outside of France. In the fall of 1972, he traveled to San Francisco to begin a new phase in his career. Advised by Girard, Stehelin selected the Bishop laboratory as the best place in the world to work on tumor virology. He and Michael Bishop agreed for him to spend a year acquainting himself with avian RNA tumor viruses such as RSV, which differed from SV40 in having an RNA rather than a DNA genome, among other things, before beginning a substantial project. He was described as a postdoc for administrative reasons, such as securing a visa, but in practice, since he had his own funding, he was more like a visiting scientist.[15]

A Probe for the Transforming Gene

Getting up to speed on avian RNA viruses, Stehelin learned that Duesberg and Vogt had shown that their transformation-defective virus was 20% smaller than a wild-type virus. They mapped mutations in transformation to one region of the genome, the part that appeared to contain one or two genes. The Bishop-Varmus group came up with a strategy to investigate whether the transforming gene, defined genetically by Vogt, Hanafusa, and Martin, was present in a similar form in normal cells. They would use the newly discovered enzyme reverse transcriptase to make a radiolabeled DNA copy of the RNA genome of RSV. They would then isolate the new DNA copy from the RNA. Finally, they would remove the parts of the copy not involved in transformation by hybridizing them to the RNA of a deletion mutant of RSV missing the gene(s) responsible for transformation. If all went well, the pieces of radiolabeled DNA left over would be copies of the oncogene that could then be used as a probe to detect the gene itself.

Varmus and his postdoc Ramareddy (Ram) Guntaka were the first to attempt to make the probe. Guntaka had joined the Bishop lab in January 1973 to work on RSV replication. He quickly had exciting results showing that the plus and minus strand of the RSV genome are made in a sequential manner. Bishop encouraged him to apply for a senior fellowship from the American Cancer Society, but because he had already supported other applicants suggested that Varmus support his application. Consequently, Guntaka won the fellowship, and Varmus become his mentor. Given his experience with DNA-RNA hybridization, Varmus asked Guntaka to make the transforming gene-specific probe in addition to his other projects. Guntaka had some limited success showing that the probe experiment was feasible, but his other projects were also bearing fruit[16] so he put the probe experiment on the back burner,[17] shifting his attention to integration of the double-stranded replicative form of avian sarcoma viruses.

The San Francisco group was not the only lab looking for the transforming gene in host cells. In 1973, Paul Neiman also obtained results that seemed to indicate that there were no cellular transforming genes. In a paper submitted to the *Journal of Virology* in November, Neiman and his colleagues demonstrated that normal Rous sarcoma virus RNA was blocked from binding to chicken DNA in the presence of RNA from transformation-defective viruses.[18] Their data suggested, or at least seemed to suggest, that the sarcoma virus-specific sequences were not present in chicken DNA.

In September 1973, about a year after his arrival in San Francisco, Stehelin went to Bishop to ask to take up the project. Bishop approached Guntaka about bringing Stehelin in. Guntaka agreed because Stehelin was his friend and because he was reassured that he, Guntaka, would remain on the project.[19] Guntaka would meet weekly with Stehelin and Varmus to talk about the project until May or June 1974, at which time he became involved with a time-intensive but ultimately successful job search.

Stehelin's background in chemistry was useful, and he stubbornly worked to perfect the oncogene purification procedure. The goal was to purify what they called "x," the "a genome" of a normal virus minus the "b genome" from a deletion mutant, following Duesberg and Vogt's terminology. Stehelin's goal was to maximize the homogeneity of the probe. Instead of taking a whole batch through each step, he took a small sample of the nucleic acids to perfect the next step in the purification procedure. Once the small sample worked, he took his entire batch though that step. The probe could be seen as a peak in the gradients, which could be made more prominent by making many small refinements in his procedures.

The Transforming Gene Probe Perfected

After six to nine months, Stehelin was happy that he had perfected the probe and significantly enhanced its purity. Stehelin liked to work at night when the laboratory was quieter and he could concentrate, and his girlfriend and future wife was still in France.[20] With unfettered access to the ultracentrifuge and chromatography column, he could run several experiments in parallel. He tried various RSV strains and discovered that a Prague strain worked best. By June 1974, Stehelin's pretty *S*-shaped curves showing thermal denaturation of RNA/DNA hybrids indicated that he had made an effective specific probe for the transforming DNA of an avian sarcoma virus. The probe annealed almost completely to RSV DNA but had no significant annealing to transformation-defective deletion mutants RSV DNA. Then came the significant result: using the probe found *src*-related sequences in normal chicken DNA. The probe for viral *src* DNA annealed to DNA from host cells.

Retrospectively, Stehelin described the night of October 26, 1974: "The intensity of the emotion I experienced and the intellectual clarity induced by the situation at that moment were very special. . . . The fantastic results . . . Normal DNA contained sequences related to the src gene of the transforming virus. . . . I suspect that few have had the privilege of enjoying such a moment when one is intensely and profoundly aware that a major step forward in science has been made, and that one has contributed to it."[21] The next morning he went to Bishop and said, "It worked!"[22] At first Bishop could not believe that cellular DNA contained a gene related to the transforming gene of the tumor virus, which was understandable as others had failed to make these experiments work. But the more he looked at Stehelin's results, the more confident he became.

Stehelin was meticulous and repeated his experiments to be sure he did make any mistakes. His procedure and initial results were sent to the *Journal of Molecular Biology* in July 1975.[23] In addition to Bishop, who was last author, and Stehelin, who was first author, Varmus and Guntaka were listed as coauthors. Stehelin thought that only he and Bishop should be the authors on the paper, but it was Bishop's call, not his. In a letter to Varmus, Guntaka pointed out that Stehelin was "reluctant to give me authorship until he talked to [Varmus]. He thought that it was all his creation and forgot completely that that was [Varmus's] idea and I worked for 5 months showing positive results."[24] In a reply to Guntaka, Varmus tried to explain Stehelin's thinking as the approach of a beginning scientist starting an independent career who is "very concerned to establish his name."[25] Additionally, Varmus thought that Peter Vogt should be included on the paper, but this did not happen. In any case, Stehelin et al. demonstrated that the transforming genes of the avian sarcoma viruses are closely related; however, they are unrelated to mouse mammary tumor virus and sarcoma-leukosis viruses of mice and cats. MMTV does not have *src*. Given that transformation-defective RSV could still replicate, a whole host of open questions remained: What was the biological role of the transforming gene? Why was it part of the viral genome? Was it part of an endogenous provirus or one of the cellular genes? What advantage, if any, did it confer to the virus? Was it related to other cancers?

The team would find answers to some of these questions using the radiolabeled probe to investigate a range of normal cells, not viruses. To look at cells rather than viruses to understand viral transformation was not completely counterintuitive. Huebner and Todaro's oncogene hypothesis held that oncogenic genes were part of endogenous viruses, fused within the genomes of the viral host cell.[26] Varmus corresponded with Huebner about the UCSF work and was funded by the Special Virus Leukemia Program that Huebner controlled, but up until this point no researchers had been able to find transforming viral genes in cells using RNA-DNA hybridization.[27] Could the transforming genes of the avian sarcoma viruses have been captured by an ancestral virus from a nonviral source? Stehelin's October evening discovery required more experiments to fully reveal its significance. It was possible that Stehelin was detecting an endogenous virus in the chicken cells he examined. Alternatively, the normal cellular genome could contain a gene closely related to the *src* gene in viruses.

To establish which possibility was correct, the natural thing to do would be to examine cells from various bird species. Varmus solicited help from evolutionary biologist Allan Wilson at UC Berkeley, who suggested they examine the cells of the ratites—emu, ostrich, kiwi, and the like—as they have the most evolutionary distance from chickens, so if they also contained the *src* gene, it would be evidence that it was a cellular gene and not a resident endogenous virus, or at least not a virus that made a home in the bird genome since the last common ancestor.[28] As a New Zealander, Wilson had a particular interest in the ratites—the kiwi is a national symbol of New Zealand. The group obtained an emu chick about the size of a chicken from the Sacramento Zoo so that Stehelin could determine whether it also had *src* sequences. The chick was so cute that no one wanted to kill it, but eventually Varmus did the deed.[29] Stehelin first took liver cells to analyze, but the emu's liver contained enzymes that interfered with his test, so he turned to muscle cells instead. Using the probe, he found *src* sequences in the emu DNA.

By March 1975, Stehelin had found *src* sequences in chicken, duck, and quail DNA. David Baltimore invited him to give a talk at MIT on his results, as did Hidesaburo Hanafusa at Rockefeller University.[30] Before the results were published, Stehelin presented his work at the Pacific Coast Tumor Virus Group Meeting at the University of Southern California School of Medicine on August 26, 1975. Bishop and Varmus also spoke about the exciting findings at several large international meetings.

Wilson estimated that the evolutionary distance between chickens and emus was 100 million years. The thermal stability of the duplexes between the *src* probe and bird DNA was proportional to the evolutionary distance from chickens. In other words, the more closely related the bird DNA was to the chicken virus probe, the more stable the double helix was that they would together form. The duck and quail *src* sequences in particular diverged in parallel with evolutionary distance. The oncogene hypothesis suggested that cancer-causing genes were present in animals because of virus DNA fused into their genomes. Stehelin et al. suggested that this was the wrong way around. It was more probable that the cancer-causing gene in avian sarcoma viruses was originally a bird gene that a virus had picked up by recombination: "We suggest that part or all of the transforming gene(s) of ASV was derived from the chicken genome or a species closely related to chicken either by a process akin to transduction or by other events, including recombination."[31] The paper explaining their argument was published in *Nature* in 1976, its title crisply capturing their insight: "DNA Related to the Transforming Gene(s) of Avian Sarcoma Viruses Is Present in Normal Avian DNA." Guntaka was not listed as an author; he was hurt but decided not to make a fuss over credit.[32]

Given the widespread conservation of the cellular sequences that hybridize with the *src* probe, the authors suggested that this protein must have an important function within the cell, possibly something to do with the normal regulation of cell growth. They could not detect *src* DNA in mouse and calf cells. Rat XC cells that were transformed and presumably contained the Prague strain of RSV "completely matched" the probe and served as a positive control.

In their *Nature* article, Stehelin and colleagues mentioned the results obtained by Paul Neiman's lab at the University of Washington that seemed to contradict the UCSF findings. Neiman had argued that only a little more than 1% of the RSV genome is present in the DNA of normal cells. He had conveyed these data to Varmus in December 1974.[33] Temin also mentioned Neiman's earlier negative findings when he commented on the paper prior to publication, "I was convinced by your paper, but previously had been convinced by [Neiman]."[34] Temin suggested they explain the discrepancy between the two labs but thought that it was a "very fine piece of work." Varmus was well aware of Neiman's work, as he had written him at the end of 1974. On the basis of competition experiments similar to Stehelin's work, Neiman concluded that 90% of the "X piece" (*src* gene[s]) are not detectable in normal chick embryo DNA.[35] His data gave rise to a "frenzy of hypothesis making" to account for the discrepancy, but in the end no good explanation was found. Stehelin and his coauthors were forced to write in the *Nature* paper, "We cannot explain the discrepancy." The implication was that the UCSF results were correct and Neiman's group had made a mistake. Later Neiman's false negative could be explained by technical aspects of the hybridization kinetics for RNA-DNA and DNA-DNA reactions.

Robin Weiss, who had worked in Peter Vogt's lab, wrote an accompanying overview for *Nature*. As evidenced by his remarks, it was clear, even at the time, that this work was important evidence of a specific gene for neoplastic transformation in the genome of cells.

By the time the *Nature* paper was published, Stehelin had moved back to France to start his own laboratory. He continued to collaborate with Thomas Graf, a virologist at the Max-Planck-Institut für Virusforschung in Tübingen. Stehelin had a gentleman's agreement not to compete with Michael Bishop on avian sarcoma viruses, so he focused on new viruses.[36] One of his goals was to replicate in Europe the successful collaboration among virology laboratories on the West Coast of the United States, which he did by developing relationships among his lab, Klaus Bister's at ICRF, and Thomas Graf's at the Max-Planck.

Extending the *src* Research

At UCSF, the job of extending Stehelin's results was given to Deborah Spector, who entered the lab in July 1975, after spending time in David Baltimore's laboratory. Her project was to continue investigating how *src* was expressed in the RNA of oncogenically transformed cells. She examined the RNA from different lines of chemically transformed quail cells, using quail, duck, and chick embryo cell RNA as controls. Surprisingly, all the cells showed comparable levels of RNA, contradicting what Stehelin found and what had been submitted in the *Nature* paper.[37] At the same time, another member of the Bishop-Varmus group, Daisy Roulland-Dussoix, found high levels of the RNA in chick brain cells. Consequently, Bishop and Varmus decided to pull the RNA data from the *Nature* paper, which was not yet published. Although when no RNA was added the negative control was clear, there was concern that the RNA detected in the samples from the various avian cells might be some artifact of a contaminant. Thus, Spector began looking for a solid negative control using sources of RNA from non-avian species, testing human and mouse RNA. These RNAs were also positive, and only yeast RNA was negative. Based on these results, Spector began to investigate whether *src* was a normal cellular gene. Within a few months, she found that not only was the gene universally present, but it was highly conserved.

Varmus corresponded with Stehelin about the new results that showed that *src* RNA was in normal cells and not only in transformed cells.[38] Stehelin replied, "I am now convinced that my earlier observations about sarc RNA not present in normal cells cannot be repeated."[39] If *src* RNA is found in both normal and transformed cells, then the oncogenic potential of Src is not just a matter of "turning on" the gene as Stehelin and others had first thought. Rather, this result suggested that *src* in transformed cells is different, possibly a mutated version of the gene found in normal cells. Indeed over the next few years this picture was confirmed, and the mutated viral gene was called *v-src* and the cellular gene *c-src*.[40] Given their potential to mutate into oncogenes, cellular genes like *v-src* are called proto-oncogenes. The general importance of this model was confirmed as other cellular proto-oncogene-oncogene pairs were consequently discovered.

Given the success of the probe, the Bishop and Varmus laboratories increased the work on *src*. Up until 1974, many RNA tumor virus scientists were concerned with viral replication, but the study of transformation now took priority. A significant open research question involved discovering the protein that the *src* gene encoded and the role it played in the cell.

In the days before molecular cloning, it was much more difficult to identify a protein from a newly discovered gene like *src*. One approach that had been successful in identifying SV40 T antigen in the 1960s was adapted to work on Src. The idea was to find an antibody specific to the gene product. This task was more difficult than finding a T antigen antibody, because there was both the cellular *src* and viral *src*, whereas the SV40 protein had no cellular homologue.[41] Nonetheless, Joan Brugge and Ray Erikson (1936–2020) at the University of Colorado Medical Center used this approach by generating tumors in rabbits by injecting newborn animals with Schmidt-Ruppin RSV.[42] After two years of work, using this antisera, they then immunoprecipitated a 60-kilodalton protein that was uniquely produced in cells that carried the genetic information of the RSV *src* gene, and it was later confirmed to be the protein product of this gene.

The Function of Src Protein

In April 1978, work at the Erikson lab identified the function of the *src* gene product: it was a protein kinase—an enzyme that attaches phosphate molecules to proteins in a chemical process called phosphorylation. With the addition of a phosphate group, the target protein can often be changed to be more or less active. Protein kinases then are regulators of protein activity and therefore regulators of cellular processes in general. Marc Collett and Raymond Erikson made this discovery using radiolabeled ATP, the source of energy and phosphate for the Src protein. They purified the 60-kilodalton protein they believed was the product of the *src* gene (figure 9.5). In the presence of this purported Src protein, the radioactive phosphate was transferred from ATP to another protein. The transfer did not take place in the presence of the purified 60-kilodalton protein from transformation-defective viruses. These results were dramatic, and Erikson had the paper published in *Proceedings of the National Academy of Science*. That Src was a kinase was consistent with the understanding of the role of kinases more generally: "The role of protein phosphorylation in the function regulation of a variety of cell processes is well documented."[43]

Independently, Art Levinson,[44] a postdoc in Michael Bishop's lab, also proved that the Src protein was a kinase, although his paper came out in *Cell* a few months later than Collett and Erikson's, so he did not get the same credit.[45] Part of the delay was due to *Cell*'s slower turnaround time. Serendipity played a big role for Levinson. He was not expecting the Src protein to be a kinase—the thought never entered his mind.[46] He had previously shown that the viral Src protein was itself a phosphoprotein and wanted to identify the kinase that phosphorylated it. So, he reasoned that if he purified the Src

9.5. Ray Erikson, Tony Purchio, Joan Brugge, and Eleanor Erickson at a lab meeting in May 1977 celebrating the identification of Src with a 60-kilodalton protein. Image courtesy of Joan Brugge

protein by immunoprecipitation with an antibody, he might coprecipitate the kinase that was phosphorylating the Src protein. But instead of finding a 60-kilodalton protein (i.e., the Src protein), he found, astonishingly, a different one. Levinson repeated the experiment many times, perplexed. He eventually realized that the protein phosphorylated in his assay was the very antibody that was binding to the Src protein; further controls established clearly that the most reasonable explanation of this was that the Src protein itself was the kinase.

The next step was to identify the natural target of the Src kinase enzyme in the cell. In late 1979, Tony Hunter and others at the Salk Institute showed that the Src kinase phosphorylated the amino acid tyrosine and not the amino acids serine or threonine as other known kinases did.[47] About the same time, David Baltimore and others at MIT were working on a protein encoded by another proto-oncogene also discovered in a tumor virus. This protein, which is related to Src, became known as ABL, and the MIT team found that it was also a tyrosine kinase.[48] This was the beginning of a deep understanding of how normal cells regulate their growth at the molecular level. Tyrosine kinases help control many different cellular processes and signal transduction pathways, including how cells communicate, how cells adhere to each other, how vesicles are trafficked within the cell, how proteins are degraded within the cell, and various aspects of immune response.

The 1989 Nobel Prize in Physiology or Medicine

The Nobel Committee in Sweden awarded the 1989 Nobel Prize in Physiology or Medicine to Bishop and Varmus for the discovery of the cellular origin of RNA tumor virus oncogenes. The press release from the Karolinska Institutet, the organization determines who wins Nobel Prizes in Physiology or Medicine, put it this way:

> Michael Bishop and Harold Varmus used an oncogenic retrovirus to identify the growth-controlling oncogenes in normal cells. In 1976 they published the remarkable conclusion that the oncogene in the virus did not represent a true viral gene but instead was a normal cellular gene, which the virus had acquired during replication in the host cell and thereafter carried along.
>
> Bishop and Varmus' discovery of the cellular origin of retroviral oncogenes has had an extensive influence on the development of our knowledge about the mechanisms of tumor development. Until now [October 1989], more than 40 different oncogenes have been demonstrated. The discovery has also widened our insight in the complicated signal systems which govern the normal growth of cells.[49]

Varmus learned the news from reporters who called him at 3 a.m., an occurrence that is not uncommon for US Nobel Prize winners since the announcement in Sweden is nine hours ahead of California. It was also a momentous day for San Francisco as the Giants won baseball's National League title, and Bishop and Varmus were given prime seats to the game. Vogt called Varmus to congratulate him. Varmus invited Vogt to come with him to the ceremony in Stockholm.[50]

Stehelin Objects to the Nobel Prize Committee Decision

Almost immediately, Stehelin objected that the Nobel Committee had made a mistake by omitting him from the prize. He appeared on French TV offering to "open his notebooks" to show how the committee had gotten it wrong. On October 11, 1989, the *New York Times* ran an article with the headline "Frenchman Says Panel Overlooked His Contribution."[51] The paper reported that Stehelin claimed that "he got the idea for the prize-winning research and carried out the experiments." Varmus and Bishop declined to comment directly to the reporter, but a UCSF spokesperson acknowledged that Stehelin had performed the experiments, but only under supervision. Deborah Spector was quoted as saying there was no substance to his claims for more credit: "Mike and Harold were the key motivating individuals." Leon Levintow, now the chair of microbiology at UCSF, who worked "10 feet from Stehelin," joined the chorus of Americans saying that Stehelin was out of line.[52]

French scientists largely took Stehelin's side, pointing out that he was the first author on the two key papers. Meanwhile, the *Chicago Tribune* also carried a story about the controversy.[53] Stehelin was inaccurately portrayed as a student under Varmus and Bishop and was quoted as saying, "I did all the work, from A to Z." Varmus was quoted as saying Peter Vogt's contribution also was important in "the brutally difficult work," but the *Tribune* mistakenly stated that Vogt had worked on the mutants in the UCSF laboratory as a postdoc, when in fact he had his own lab. The controversy made the pages of *Nature*. On October 19, it was reported that Stehelin had softened his position from the early comment—"I did all the work from A to Z"—but still thought that the Nobel Committee should have included him as a third person for the prize. Nobel Prizes can be split three ways, but no more. His role was significant, and his scientific research was awarded the Rosen Prize in 1980, the Leopold Griffuel Prize in 1982, the Louis-Jeantet Prize in 1987, and a Silver Medal from the Centre national de la recherche scientifique (CNRS).[54]

Stehelin laid out his reasoning in an open letter to the Nobel Committee. He declared "that the Nobel Committee, by excluding from the original three-person team the very person who facilitated the experiments crucial to the project's success, committed an injustice, the importance of which could not have eluded them had they been in possession of all of the requisite information."[55]

A November editorial in *Nature* argued that Stehelin's vocal approach was "ungentlemanly" and constituted "conduct unbecoming,"[56] noting that Alfred Nobel's will stated that there was no appeal of the committee's decision. However, the editorial suggested that the Nobel Prize as a winner-take-all contest was in need of revision. The author speculated that Stehelin was continuing to make this an issue in response to a claim levied by someone on the Nobel Committee that he had done nothing of note since leaving San Francisco. The Swedish virologist Erling Norrby, whose article was cited as a reference in the original press release, was quoted as saying as much in the newspapers.

Norrby's claim was clearly false. Stehelin was one of the most notable European tumor virologists, achieving significant results beyond Bishop and Varmus's research program. As early as September 1974, Stehelin had tested viruses sent to him by Thomas Graf and had written to Graf with exciting results. The MC-29, ES-4, and R-strain viruses, he said, did not hybridize with Stehelin's probe. Considering that these viruses caused transformation, Stehelin inferred that they must contain other transforming sequences. He highlighted this last point by typing it in all caps.[57] The gene *src*, then, was possibly the first of many oncogenes found in tumor viruses. Graf was surprised by the results and suggested additional experiments to be sure. These were done, and Stehelin and Graf published the results in *Cell* and in *Nature*.[58] His work over many years cemented Stehelin as one of the best French scientists working on the molecular basis of cancer (figure 9.6). Stehelin's letter and appeal, however, had no effect on the decision of the Nobel Committee.

In Stockholm receiving the prize, Varmus read a passage from *Beowulf* when he took the podium and followed with the remark: "We recognize that, unlike Beowulf at the hall of Hrothgar, we have not slain our enemy, the cancer cell. . . . In our adventures, we have only seen our monster more clearly and described his scales and fangs in new ways—ways that reveal a cancer cell to be, like Grendel, a distorted version of our normal selves."[59]

Following the Nobel Prize, Stehelin, Varmus, and Bishop all had successful careers in science and science administration. Of the three, Varmus flew the highest. He was nominated to be the director of the NIH by President

9.6. Dominique Stehelin in 2016 with the lab notebooks he used when making the *src* probe. Image in the author's collection

Bill Clinton in 1993. He assumed the position and was successful in nearly doubling the NIH budget during his six-year tenure. He went on to become president of the Sloan-Kettering Cancer Center in New York City for nearly 10 years. He was a member of President Barack Obama's Council of Advisors on Science and Technology. Obama appointed him to lead the National Cancer Institute in 2010.

Michael Bishop remained in San Francisco. He ran his lab and was involved in the administration of the university, becoming chancellor of UCSF in 1998. In 2003, he was awarded the National Medal of Science, as Varmus had been in 2002.

Dominique Stehelin has spent the remainder of his scientific career at Lille, France, heading a cancer laboratory. He was instrumental in the creation of a major center for biology in Lille in 1996. France made him a knight (*chevalier*) in the Order of Merit. He has remained steadfast in his view that he should have received the Nobel Prize with the American researchers.

10

Mecca for Tumor Virology

James Watson, Joseph Sambrook, SV40, and the Growth of Tumor Virology at Cold Spring Harbor Laboratory

James (Jim) Watson is rightly famous for his monumental elucidation of the structure of DNA with Francis Crick in 1953, signaling the beginnings of a revolutionary new biology at the molecular level. What is lesser known is Watson's enduring interest in viruses, tumor viruses in particular. This interest explains important career decisions he made in his professional life and gives a fuller picture of him as a scientist and science administrator. Watson's relationship with Joe Sambrook, arguably the most important scientific pairing in Watson's post–double helix career, led to an additional significant impact on the history of molecular biology. With Sambrook and other virologists, Watson built Cold Spring Harbor Laboratory into a molecular biology powerhouse largely because of its focus on using viruses to study cancer. The science developed at CSHL, including the many advances by Sambrook's group, cemented Watson as one of the central figures in biology for decades.

James Watson and Salvador Luria

The young Watson was recognized as being exceptionally smart. At the precocious age of 15, he entered the University of Chicago, where he would eventually trade in his passion for ornithology and bird watching for one in genetics. After graduating from Chicago, he began graduate school at Indiana University, where he was deeply influenced by his mentor Salvador Luria, an Italian immigrant who worked on viruses that attack bacteria and who was mentioned in chapter 3. In a seminar in virology, Luria taught Watson about tumor viruses, an interest he promoted in part because he was supported by one of the first research grants of the American Cancer Society

(ACS). These viruses interested Watson's in part because one of his uncles on his father's side had died from cancer at a young age. Luria's approach to the genetics of bacteriophages, for which he would later be awarded the Nobel Prize, drew Watson, in the spring of 1949, to begin his PhD research in Luria's bacteriophage laboratory as his first graduate student.

Luria's 1953 textbook *General Virology* was an "outgrowth" of the course Watson took and gives a flavor of what would have been discussed in the Luria lab. Devoting an entire chapter exclusively to tumor viruses, Luria described them as carcinogenic agents that accompany neoplastic cells in their growth period. He summarized Rous's work and claimed that the agent he discovered "was indeed a virus in the accepted sense of the word."[1] He called rabbit papilloma a "very interesting virus tumor" even though it is often difficult to isolate. Further, Luria mentioned serological data that suggest that this particular virus persists in a "masked noninfectious form," and then draws an analogy with a bacterial prophage that inserts its DNA into the genome of the host. In Luria's eyes, there were challenging and important questions. How could such a provirus be detected? What is the role of the virus in transforming normal cells into tumor cells? Does it act only at the beginning of the transformation or throughout the growth of the tumor? He thought the evidence suggested an "intimate fusion of cellular and viral properties"[2] and took it as a "fruitful working hypothesis" that latent and masked viruses are the cause of most cancers, acknowledging that at the time science did not have sufficient evidence to accept or reject the hypothesis.

The challenge for the theory that latent viruses cause most cancers—which strike practically all animal species—was to show how an almost ubiquitous distribution of latent viruses could exist. Additionally, one would expect a latent virus such as the Rous sarcoma virus to cause tumors "from the inside" in addition to external infections. Ultimately, Luria thought that determining whether the latent viruses cause cancer would require a better evolutionary understanding of the origin of viruses. His view was that the virus "is nothing but a part of the cell."[3] Over time, a virus could be a regressed parasite or cellular component. One question remained unanswered: Did the merging of a virus and a cell mainly lead to disease, or did it also have a more positive role on the evolution of the host?

In the same year Luria published his *General Virology*, Watson had his remarkable year with Crick. Their foray into the structure of DNA was closely connected to Watson's interest in viruses. Before and after the DNA discovery, Watson worked on the rod-shaped tobacco mosaic virus, first in Cambridge and then at Caltech with Don Caspar in 1955, but they did not publish their speculative, theorized TMV structure.[4] In the next two years, Watson paired up with Crick again, and together they published two papers on the structure of small viruses in general, using a mixture of structural and information-theoretic reasoning. They argued that, given the small size of the viral genome, the number of types of viral coat protein too must be small, perhaps one type repeated numerous times, to create a closed shell to protect the viral DNA. This was likely the first time information theory was used to argue about the design of biological structures. The most efficient such shell has icosahedral symmetry, much like an old-fashioned black-and-white soccer ball.[5]

In 1958, while visiting Luria's laboratory, Watson attended a talk by Wisconsin professor of oncology Van Rensselaer Potter, on the biochemistry of cancer. This talk helped crystallize Watson's idea that eukaryotic cells need specific signals to divide. He speculated that animal viruses coded for enzymes that force host cells into the S phase, the time in the cell cycle when new DNA is synthesized. In the following year, while preparing a freshmen lecture on cancer at Harvard, Watson developed the idea that viral genomes insert themselves into host chromosomes, thereby converting healthy cells into cells that permanently have the signal to divide switched on.

It would not be long before Watson had the opportunity to present these ideas publicly as a recipient of the Warren Triennial Prize at Massachusetts General Hospital, an award that is, as the name suggests, given every three years. Watson and Crick jointly won in 1959 on the basis of their DNA discovery, and they delivered two talks to a crowd of medical professionals. Watson's talk in May was titled "The Role of Nucleic Acids in the Viral Induction of Cancer." He suggested that DNA from tumor viruses can improperly move cells into the S phase. The weakness of Watson's hypothesis was that it did not explain how RNA viruses, such as Rous sarcoma virus, caused cancer: cells do not need to be moved into S phase to make RNA. Overall, he thought that his talk was "much less convincing" than Crick's,[6] which was titled "The Structure and Replication of Deoxyribonucleic Acid." Watson found Crick more convincing because his talk was more data driven and less speculative, and it appears that he felt some competition with his former collaborator. It was clear to Watson that he needed more experimental data to bolster his speculation.

James Watson, John Littlefield, and the Shope Papilloma Virus

Watson started working on Shope papilloma virus (SPV) with John Littlefield in the spring of 1959. Littlefield was purifying the virus from warts harvested from wild rabbits caught by Kansas trappers. SPV was the smallest known tumor virus, fitting Watson's methodological heuristic of working on a simple system. In learning that he could purify two different "homogeneous components" of virus DNA, Watson showed that they had roughly equal proportions of A, T, G, and C. However, it was hard to come by large quantities of SPV, and Watson and Littlefield eventually ran out of samples, halting their work. Watson was still able to write a short note for the *Journal of Molecular Biology* about his results. Unfortunately, however, Watson and Littlefield did not look at the SPV DNA through an electron microscope. If they had, they would have seen that the viral DNA was not linear but circular with the ability to supercoil like a rubber band. The supercoiled version of the genome sedimented at a different rate, which is why there were two different components to the viral DNA.

In his influential 1965 *Molecular Biology of the Gene*, the first textbook on molecular biology, Watson included a chapter on tumor viruses. It was titled "The Problem of Cancer to a Geneticist" (Watson indeed thought of himself as a geneticist), and it focused on the problem of the control of cell division.[7] In a letter to Howard Temin, Watson described chapter this way: "Some may regard it as propaganda; on the other hand, I always find the introductory students quite interested in hearing about the connections between pure science and the cancer problem."[8] After a discussion of cancer in general, Watson turned to tumor virology: "The relevant question is thus not whether viruses cause cancer, but whether a sizable fraction of cancers are virus-induced."[9] Watson focused on two viruses, RSV, an RNA virus; and polyoma virus, the DNA virus discovered in 1953 by Ludwik Gross but renamed polyoma in 1958 after it was shown that it could cause many types of cancer in many species (see chapter 2). Watson speculated that since no infectious polyoma virus could be detected in transformed cells, polyoma might become a provirus in infected cells. Using an argument based on coding similar to that which he and Crick had used in 1956, he wondered whether the antigenic properties of cells transformed by polyoma might be the result of an enzyme responsible for viral nucleic acid synthesis.[10]

Unlike polyoma, which contains double-stranded DNA, RSV contains single-stranded RNA. In this one respect it is like TMV, the plant virus that Watson had worked on in the 1950s.[11] RSV is also a larger virus than polyoma, and, as Watson pointed out, its genome could potentially code for 25 proteins. He also noted about RSV that, remarkably, a single virus particle infecting a susceptible cell can create cancer, and that RSV is a defective virus requiring a helper virus, Rous associated virus, to multiply. In other words, a cell would need to be coinfected with another virus to produce coat protein for RSV. Watson speculated that the RSV coat protein does not aggregate properly to make the virus but moves to the cell surface and disrupts contact inhibition. He also considered the "radical proposal" that RSV has a DNA provirus stage, but he concluded that much more evidence was needed. This proposal was radical because the central dogma of molecular biology states that DNA can make RNA but not vice versa (see chapter 8). In generalizing to other cancers, Watson was open to the possibility that many of what were called "spontaneous" tumors are in fact caused by defective viruses. He ended his discussion with the following remark: "Naturally we should not underestimate the difficulties ahead. These viruses multiply in cells that we are only beginning to understand at the molecular level. Nonetheless, most important is the fact that at last the biochemistry of cancer can be approached in a straightforward, rational manner."[12] The placement of this roadmap for further study at the end of Watson's highly successful text for the new field of molecular biology highlighted the promise Watson foresaw in the new field of molecular tumor virology.

Watson as Director of Cold Spring Harbor Laboratory

Later in the decade, Watson would get a chance to push the idea that researchers could better understand cancer by studying tumor viruses. On February 1, 1968, Watson officially became director of CSHL, "the emotional home of American genetics."[13] A biological laboratory built at Cold Spring Harbor in 1890 would be the source of many twentieth-century discoveries in genetics and quantitative biology.[14]

> My decision to take on Cold Spring Harbor's troubles was in no small part sentimental. When I was there I was home. To me it was science at its best, where finding deep truths mattered more than personal advancement. . . . As director, moreover, I could test my 1958 hypothesis that the cancerous potential of DNA

> tumor viruses is owing to the presence in their genomes of genes encoding enzymes that turn on DNA synthesis. It was too good an idea not to have a high chance of being correct, but because of space and funding limitations it had no chance of being tested at Harvard.[15]

Taking this position was not without risk. The laboratory's finances were not good. Traditionally, the lab had been funded largely by the Carnegie Institute of Washington, but the institute withdrew most of it its financial support in 1963 after their leadership changed their financial commitment to genetics. The condition of the buildings and grounds were poor, and to survive CSHL would need new sources of money. In his early years as director, Watson also kept his academic position at Harvard.

In his first *Director's Report*, he wrote that CSHL "must be a place where trends of the future are anticipated."[16] With this in mind, Watson established a new summer course on tumor virology, which would be repeated for the next five years and which supplemented the established courses in animal cell culture and animal viruses.[17] Another important element to fulfilling his vision for Cold Spring Harbor as a center for tumor virology was finding the right people to perform research. Watson needed to hire someone smart and able to lead a team of scientists. He wanted a "first class Assistant Director"[18] so that he could remain at Harvard most of the time and simultaneously change the focus of CSHL from bacteria and bacteriophages to animal cells and tumor viruses. Joseph Sambrook was that person.

The Early Career of Joseph Sambrook

Joe Sambrook grew up in Liverpool, England, in a working-class family. His father and grandfather operated the family's plumbing business, Sambrook and Son. Joe did well at school. His grammar school teachers saw his academic potential and advocated a track of studying French, German, and history to prepare for Cambridge or Oxford University. However, early in 1955, while on a cycling holiday, he happened to visit Oxford when the May balls were in full swing. He quickly realized that this was not the place for a working-class student like him. And yet, a career in plumbing was even less appealing.

A way out of his dilemma appeared almost by accident. Late in secondary school, a girl he liked attended Science Club, so he did too. No relationship materialized between Sambrook and the girl, but he got hooked on science, which seemed fresh and exciting and much different from the traditional humanities he was studying with increasing reluctance.[19] Sambrook dropped out of school, took a job at a pharmacy, and enrolled in science classes at night. After two and a half years of science classes, Sambrook took the university entrance examinations and did so well that he won a scholarship to the University of Liverpool, where majored in microbiology. In addition to classes, Sambrook attended talks by visiting scientists. One such speaker was Frank Fenner, an eminent Australian virologist. Upon speaking with Sambrook after the presentation, Fenner was so impressed that he invited the young man to enroll in the PhD program at the Australian National University. Fenner even offered to pay for his family's travel to Australia on the condition that Sambrook graduate with first-class honors at Liverpool. In addition to his passion for microbiology, Sambrook by now had a pregnant wife, a two-year-old son, and no money, which made Fenner's offer even more attractive. In December 1962, Sambrook and his young family boarded the SS *Oronsay* for a five-week voyage to begin his postgraduate career on the opposite side of the world.

At that time, the Australian National University was world renowned for virus research. There, Fenner had collaborated with the great Australian virologist Macfarlane Burnet on pox viruses, among other viruses.[20] More practically, Fenner had worked on the myxoma virus as a way of controlling the plagues of invasive, wild rabbits that ravaged rural Australia in the 1940s and 1950s.[21] Once he arrived, Sambrook began working on the genetics of vaccinia, a large DNA virus; and Semliki Forest virus, a small RNA virus. Additionally, Fenner and Sambrook generated a set of conditional lethal mutants of rabbitpox virus, which they used to measure rates of recombination and construct a rough genetic map of the viral genome. In three happy years, Sambrook learned enough viral genetics to write an authoritative overview with Fenner.[22] He returned to England for a postdoctoral position at the MRC Laboratory of Molecular Biology (LMB) in Cambridge with Sydney Brenner. Despite his earlier aversion to Oxbridge, Sambrook became the Harrison Watson Fellow at Clare College, the second-oldest college in Cambridge. In his year with Brenner, and with "tenacity" and "brute force,"[23] as Brenner would describe it, Sambrook worked with *E. coli* and isolated and characterized a strong suppressor of UGA-chain-terminating mutants. (UGA is the signal for "stop" on a messenger RNA sequence and one piece of the genetic code.) Brenner saw Sambrook as an "extremely competent and original researcher" but unofficially thought he was "a rough diamond," someone who was unsure about his career or even where to live.[24] It was certainly true: Sambrook was not happy in Brenner's lab, not for sociological reasons but because the golden age of prokaryotic genetics at LMB was fast coming to a close.[25] Rather than linger at Cambridge and eventually take a lecturing job at a provincial university, he began to think about the United States.

Joseph Sambrook and Renato Dulbecco

In 1967, Sambrook visited his graduate school acquaintance John Cairns in the United States. Cairns, now the director of CSHL, told him about the changes afoot at the laboratory and helped him arrange a second postdoc at the Salk Institute with Renato Dulbecco, whose lab was arguably one of the best places in the world to study the molecular biology of animal tumor viruses (see chapter 3).

Dulbecco had mostly worked on poliovirus and polyoma viruses, but Sambrook preferred SV40. SV40 and polyoma virus are especially closely related; they both have tiny genomes. But among their important differences, SV40 is easier to work with and can be "rescued" from transformed cells in a way that polyoma cannot. By fusing infected cells with healthy cells, researchers can purify SV40 particles again.

SV40, also called simian virus 40 or simian vacuolating virus 40, has an interesting history. In 1960 Ben Sweet and Maurice Hilleman discovered it in cultures of Rhesus monkey kidney cells being used to produce Sabin's live poliovirus vaccines,[26] and Bernice Eddy (see chapter 1) further showed that it was oncogenic in newborn hamsters. Injecting cell-free fluids from Rhesus monkey kidney cell culture into the check pouch of hamsters in June 1959 produced transmissible tumors in 3–15 months.[27] These results did prevent SV40 being injected into many thousands of people who got poliovirus vaccinations between 1955 and 1963.[28] Although some think these injections might have increased the cancer rate, molecular biologists are not convinced, and it would appear that the choice to eradicate polio as quickly as possible, betting that the carcinogenic results in newborn hamsters would not be replicated in humans, was a reasonable gamble.

Given the history of SV40 and its potential as a research tool, this virus was a "hot item" in the field.[29] A key motivating area of research was to determine why and how it transformed cells and why it sometimes entered the lytic cycle and destroyed the host cells.

While in Dulbecco's laboratory, Sambrook and his colleagues performed a particularly elegant experiment on SV40 virus. Sambrook, Dulbecco, and fellow Dulbecco lab scientists Heiner Westphal and P. R. Srinivasan considered the physical state of SV40. By looking at a cell line SV3T3, which is the well-studied 3T3 cell line permanently transformed by an SV40 infection, they showed that the viral DNA does not exist freely in the cells; rather, during the infection process, viral DNA is integrated into the DNA of transformed cells.[30] Their experiments showed that SV40 complementary RNA hybridized with the chromosomes extracted from the transformed cell line. To keep track of the viral DNA, they used radioactive nucleotides as a "primitive sort of labeling."[31] The resulting paper summarized their conclusions: "The results presented here provide evidence that the viral DNA molecules in SV3T3 are integrated with cell DNA by alkali-stable covalent linkages."[32] This integration of viral and host DNA would explain how transformation could be maintained in a cultured cell line—it was likely that viral proteins did not merely trigger transformation but continued to influence the cell over time. This work was one of the high points for the Dulbecco laboratory and was part of the justification for Dulbecco's Nobel Prize in 1975.

Sambrook's excellent work in Dulbecco's laboratory impressed Watson, who had been friends with Dulbecco since their days working together in Luria's laboratory in Indiana, when Watson was pursuing his PhD. John Cairns also advised Watson to recruit Sambrook.[33] So, about a year into Sambrook's postdoc with Dulbecco, Watson "asked in his oblique way whether [Sambrook] might be interested in a job" at Cold Spring Harbor.[34] Given the successes of Dulbecco's laboratory, Watson was not taking a huge gamble by moving into tumor virology and hiring Dulbecco's laboratory personnel.[35] At the end of the 1960s, many molecular biologists were ready to move from studying bacteria and bacteriophage to eukaryotic systems. Specifically, well-funded animal virus research was the natural next step, given that the successes of early biology in understanding the prokaryotic cell were due to a focus on bacterial viruses. To begin, Sambrook would work with Lionel Crawford to help run the 1968 summer course in animal virology. Although it was somewhat isolated on Long Island, Sambrook enjoyed the summer class and its social dimension. Not all the appeal was strictly scientific: as he put it, "You have to remember that this was in the late 1960s when love was free and drugs soft and plentiful." At the end of the course, Sambrook received a formal offer from CSHL. He moved his family from California to New York in

May 1969. He was not sure how long he would last at CSHL, as Watson had expressed the view that scientists were "over the hill" by age 30 and he was close to that threshold when he arrived.

One of the first things Sambrook did after accepting the Cold Spring Harbor job was help complete a significant NIH grant application. Lionel Crawford and Watson had drawn up plans for turning the James Laboratory into a tumor virus lab. To fund the new push into tumor virology, CSHL was awarded a $1.6 million grant from the National Institutes of Health. The team described the research plan: "We intend to study how the SV40 (polyoma) viruses replicate, hoping that our findings will be relevant to the mechanism by which these viruses induce malignant cellular transformations. Our approach will be that of the molecular biologist, visualizing SV40 (polyoma) as a very small virus whose DNA content suggests a genome of at most 8–10 genes. We hope to obtain evidence to tell us what those genes do, both in the lytic cycle and when they are integrated into the host genome."[36] In practice, they needed funds to establish methods for large-scale virus, DNA, and protein purification. They hoped to find temperature-sensitive conditional lethal mutations in order to investigate the SV40 viral machinery. The grant application included salaries for Watson, four scientists (Sambrook, Bernhard Hirt, Carel Mulder, Crawford), postdocs, and technicians.

Their hope was that one of the SV40 genes was responsible for transforming cells into cancer cells and that the Cold Spring Harbor team would identify which one. Watson, Sambrook, and coworkers proposed a number of experiments, a "dog's breakfast" of various approaches, as Sambrook put it.[37] Before the age of restriction enzymes and recombinant DNA technology, no direct method existed for locating the specific function on a specific gene.

Watson was happy investigating the DNA tumor virus SV40. Given previous work in the field, the natural alternative was an RNA tumor virus like Rous sarcoma virus. However, Watson thought that RNA viruses were more dangerous to have in the laboratory. As described in chapter 4, a feline leukemia virus contagious among cats was discovered by Jarrett in Scotland in 1964. It was an RNA virus that easily transferred from one cat to another, as could be observed among the many cats owned by "cat ladies." Although there was no specific evidence the virus could infect humans, Watson went so far as to try to rid Cold Spring Harbor Laboratory of cats because of his concerns about safety. Sambrook did not share Watson's worries; he reasoned that Rous sarcoma virus had been studied for 60 years without a single documented case of a scientist developing cancer from it. In any case, CSHL would not work on RNA tumor viruses, or retroviruses, as they came to be called.

During the next several years, CSHL boomed. After obtaining the grant money, the James Laboratory was extensively renovated and expanded. The lower floor was turned into three biochemistry laboratories. This work took several months and was followed by renovations to the upper floor to create suites where cultured cells and viruses such as SV40 could be safely handled. Robert Pollack, who had shown that one could select for normal noncancerous cells from a population of transformed cells, that is, select for healthy reverse mutations in transformed cells, was hired from NYU Medical School.[38] In work that went beyond tumor virology, Pollack would work with Nancy Hopkins on the cytoskeleton of the cell to show that the cancerous phenotype does not depend entirely on the presence of a particular gene.[39]

In 1971, the National Cancer Institute awarded CSHL $1 million per year for five years to fund a series of interconnected research proposals—an award that would be renewed twice. The buildup of staff and facilities had paid off financially, and CSHL began to benefit significantly from President Nixon's newly announced War on Cancer. The first phase of CSHL's redevelopment was to reequip the old Demerec Laboratory as a modern animal cells culture lab with chromographic facilities and a "hot lab" for Pollack.[40]

By the time the research assistants were hired and trained, a full year had passed since Sambrook's arrival at Cold Spring Harbor. He wondered whether the loss of year at a crucial stage of his research career would prove fatal.

Watson wanted Sambrook to be the nucleus of a new group devoted to tumor virology. Along with Sambrook, Westphal and then Mulder were also hired from Dulbecco's lab at the Salk Institute. Bernhard Hirt came from Lausanne, Switzerland. Lionel Crawford came with experience from Michael Stoker's lab in England. Watson also brought students from Harvard to work in the summers, including Hopkins and William (Bill) Sugden. Some students came to the lab knowing that Watson would write the draft board to secure deferments.

In the beginning there was some tension among the transfers from the Dulbecco lab. Watson had not made it clear that Sambrook would be the leader, whereas Westphal and Mulder thought the three of them would be coleaders.[41] This lack of articulated structure at the beginning fostered some

10.1. Jane Flint, Terri Grodzicker, Phillip Sharp, and Joseph Sambrook in 1973. Courtesy of Cold Spring Harbor Laboratory Library and Archives

mistrust and fueled a competitive atmosphere among members of the tumor group. Sambrook valued the competition and thought it accelerated and honed the research. Others disliked it; eventually, the stringent environment would be one reason Mulder left CSHL after a couple of years. Westphal, too, would stay only two and a half years, leaving to join Phil Leder at the NIH. Watson's biggest fans were drawn to him because he was irreverent and seemed "like one of us": young, ambitious, and brash.[42] The research atmosphere was schizoid: many summer visitors for courses and meetings, in contrast with a monastic existence in the winters with few to socialize with outside of CSHL and little to do beyond work long hours in the lab. The environment was hard on spouses and marriages, Sambrook's included.

By January 1973, CSHL had 14 staff scientists, 2 visiting scientists, 16 postdoctoral fellows, and 6 graduate students busy for hours on end with their research work (figure 10.1). Some of the lab's results were published successfully, and the work seemed to be moving in the right direction. However, Watson was frustrated that so much time had passed since the formation of the lab without any "major breakthroughs."

Despite Watson's frustrations, much progress had been made. In 1971, the SV40 research group began to use newly discovered restriction enzymes to cut the SV40 DNA, inspired by Daniel Nathans and Kathleen Danna at Johns Hopkins, who cleaved the genome of SV40 into 11 different-sized pieces using a restriction enzyme purified from *Haemophilus influenzae*.[43] Restriction enzymes appeared to be part of the bacterial defense system—they cut up invading DNA into small pieces. Different restriction enzymes cut the genome in different places and so made different-sized fragments. In 1973, Phil Sharp, Sugden, and Sambrook developed an assay for identifying restriction fragments. Using electric current, they ran pieces of DNA in an agarose gel that separated the pieces according to size: larger pieces of DNA took longer to migrate through the gel. They also stained the DNA using ethidium bromide. Others had used ethidium bromide before, but the CSHL team made the process more efficient by finding the optimal amount of the stain. This technique was widely adopted and used for decades after its invention to separate and purify DNA fragments. By using two or more enzymes, one could construct a physical map of the genome of SV40 that showed the relative locations of the specific cut sites in the viral genome.

The arrival of a Swedish adenovirus expert, Ulf Pettersson from Uppsala, allowed the Cold Spring Harbor researchers to work on a second tumor virus. Adenovirus was 5–10 times bigger than SV40, but it was easier to grow. It had been discovered in cells taken from the adenoids of young children by Wallace (Wally) Rowe and NIH researchers in 1952–53 and during the same winter from military recruits at Fort Leonard Wood in Missouri by Maurice Hilleman.[44]

In late 1972, Watson hired Richard Roberts, a British RNA sequencing expert from Harvard, to sequence the genome of SV40. However, since two other labs—Walter Fiers's at Ghent University, Belgium,[45] and Sherman Weissman's at Yale—were attempting to sequence SV40, Roberts thought it more productive to purify additional restriction enzymes instead. This was not a complete change of focus because to perform successful DNA sequencing, small pieces of DNA were needed; restriction enzymes create smaller pieces of DNA from larger ones. The change of plans did lead to tension with Watson, however. Roberts's group would eventually find half the commonly used restriction enzymes and assume the time-consuming responsibility of supplying all the investigators in CSHL, and eventually investigators all around the world, with restriction enzymes that were becoming increasingly important tools for new experiments in molecular biology. In the mid-1970s there was no comprehensive commercial source for the restriction enzymes that molecular biologists craved. New England Biolabs (NEB) was one of the first companies to move into this space. NEB hired Roberts to be their chief consultant in 1975, but he remained a staff member at CSHL until 1992. (See chapter 11 for more about Roberts's career and his Nobel Prize.)

The primary focus of Sambrook's lab was the nature of tumor virus transcription: the team at first studied transcription of SV40 RNA from SV40 DNA and later turned its attention to adenovirus transcription from adenovirus DNA. Understanding how the virus controlled the expression of its genes would provide some insight into how the genes functioned and which gene(s) were cancer causing. By 1972, it was known that some viral RNA sequences are expressed early in infection while others are expressed later. Westfall had shown that he could transcribe some SV40 DNA into RNA in vitro using an *E. coli* polymerase and the RNA came preferentially from one stand of the DNA double helix.[46] Sambrook's group used this insight to develop a tool to separate the two strands of the double helix of SV40 DNA and compare their expressions in two cases: during a lytic infection, in which the virus replicates and destroys the host cells, and also in transformed cells, where the virus integrates into the host DNA and does not kill the host cell.[47] Sambrook's hybridization experiments suggested that the genes expressed early in an infection were on the opposite strand of DNA from those expressed late. The early and the late genes, as they were known, each took up roughly half of the genome. The late genes were expressed only when the SV40 DNA was replicating and appeared to be genes that were needed to make new viral particles. However, a key question was still very much open: Sambrook, Sharp and Keller wrote, "What the nature of the accompanying switch from early to the late pattern of transcription might be is entirely a matter for speculation."[48]

A project four years in the making came to fruition in 1973, when Cold Spring Harbor published a 743-page book summarizing the state of the field. It was titled *The Molecular Biology of Tumor Viruses* and edited by John Tooze of the ICRF. It was an unusual book for a number of reasons. Namely, although it was an edited volume, the individual chapters were not identified by author but rather rewritten by Tooze and Sambrook during a frantic month in London to give a uniform voice. It was a group effort, with 22 contributors including current and former CSHL scientists Watson, Sambrook, Mulder,

B. Ozanne, Pollack, Hirt, and Crawford. Second, although it was the first book published on the molecular biology of tumor viruses, Tooze called it a second edition. The first edition was completed in November 1970 and although set to type, did not go into production. Nonetheless, the galleys of the "first edition" were copied for limited circulation. As Watson noted in the foreword, the frenetic pace at which new data were being generated about tumor viruses made it difficult to write a book that was up to date. Nonetheless, the "second edition" was testament to how much new data a molecular approach to tumor viruses was yielding. Of the thirteen chapters, four covered the structure, lytic cycle, transforming ability, and genetics of SV40 and polyoma virus.

The 1974 Tumor Virus Symposium

The year 1974 was important for tumor virology at CSHL. Although Watson still split his time between Harvard University and CSHL, he had time to organize an important, large meeting on tumor viruses. A multitude of scientists traveled to Cold Spring Harbor to listen to talks on the latest experiments. Among the attendees, seven future Nobel Laureates: Baltimore, Dulbecco, and Temin (1975); Nathans (1978); Varmus (1989); and Roberts and Sharp (1993) presented material. (See chapters 3 and 8 for Dulbecco and Temin, chapter 9 for Varmus, and chapter 11 for Roberts and Sharp.)

Watson called the timing of the meeting "propitious" because

> tumor virus research has a uniquely favorable position in contemporary biology. It affords the opportunity to probe both the fundamental biology and chemistry of higher animal cells, as well as the nature and origins of cancer. So it combines exciting science with only a marginal concern for next year's research dollars. Not surprisingly, masses of scientists have moved into this arena, and the pace of good research has increased almost overwhelmingly. Rapid accumulation of new facts, however, should not be confused with deep understanding, and all too many basic principles about how tumor viruses act remain to be worked out.[49]

Tumor virologists from around the world converged on Long Island to present papers and enjoy informal discussions after the talks. Renato Dulbecco gave the introductory remarks: as he saw it, the field had experienced a "vast development" since 1962 when CSHL last devoted a symposium to virology. Sambrook and his coworkers contributed four papers, the maximum allowed from any one laboratory. As Sambrook put it, CSHL "dominated" the proceedings.[50] One of the more significant papers came from Terri Grodzicker, J. Williams, Sharp, and Sambrook.[51] They had managed to locate mutations on the physical map of the adenovirus genome. By recombining different mutants, using restriction enzymes to cut the recombinants into pieces, and analyzing the pattern of fragments, the CSHL team was able to infer the position of the mutations in the genome.

The inspiration for this approach to mapping is worth recounting for the way it demonstrates the value of interpersonal connections in science. Six months earlier, the Scottish adenovirus geneticist Jim Williams visited CSHL. After a late night of welcoming Williams to the lab, Sambrook and others were sitting on the floor in an office the next morning, nursing hangovers. Williams asked Sambrook what he should work on. An idea came to Sambrook: that recombinants could be made between adenovirus 2 and 5 to aid in mapping the mutations using restriction enzymes. This insight spawned a new project, on which they worked as a team for six months, producing a "most satisfying" piece of work. Eventually, sequencing technology would eclipse mapping mutations using restriction enzymes, but at the time restriction enzymes allowed for the mutations found in various mutant viruses to be mapped to a location on the genome.

The Function of T antigen

One aspect of the symposium centered on what seemed to be the central element in the transformation of healthy cells: something called SV40 tumor-antigen, or T antigen.[52] It had been known since the early 1960s that injecting a live virus such as polyoma virus into mice or hamsters protects the animals from transplantation of tumor cells, while injecting a killed virus has no effect. This discrepancy suggested that the live, growing virus creates a new antigen that is not part of the virus particle itself.[53] Researchers from the NIH, using antibodies, showed that T antigen was present in SV40 transformed cells of various species, suggesting that the T antigen had a viral origin.[54] They also showed that the T antigen does not exist in normal cells or in tumors that are not caused by viruses. Thus, T antigen was considered to be a signature of transformation. David Baltimore summarized the state of knowledge at the meeting: "The present evidence, while not totally compelling on this point, argues that the early region of the SV40 genome encodes one protein which is at once T antigen."[55] It was possible for transformed cells to only have mRNAs expressed from the early region of the SV40 genome. The interesting research question was to determine the function of T antigen. Specifically, how did this protein transform the host cells?

Better understanding of T antigen came via a discovery in 1977 that transformed eukaryotic biology: mRNAs in higher cells may be spliced or composed of segments that were encoded by separate regions along the gene.[56] Two teams of scientists, Richard Roberts, Richard Gelinas, Louise Chow, and Tom Broker at CSHL; and Phillip Sharp, who had left CSHL for MIT, and his postdoc Susan Berget, independently discovered RNA splicing in their studies of adenovirus. Unlike bacterial genes, in which the messenger RNA transcript is uninterrupted and collinear with the DNA sequence, eukaryotic RNA transcripts are modified: RNA sections called introns are removed, and exons are joined or spliced together to form the informational message. Following a suggestion by Roberts and using reagents prepared by Gelinas, Chow, an electron microscopist at CSHL, observed the results of splicing by looking at micrographs of adenovirus RNA. (See chapter 6 for more on this discovery.) This was a remarkable research result, one that Watson considered comparable in significance to the discovery of messenger RNA.[57] One of the first implications of this groundbreaking research was the realization that the SV40 T antigen gene encoded two proteins that were named large T and small t. The large T appeared to be the cancer-causing gene.

The Moratorium on Recombinant Technology

Further progress on understanding the function of T antigen would depend on recombinant methods: researchers needed to clone the gene for T antigen by transferring it into bacteria to purify and study larger amounts of the protein. These methods had been developed in 1973 but were prohibited for NIH-funded work, over safety concerns, in 1976.[58] Ironically, the person who began the movement toward a ban was CSHL cancer researcher Robert Pollack. While teaching the tumor virus course in 1971, Pollack heard of plans to clone SV40 genes into *E. coli* bacteria in the biochemist Paul Berg's lab in California.[59] Pollack was worried that there were significant risks of transmitting cancer to humans by introducing oncogenes into the bacteria that live in the human gut and wanted safeguards and guidelines to address them. For his part, Paul Berg was not worried about the risk, although he acknowledged it was not zero.

One effect of Pollack's worry about transmitting cancer was the formation of a 1973 conference on the biohazards in biological research in Asilomar,

California.[60] Both Pollack and Watson attended the conference, and Cold Spring Harbor Laboratory published the proceedings. A number of virologists were present. Maurice Hilleman, who discovered SV40 in the poliovirus vaccine, delivered a talk and mentioned during the discussion that there was no evidence of short- or long-term adverse effects in people who had received SV40 in the poliovirus vaccine. Watson, who spoke in the last discussion period, argued that the National Cancer Institute had a moral responsibility to protect cancer scientists by providing more funds to build "safe labs." He suggested that, when coming to any decision about the safety and risk of recombinant DNA, they should adopt an approach that mirrored the Atomic Energy Commission's reasoning concerning nuclear power plants.[61] Paul Berg had the last word: "I am persuaded by what I have heard that prudence demands caution and some serious effort to define the limits of whatever potential hazards exist."[62] Following the conference, Berg, Baltimore, Watson, and others drafted a letter to *Science*.[63] They proposed voluntarily deferring certain experiments—including experiments linking oncogenic viruses or pieces of oncogenic viruses with autonomously replicating DNA such as plasmids—until the hazards could be better evaluated. In retrospect, Watson regretted the letter for unnecessarily raising public alarm.[64]

The 1973 conference in Asilomar was followed up by another in the same location on the same topic two years later. This 1975 conference is better known, in part because by design a number of journalists were invited.[65] By this point, virtually all recombinant DNA technology was seen as potentially dangerous, and it was agreed that certain experiments were to be deferred, especially cloning DNA from pathogenic organisms. The NIH adopted the recommendations of this meeting. In particular, experiments with higher levels of risk were permitted only in more secure facilities.

Potential experiments fell within different general classes ("shotgun" experiments with *E. coli*; use of recombinants to insert genes from viruses, plasmids, and organelles into *E. coli*; and use of animal virus vectors), and the NIH guidelines assigned a physical level of containment, designated P1 to P4, to be used in each class. Physical containment P1 consisted of standard microbiological practice, P2 required extra precautions such as not creating aerosols, and P3 called for having the lab itself under negative air pressure. The highest category, P4, involved techniques such as airlocks, protective clothing, and showering on exit for handling the most dangerous pathogens.[66]

Watson was not happy about the outcome of the second meeting. He went in intending to address whether the voluntary moratorium should be lifted, not whether it should be extended and expanded, which was exactly what was discussed.[67] Cloning oncogenes would now require a P3 or P4 level containment, which did not exist at CSHL. His view was that "the whole affair was a hasty rush into unjustified bureaucratic rules that would set back the course of legitimate science." He was aware that such a view made him seem eccentric and irresponsible. He acknowledged that "much too late" he changed his mind and came to the view that the experiments that Asilomar participants urged to halt were not novel (and so not dangerous) because bacteria in nature are continually taking up foreign DNA. To explain why many scientists disagreed with him, the governor of California Jerry Brown suggested to Watson that it might be a case of "liberal guilt": "In private conversations with the vast majority of molecular biologists, I find almost universal agreement that the guidelines are a total farce but no one feels equal to the task of telling the Senior Senator from Massachusetts that the Emperor has no clothes."[68] Senator Edward Kennedy had publicized the controversy by holding congressional hearings on the potential dangers of recombinant technology. The issue clearly concerned Watson, since the new restrictions on cloning tumor virus genes would severely hamper the progress of tumor virus research at Cold Spring Harbor Laboratory, and he devoted five pages of his *CSHL Annual Report* to the issue.

Robert Tjian and the Function of T antigen

Despite the new restrictions on cloning, there was still progress in understanding T antigen. In 1978, it was shown that T antigen binds to specific DNA sequences at the SV40 origin of replication, the place on the genome where the SV40 genome first starts to replicate. The biochemist Robert Tjian, who had been recruited by Watson to join CSHL in the spring of 1976 as a postdoctoral fellow, determined some of the functions of T antigen (figure 10.2). "Tij," as he was known, was raised in Hong Kong, São Paulo, Brazil, and New Jersey. He completed an undergraduate degree at University of California, Berkeley, and then a PhD at Harvard. Like many others entering the tumor virus field, Tjian had a background in bacterial systems but was interested in transitioning into eukaryotic systems.

At CSHL, Tjian developed antibodies specific to T antigen by injecting SV40 into hamsters and waiting for their immune response to the SV40 proteins. These antibodies, once purified, allowed him to track T antigen through

10.2. Robert Tjian and James Watson in 1998. Courtesy of Cold Spring Harbor Laboratory Library and Archives

various biochemical purification techniques. Because it was difficult to purify large quantities of T antigen, he used a "rather exotic" adenovirus-SV40 hybrid virus, Ad2+D2, discovered at CSHL, that overproduced a hybrid protein highly similar to SV40 T antigen.[69] Tjian did not create the hybrid virus, but with his antibodies demonstrated that it overexpressed T antigen. He investigated the hypothesis that T antigen binds a specific sequence of DNA using the enzyme DNase that degrades DNA. The hybrid protein protected the DNA binding site from degradation by DNase. (The pancreatic DNase I enzyme will degrade naked DNA but not DNA bound to a protein like T antigen.) Watson took much interest in Tjian's results, stopping by to talk to him "every other day."[70] Joe Sambrook was in England when Tjian arrived, so Tjian had a good deal of autonomy in his approach to investigating T antigen. The binding sites were actually tandem repeats of DNA that contained both the origin of viral replication and key elements of the SV40 early promoter, and T antigen turned out to be the first eukaryotic sequence-specific DNA binding factor discovered. In addition to increasing biologists' knowledge of SV40, this work gave biologists new insights into the regulation of eukaryotic transcription. This result was the first of many for Tjian, who made his name discovering a number of sequence-specific DNA binding proteins and core components of the molecular machinery that regulate RNA transcription in humans.

It was known that certain bacterial enzymes that facilitated the initiation of DNA replication were ATPases. Tjian showed that early speculations that T antigen was a kinase were incorrect, and that it was in fact an ATPase. There was a lot of scrutiny of Tjian on this point, because the oncogene *src* (pronounced "sark") had been discovered in Rous sarcoma virus (see chapter 9), and it was a kinase. Many thought that T antigen would also be a kinase, and Tjian was challenged by "heavy duty people" on his data, but he stood his ground.[71] The ATPase activity supported the view that T antigen was a sequence-specific DNA binding protein that bound at the origin of replication and so regulated SV40 DNA replication.

By this point in 1978, Sambrook had left the United States for London's Imperial Cancer Research Fund, the "spiritual sister lab" of CSHL,[72] so that he could do recombinant work that was under moratorium in the States but not in England. There, he was able to clone adenovirus cDNA into an *E. coli* plasmid. Sayeeda Zain and Richard Roberts then sequenced the clone using techniques pioneered by Fred Sanger. The splicing points in the adenovirus mRNA were the same as splicing points on other mRNAs, suggesting that similar enzymes splice different mRNAs. For Sambrook, the appeal of England was not only scientific. After work he would often go to a pub behind the lab where many unattached scientists would socialize.[73] While there, he got to know Mary-Jane Gething, a protein chemist. Romance blossomed, and she would become his second wife.

The Rise of Cloning at CSHL

In February 1979, the NIH rules on recombinant work were relaxed, clearing the way for the CSHL tumor virus group to clone SV40 and adenovirus genes. Within a year, the sequence of SV40 was determined. Given their rapid successes, Watson decided to devote the 1979 CSHL Symposium to viral oncogenes. The focus on genes rather than viruses reflected the explosive growth in understanding of tumor viruses that had occurred over the past five years. It was a large meeting—as Watson put it, with almost too many papers to absorb—and CSHL used closed circuit TV to telecast lectures because there were 200 more researchers than seats in the largest auditorium. The distinctive red volumes of the proceedings included 141 papers.

Tjian had moved to UC Berkeley in 1979 and continued to work with SV40 and adenovirus hybrids. With his student Carl Thummel and CSHL's Terri Grodzicker, Tjian created a hybrid virus that placed the large T and small t proteins under an adenovirus promoter. With this designer virus, Tjian was able to purify native T antigen in larger quantities. The approach created a new vector system for overexpressing eukaryotic proteins.[74] In this way, a recombinant virus was turned into a general tool.

With the lifting of the NIH moratorium on many recombinant biological techniques, Cold Spring Harbor devoted the 1980 summer course to molecular cloning. The instructors were Tom Maniatis and Edward Fritsch from Caltech. Nancy Hopkins, who had moved from CSHL to MIT, was part of the previous year's tumor virus course and a motivating force for its organization and continuity. Maniatis had been at Harvard, but Watson invited him to CSHL in 1974 when the city of Cambridge instituted an outright ban on recombinant experimentation, which precluded Maniatis's planned cloning of full-length cDNA copies of b-globin mRNA. Sambrook was upset that he was not consulted on the plan to bring Maniatis to CSHL, but Maniatis came to CSHL, nonetheless. In 1977 he was offered an attractive job at Caltech, which was building a P3/P4 containment facility that would allow cloning of human genes. When Watson asked him to instruct the CSHL summer course, he was somewhat reluctant as he was only a third-year faculty member, but he brought his postdoc Ed Fritsch to help.[75] Fritsch had trained in Howard Temin's lab and assembled all of the protocols to be used in the course into a loose-leaf binder. The course was a success and became a staple of the curriculum, always offered as one of the three or four summer courses taught every year at CSHL in the 1980s.

For this second run of the course, the group assembled and refined a set of "consensus protocols" based on protocols in use at Harvard, Caltech, and CSHL. This set of protocols was written in the winter of 1981–82 and published by Cold Spring Harbor Laboratory as *Molecular Cloning: A Laboratory Manual*.[76] Watson saw the utility of and need for such a manual in the wider biological community and pushed the at first reluctant Maniatis to polish the set of protocols into something publishable.[77] Watson envisioned the proposed manual as a "modern Jeffrey Miller manual," which at the time was the definitive laboratory manual for microbial genetics.[78] "Anyone who has worked closely with Jim knows that you cannot say no to him,"[79] Maniatis recalled. Watson sold Maniatis on the project by emphasizing Sambrook's writing ability as a coeditor. They started with the loose-leaf binder that Ed Fritsch had assembled. Conceptual information was added so that practitioners could troubleshoot problems themselves when the protocols did not work as expected. The volume ended up having three editors—Maniatis, Fritsch, and Sambrook—and went through multiple drafts. The spiral-bound, 545-page manual was an instant hit when it was published in 1982. It became the bible of cutting-edge laboratory techniques, and by 1985 it sold 38,000 copies.[80] It contained material on how to clone into plasmids, how to propagate bacterial strains, how to use a variety of restriction enzymes, how to work with gel electrophoresis, how to create and clone cDNA, and many other topics. It was translated into multiple languages, and pirated copies were made in Taiwan, increasing its influence in Chinese labs. This book, which spread the know-how and procedures of the CSHL summer class to a broad biological audience, is one of the enduring products of the 1970s Watson era of CSHL. It was not merely a collection of recipes but rather the consequence of intense development at Harvard, Caltech, and CSHL.[81] It left a lasting impact on the development of molecular biology.

The 1970s could be thought of as the decade of the tumor virus, the 1980s as the decade of the cloned gene, and the 1990s as the decade of the human genome. By the early 1980s, Sambrook understood SV40 and adenovirus at the level he was aiming for, and consequently, he disbanded his lab. He left for the University of Texas Southwestern Medical Center in Dallas in July 1985. Watson spent much of his energy in the 1990s as head of the Human Genome Project and remained at CSHL until 2019. Many other tumor virologists at CSHL and elsewhere moved into more general cancer research. With the ability to clone genes including oncogenes, the experimental utility of tumor virology diminished. Two months after Sambrook left CSHL, he returned to celebrate the completion of a new laboratory building named in his honor. The Joseph Sambrook Laboratory extends from the north end of the James Laboratory where he had worked for 16 years. Over that time period, 97 people were associated with the James Laboratory, either as graduate students, postdocs, visiting scientists, or scientists.[82] In addition to Watson's remarks, Renato Dulbecco, Sambrook's postdoctoral advisor, spoke, as did Mickael Botchan, who, like Tjian, left CSHL for UC Berkeley.

The Sambrook-Watson Relationship

The relationship between Joseph Sambrook and James Watson is one of the more important in the development of 1970s molecular biology. It is, however, unusual in a number of respects. Famous pairings in the history of molecular biology most commonly occur between two scientists whose skills complement each other as they work together a project.[83] Watson and Crick working together to solve the structure of DNA is a case in point. In this instance, however, it was a relationship between a forward-thinking, hands-off administrator and an extremely capable chief scientist. Watson provided Sambrook a place where science of the highest order could be performed. Sambrook allowed Watson to realize his dream that the development of the molecular biology of tumor viruses would shed light the mechanisms of cancer. While Watson did not direct Sambrook, he did steer the direction of research at CSHL by hiring Sambrook in the first place. They agreed that competition fueled the pace of science and encouraged competition with other labs as well as within CSHL. However, the relationship was not without friction, in part because Sambrook was a "tough guy."[84] Watson "fired" Sambrook a number of times, but once emotions waned, it was clear that Watson had not meant what he said in those disagreements. Sambrook assumed directorship of CSHL when Watson took a sabbatical. Sambrook's second wife, Mary-Jane Gething, put it well: beyond the large personality, Sambrook "is often a selfless person—throughout his academic career he has built things for other people."[85] Sambrook was certainly instrumental in building CSHL into a world-class center for molecular biology, the institution that James Watson had dreamed of in the late 1960s. Sambrook reflected on his relationship with Watson thus: "Jim is a complicated man, but his heart is in the right place. He loves good science. He loves to be surrounded by good scientists."[86]

Watson remained in the leadership of CSHL, surrounded by good scientists for the remainder of his career. He headed the Human Genome Project in the early 1990s. His affiliation with CSHL ended in January 2019 when a PBS television documentary aired in which he failed to denounce his earlier racist remarks that Africans have lower IQs than Europeans. He did not respond to the controversy as he had been seriously ill since October 2018 following a car accident.

Sambrook also became a science administrator. After CSHL he worked at the University of Texas Southwestern Medical Center and then as director of research at the Peter MacCallum Cancer Center in Melbourne, Australia. He was elected as a fellow of the Royal Society and also a fellow of the Australian Academy of Science. He passed away June 14, 2019.

Further progress on understanding the mechanism of T antigen would hinge on understanding the host cellular machinery that interacts with T antigen; this topic will be explored in chapter 13.

11

Control Mechanisms beyond Viruses

Louise Chow, Phillip Sharp, and Richard Roberts, and the Discovery of RNA Splicing in Adenovirus

Phillip (Phil) Sharp was born in rural Kentucky, near Falmouth. Like many in the area, his parents started out as tenant farmers; they tried running a grocery store, but it failed in the Depression. They bought a farm when Sharp was seven, and there he grew up. Although they did not attend college, they understood the importance of education and encouraged Phillip to attend. They paid for college by raising Hereford calves and growing tobacco. He attended Union College, a small liberal arts school run by the Methodist Church, and majored in mathematics and chemistry. A young professor at Union, Dan Foote, encouraged Sharp to consider graduate school in chemistry. Despite failing every entrance exam except organic chemistry, Sharp, with support from Foote, was admitted to the University of Illinois.

After completing a PhD in theoretical chemistry, Sharp left for a postdoc at Caltech with Norman Davidson, who had transitioned from chemistry to biochemistry and made a name for himself studying nucleic acids. Jerry Vinograd's lab was next to Davidson's, and Sharp attended both lab meetings. Like Davidson, Vinograd was interested in questions that straddled the border between chemistry and biology. At the end of his postdoc in 1971, Sharp decided he wanted to study DNA tumor viruses. He approached David Baltimore, who said he would have space for him if he could wait two years, and James Watson, who called him back and offered him a job and gave him two weeks to decide. Sharp discussed the position with Vinograd, who encouraged him to take it. "Jim has been an outstanding leader," Vinograd said. "If he is putting something together [at Cold Spring Harbor], it's going to be excellent." He also told Sharp about a scientist at Cold Spring Harbor named Joe Sambrook, "who is really good young guy."[1]

Sharp began his time at Cold Spring Harbor as a postdoc in 1971 but was made a staff member in 1972. He worked with Joe Sambrook on SV40 to purify the viral genome and separate the two strands of DNA using electrophoresis. Sharp and the CSHL group pioneered the use of ethidium bromide to stain the DNA and make otherwise invisible bands in gels visible under UV light. Other members of Sambrook's group were interested in replication of SV40 and adenovirus, but Sharp was more interested in transcription of mRNA from viral DNA. As detailed in chapter 10, the Cold Spring Harbor group used restriction enzymes to construct a genetic map of the genome. Different strains of the virus could be compared, and linkage maps could be created.

Sharp's Work at MIT

In his third year at CSHL, Sharp decided to return to a university position and have his own lab. He was happy to be offered a position at MIT, where a new cancer center was getting off the ground, headed by Salvador Luria. David Baltimore, Robert Weinberg, Dave Hausman, and Nancy Hopkins made up the molecularly focused cancer group. They held weekly joint lab meetings, so information flowed among them.

Sharp would bring physical and chemical expertise to the new center. Once there, he continued to focus on the transcriptional cycle of adenovirus. James Darnell and others had identified a high molecular weight long RNA in the nucleus of infected cells. To see how this RNA moved into the cytoplasm was one of Sharp's motivating research questions. Pulse chase experiments, designed to track the large RNA into the cytoplasm with radioactive labels, were challenging because the population of radioactive triphosphates could not be changed fast enough to get useful data. Sharp began to investigate adenovirus RNA using other methods he had studied in the Caltech laboratory, including electron microscopy and DNA-RNA hybridization. He trained Claire Moore, who was the electron microscope (EM) technician for the MIT cancer center, to make micrographs of single molecules of DNA and RNA.[2] His postdoc Susan Berget was working on mapping adenovirus mRNA to the corresponding positions in the adenovirus DNA genome. Berget and Moore would make the RNA and DNA preparations and explore the conditions for hybridization before producing grids to be examined in the EM. Once the images were made, the three participated in examining and interpreting the results.

After starting to look at adenovirus hexon mRNA in September 1976, the MIT group had, by January 1977, found a small piece of the 5′ end of the mRNA that unexpectedly would not anneal with DNA at that position. (Hexon protein makes up the protective coat, or capsid, of the adenovirus particle.) This "tail" did not appear to be an experimental artifact due to a particular salt concentration, as it remained after changing salt conditions and it was repeatedly found in many instances of hexon mRNA. The length of the tail was the same each time and did not exhibit the variation one would expect if it were due to branch migration. Berget presented this result at the weekly seminar group that included the labs of David Baltimore, Robert Weinberg, Nancy Hopkins, and David Housman. These were significant results, so Sharp and Berget spoke with Baltimore about communicating a paper to the *Proceedings of the National Academy of Sciences*. Until 1995, only members of the academy could submit papers, either their own or others', to the journal. In early February, they discussed the several-hundred-nucleotide "tail" segment.[3] Since it would not hybridize to adjacent viral DNA, it must have a different origin. They considered three possibilities:

1. They were random sequences created by a nonspecific polymerase.
2. They were transcribed from cellular sequences and added to the viral mRNA.
3. They were transcribed from viral sequences and added to the viral mRNA.[4]

Baltimore suggested a way of testing option 3 using electron microscopy and hybridization.[5] (The other two options would be difficult to test given the state of DNA technology in 1977.) But in the meantime, Baltimore sent a paper with the preliminary finding of the tail to *PNAS* reviewers. One reviewer was James Darnell.[6] While the paper was under review, Sharp, Berget, and Moore used the R-loop procedure—making DNA-RNA hybrids where the sequences match and observing the displaced single-stranded DNA loops—to observe complementarity and lack of complementarity. (DNA-RNA hybrids are more stable than double-stranded DNA under certain conditions.) When they examined the upstream sequences of the hexon mRNA they were in for a surprise. While looking at the micrographs, the usually soft-spoken Sharp was so excited, he exclaimed, "There are three f__ing loops!"[7] (figure 11.1). These electron micrographs would be incorporated into the paper once the reviews of the first version came back. An invitation from adeno-

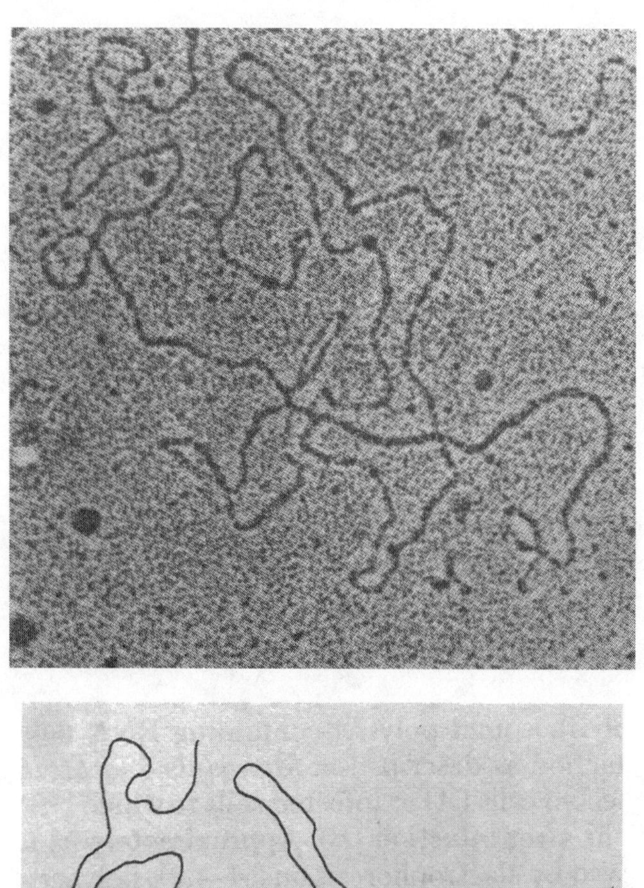

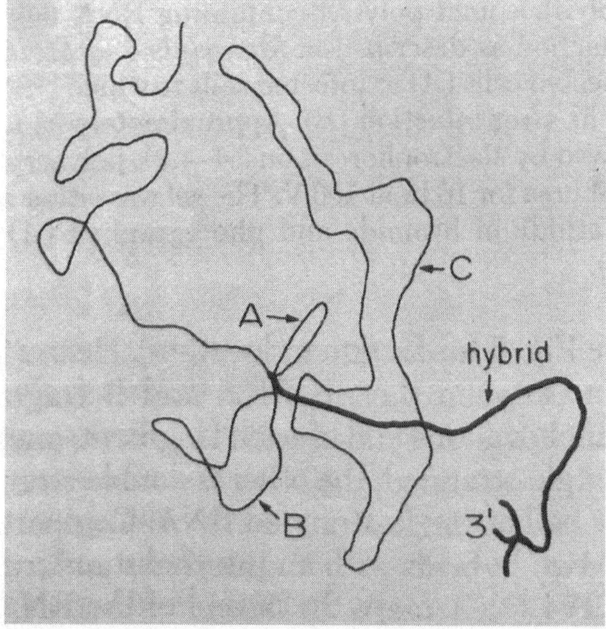

11.1. Adenovirus splicing micrograph from MIT group showing three loops. Image courtesy of Phillip Sharp

virus expert Joe Weber to present a paper at Université de Sherbrooke in Quebec, Canada, on April 15 allowed Sharp to present his group's results.[8] The Canadians were "astonished" and "shocked" at the revolutionary results, and it took a while for the implications to sink in.[9] The resulting revised paper was sent out to the reviewers again, and the final version was communicated to *PNAS* by Baltimore on May 8, 1977.

The English Chemist Richard Roberts

Sharp was not the only one thinking about how to investigate mRNA production. Richard (Rich) Roberts grew up in England. His early interest in science was nurtured by his father who bought him a chemistry set, which began his career path. He majored in chemistry as an undergraduate at the University of Sheffield, and, influenced by a charismatic chemistry professor, David Ollis, continued on to get his PhD. His research proceeded quickly, giving him time to read outside of his area. John Kendrew's 1966 book, *The Thread of Life*, an introduction to molecular biology based on a BBC television show, re-reoriented his focus toward molecular biology. Knowing he wanted to be a molecular biologist, he applied to a number of labs in the United States and was accepted into Jack Strominger's at the University of Wisconsin. However, before Roberts arrived, Strominger moved to Harvard, so Roberts ended up doing his postdoc there instead.

Strominger, while supportive, was not particularly interested in Roberts's project—investigating transfer ribonucleic acid (tRNA) and left him alone for the most part. This hands-off approach suited Roberts. To learn the newest RNA sequencing techniques based on radiolabeling, Roberts spent six weeks with Fred Sanger at the MRC in Cambridge, England. Returning to Cambridge, Massachusetts, Roberts began sequencing tRNAs and became the go-to guy for RNA sequencing.

At the end of his postdoc, Roberts was considering two jobs, one at Harvard and one at CSHL. Harvard biologist Mark Ptashne had told Roberts about the CSHL job. James Watson wanted someone to sequence the genome of SV40. The interview in Watson's office "lasted all of five minutes" before Watson offered Roberts the job.[10] Ptashne, behind the scenes, had put in a good word. However, when Roberts turned up at CSHL in 1972, one of the first things he said to Watson was that he was not going to sequence SV40. There were two other labs racing to sequence the SV40 genome, and Roberts thought it was redundant for CSHL to join the race. This caused some friction between the two of them, but Watson allowed Roberts to do what he wanted. Instead of sequencing, Roberts was interested in exploring a new type of enzyme, called restriction enzymes, that he had heard about from Dan Nathans at a talk at Harvard.

The Rise of Restriction Enzymes

The restriction enzyme endonuclease R could cleave DNA into pieces if the DNA had a specific short string of bases. Roberts saw the potential for this enzyme to help with DNA sequencing. Short pieces of DNA are much easier to work with than long pieces. At the time, short sequences of DNA could be sequenced via RNA—that is, transcribe the DNA into RNA and then analyze the RNA. Roberts began purifying restriction enzymes. He found many of them, each one differing in its cutting specificity (figure 11.2). Each newly discovered enzyme cut DNA sequences in a different place. Biologists visiting CSHL would often ask Roberts to see whether he had a restriction enzyme that would cut the DNA they were researching. Given the growing popularity of restriction enzymes as general tools in molecular biology, Roberts suggested to Watson that they commercialize the sale of restriction enzymes, but Watson was not interested in mixing this type of commercial business with the pure research at CSHL.[11] In retrospect, CSHL could have made a lot of money, as this profitable space was filled by private companies like New England Biolabs.[12] Watson later admitted he made a mistake in rejecting Roberts's proposal.[13]

Roberts hired his third postdoc in 1974. Richard Gelinas, who had completed his PhD at Harvard, joined the Roberts laboratory as a National Cancer Institute postdoctoral fellow. His project was to study the adenovirus sequences associated with initiating transcription and to see whether eukaryotic promoters were identical to prokaryotic promoters.

In February 1975, Cold Spring Harbor Laboratory welcomed two more scientists trained in Norman Davidson's laboratory at Caltech: the husband-and-wife team of Tom Broker and Louise Chow (figure 11.3). Chow was born in Hunan Province in China and completed her undergraduate studies at National Taiwan University before coming to the United States for her graduate studies at Caltech. Broker was appointed as a staff scientist and promoted to chief of the electron microscopy section after the previous microscopist left CSHL.[14] Together Broker and Chow made up the EM section.

Chow worked on adenovirus transcription and also on bacteriophage Mu,

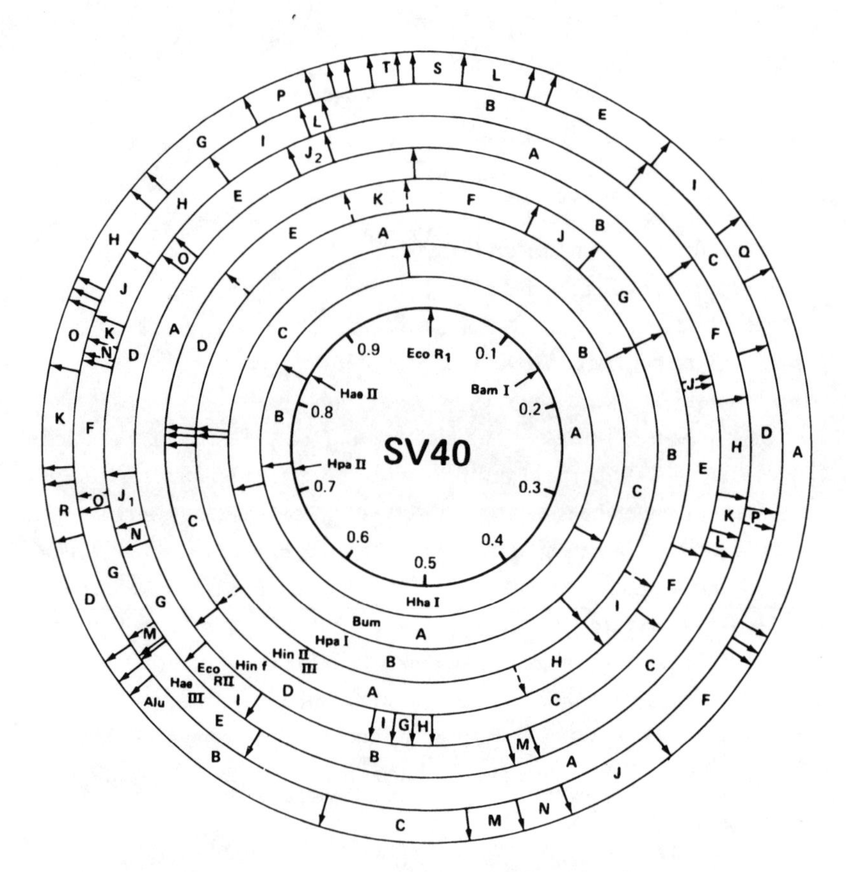

11.2. SV40 restriction map published for the 1975 Tumor Virus Meeting. Arrows represent restriction enzyme cut sites. Image courtesy of Cold Spring Harbor Laboratory Library and Archives

which could invert a piece of its genome and so its gene expression, to change its host range, the species of bacteria it can infect. This early work led to an NIH research grant with Chow as principal investigator.

She and Broker worked to adapt the R-loop hybridization, a method from the Norman Davidson laboratory, to mapping adenovirus mRNAs to their position in the DNA adenovirus genome. By February 1977, they had a paper ready to send to *Cell*, a prestigious journal for molecular biology.[15] The paper identified the coding regions for the viral proteins expressed late in the viral replication cycle. Also visible in the electron micrographs were unhybridized tails on the ends of the R-loops. A poly(A) sequence probably explained the

11.3. Louise Chow and Thomas Broker in 1974. Image courtesy of Cold Spring Harbor Laboratory Library and Archives

3′ tail, but the tail at the other end of the loop—the 5′ end—was more mysterious, and a number of hypotheses might explain it.

Richard Roberts's postdoc Richard Gelinas discovered a common capped 11-base-long sequence on the 5′ end of adenovirus mRNA that was the same in multiple late mRNAs.[16] This finding was unexpected—Roberts and Gelinas expected to find different 5′ sequences for each different mRNA—and its meaning was unclear. Roberts and Gelinas played with various hypotheses to explain the data, with Roberts speculating that short VA (virus associated) RNA might serve as a primer for late mRNA synthesis in adenovirus.

Around this time, Roberts heard a rumor that Phil Sharp at MIT was on the verge of a big discovery. He inferred that it must be something to do with EM of adenovirus RNAs, since this was the project Sharp had taken with him to MIT from CSHL. This possibility made testing his ideas with EM more urgent, so Roberts and Gelinas proposed a collaboration with Chow and Broker. They would use the R-loop method invented by Raymond (Ray) White and then further refined by Norman Davidson's laboratory at Caltech. This was the "perfect" method, accurate within 50 nucleotides.[17] Chow and Broker refined the thermodynamic and kinetic parameters to make their experiment work,[18] while Gelinas worked to prepare the pure RNA and DNA specimens needed.

The Electron Microscopy of DNA-RNA Hybrids

The CSHL team realized that single-stranded DNA would work best and set to work on the testing. They began by looking at an adenovirus DNA fragment excised by the restriction enzyme HindIII that included VA genes. The experiment almost immediately gave clear results in the first week of April. Unexpectedly, the short DNA hybridized to many late mRNAs. It was soon apparent that VA RNA did not have the role Roberts had hypothesized. Chow and Broker worked "feverishly," examining additional restriction fragments, and by the third week of April they discovered that all of the adenovirus late mRNAs had the same three-part leader sequence at the 5′ end. They came from upstream on the DNA molecule and were joined to downstream regions. Messenger RNA then need not simply be a sequence of bases colinear with the bases found in a strand of DNA, as molecular biologists had believed. Colinear sequences of DNA and RNA share the same linear order of codons. Chow and Broker repeatedly relayed the newest EM results to the team. Roberts later recalled, "The best thing about it was that no one had predicted it, I mean it was just totally out of the blue."[19] Initially Watson was skeptical of the discovery of split genes but soon became "enchanted" with the idea.[20] It was a major discovery that vindicated Watson's decision to move CSHL into tumor virology.

The new interpretation of split genes made sense of other data produced at CSHL and generated four papers, which were prepared for *Cell* and submitted in early June. The title of the most important paper, "An Amazing Sequence Arrangement at the 5′ Ends of Adenovirus 2 Messenger RNA," was remarkable in having a word like "amazing" in it—most scientific presentations avoid hyperbole. The editor of *Cell*, Benjamin Lewin (and the journal's referees) initially objected but backed down after agreeing with Roberts that the result was in fact amazing.[21] What made it amazing was that it overturned a long-held, central assumption of molecular biology, that a protein-coding DNA sequence was colinear with its mRNA sequence. Chow was first author of the paper, reflecting the fact that she had done most of the experimental work, and Roberts was the last author, reflecting that he was the motivating force behind the experiments.

In mid-April, Roberts received a manuscript from *PNAS* that was under review. It described the experiments of Phil Sharp, Claire Moore, and Sue Berget. The CSHL team was relieved to find that the paper described unhybridized tails on adenovirus hexon mRNA but did not identify the source of the unhybridized sequence. Chow and Broker worried that the CSHL results would leak before they got published. At a late April meeting on RNA at Rockefeller University, Roberts requested to talk briefly about some exciting new results to the small group that James Darnell had assembled. He discussed the EM observations made at CSHL. But their worries about leaks were perhaps unfounded, as by late April the MIT group had obtained similar results.

The CSHL results were formally presented at the 1977 Cold Spring Harbor summer symposium in June. Everyone wanted to be the one to present the groundbreaking results, but Watson, who viewed the discovery as a group effort, selected Tom Broker to deliver the talk.[22] Broker wrote the paper, incorporating other results from CSHL that now made sense in light of split genes. The authors of this paper were listed alphabetically, which temporarily solved the problem of assigning relative credit. Phil Sharp also presented the MIT work at the "adrenaline filled" meeting.[23] Walter (Wally) Gilbert chaired the session, and at the meeting Michael Bishop reinterpreted some of his results concerning mRNA and found the new ideas made sense of his otherwise confusing data. The results also suggested a solution to an ongoing mystery: RNA in the cytoplasm was often shorter than that in the nucleus even though the ends of the RNA could be the same in both. Sambrook wrote a review of the meeting for *Nature*: the audience at the symposium was "amazed, fascinated, and not a little bewildered to learn that the late mRNAs are mosaic molecules consisting of sequences complementary to several noncontiguous segments in the viral genome."[24] The idea of split genes caught on quickly and showed that eukaryotic RNA processing was quite different from the simpler prokaryotic system. Sharp's terminology, "RNA splicing" caught on, as did Gilbert's terms "exon" to refer to the parts of the RNA that are spliced together and are *ex*pressed cistr*on*ic regions and "intron" to refer to the *intr*a-cistr*on*ic regions that are not exported. (Coined by Seymour Benzer, "cistron" is a precise way to refer to a genetic region.) Subsequent work showed that RNA splicing was not merely a viral phenomenon but a feature of practically all eukaryotic cells. Gilbert speculated that introns could speed up eukaryotic evolution by increasing the probability of recombination among functional pieces of DNA—exon shuffling it was called.[25] Influenced by Carl Woese, who thought eukaryotic cells did not evolve from prokaryotic cells, Ford Doolittle argued that the earliest cells had gene splicing.[26] Subsequent research found that gene shuffling was particularly important in generating the diversity of antibodies needed in a healthy immune system.

The Nobel Prize for RNA Splicing

The discovery of RNA splicing was clearly worthy of a Nobel Prize. James (Bob) Darnell at Rockefeller University put it this way: "Because of their explanatory power and total surprise of the results, the EM experiments with adenovirus rank among the most informative biological experiments ever performed."[27] But the path to a Nobel is often not straightforward. Sharp was nominated for the prize relatively early on and won many other prizes. CSHL also deserved credit, though, and because Nobel Prizes can be awarded to a maximum of three people, some CSHL scientists involved with the work would miss out. In 1989, the Nobel Prize was awarded to Thomas Cech and Sidney Altman for their discoveries of the catalytic properties of RNA, work that depended on the discovery of RNA splicing. Some biologists wrote Roberts and Sharp that they thought the earlier RNA splicing discovery should have been recognized as well.[28] Susumu Tonegawa, an

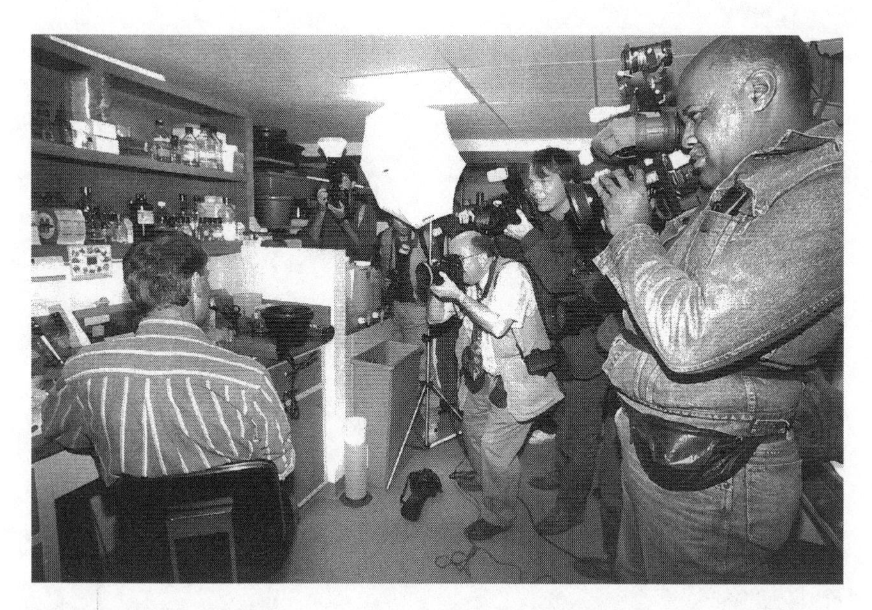

11.4. Richard Roberts with photographers after the Nobel Prize was announced. Image courtesy of Cold Spring Harbor Laboratory Library and Archives

MIT Nobel laureate who thus could nominate people to be considered for the prize, called Watson to push him to select someone from CSHL so that a joint prize could be awarded and finally Phil Sharp would get his Nobel.[29] To help him with this task, Watson asked Roberts to write an "intellectual history" of the discovery, which he did. Broker objected to Roberts's account, however, and wrote his own version of events to capture the electron microscopists' perspective. Both versions were eventually leaked to the press. With two conflicting accounts of the discovery, Watson resisted deciding who in particular at CSHL should get the most credit.

The nomination process for Nobel Prizes is confidential for 50 years, so from the outside it is still unclear how the Nobel committee made its decision. (But tune back in 2043.) In the end, Roberts and Sharp were awarded the prize in 1993 for "the discovery that genes in eukaryotes are not contiguous strings but contain introns and that the splicing of messenger RNA to delete those introns can occur in different ways, yielding different proteins from the same DNA sequence." Chow and Broker were "deeply disappointed" at being left out.[30] (Roberts was disappointed that Gelinas was left out.) Some speculated that the nominators did not include Chow because then

Susan Berget, who had done most of the EM work for Sharp, would have to be included, too, exceeding the maximum of three people who can share the prize. James Watson was quoted saying just this in the *Boston Globe*.[31] However, one difference between the two women was that Berget was a postdoc working under Sharp, whereas Chow was an investigator in her own right and not working under Roberts. Chow's advisor at Caltech, Norman Davidson, expressed a different view in the same *Globe* article: "She's a woman, an Asian woman who's a little quiet. Sometimes they get ignored." This perspective might be closer to the truth.[32] In a rare retrospective piece on the prize, Chow wrote about this aspect of her life: "Because I was brought up in Confucian teaching, I found overt self-promotion distasteful. Furthermore, being young and naive, I believed that published work would speak for itself and that we would be accorded appropriate recognition without having to campaign for it."[33]

Roberts in his Nobel Prize acceptance speech acknowledged the work of others: "We view this award as a tribute to those colleagues and especially to our co-workers, Richard Gelinas, Louise Chow and Tom Broker from Cold Spring Harbor and Susan Berget and Claire Moore from MIT."[34] Chow and Broker did not publicly discuss their views, taking an approach very different from that of the vocal Dominique Stehelin, who also felt slighted by the Nobel Committee (see chapter 9), but with the same result. The Nobel Committee does not reconsider its decisions.

Louise Chow and Tom Broker remained at CSHL until 1984. They moved to the Rochester School of Medicine and then to the University of Alabama at Birmingham, devoting their careers to studying pathogenesis, genetics, and biochemistry of human papillomaviruses. Chow is an associate editor of the journal *Virology* and organized the 17th International Papillomavirus Conference in 2000. Broker was president of the International Papillomavirus Society.

Phil Sharp has remained at MIT for the remainder of his career. He was the head of the MIT Biology Department throughout most of the 1990s. He received a large number of awards, including the National Medal of Science in 2004 and election to the Royal Society in 2011. He also cofounded successful biotech companies: Biogen[35] in 1978 and Alnylam Pharmaceuticals in 2002. Biogen had a market capitalization of more than $44 billion in 2021, and Alnylam over $22 billion.

Richard Roberts remained at CSHL until 1992, serving as assistant director of research in his last six years. He compiled an extensive database of restriction enzymes called REBASE, and his focus on these important tools of molecular biology has continued throughout his career.[36] In 1992, he joined New England Biolabs after consulting for them for decades. NEB is a leading manufacturer of restriction enzymes and other reagents. It has a strong focus on research, and in his position there Roberts helps set the company's scientific goals. In 2008 he was knighted by Queen Elizabeth II for his services to molecular biology.

In October 2017, CSHL held a retrospective conference on the 40-year history of research on RNA splicing. Sharp, Moore, and Roberts were there. Berget did not attend. Chow and Broker decided to break their relative silence about the Nobel episode. Sharp gave a 20-minute talk, but Roberts only spoke for a few minutes, introducing Gelinas. He seemed unwilling to say much about the controversy in the early history. In her remarks, Chow emphasized the importance of her work. "Had Roberts's VA hypothesis been correct, then there would not have been a Nobel Prize," she said. There were no questions following her talk, and there was no public confrontation between her and Roberts. It became apparent that the participants did not want to rehash the past disputes. Tom Broker presented a poster, but unfortunately for those interested in the early history of RNA splicing, the center of gravity of the meeting was future developments in biologists' understanding of how splicing was carried out in the cell by the complex molecular machine called the spliceosome.

12

A Second Cancer Gene

Edward Scolnick, Robert Weinberg, Geoffrey Cooper, Michael Wigler, and the Oncogene *ras*

Edward Scolnick's career illustrates the connections between private and public science (figure 12.1). He was educated at Harvard for both his undergraduate degree and his medical degree, which was awarded in 1965. Knowing that he might be drafted into the Vietnam War after he graduated, Scolnick looked for other options. As it was for Varmus and Bishop, the Public Health Service at the National Institutes of Health was attractive to Scolnick. It allowed young scientists the opportunity to do research and become commissioned officers who would not be sent to Vietnam. He applied and was accepted into the National Heart Institute, where he was matched with Marshall Nirenberg, future winner of the 1968 Nobel Prize for his work unraveling the genetic code.

Scolnick arrived in Nirenberg's laboratory in July 1967, but in the interim, Nirenberg had decided to move his research into neurobiology, so Scolnick joined a subgroup run by Tom Caskey, who worked on the genetic code. The members of this subgroup were attempting to purify a translation termination factor, a protein whose function is to trigger the end of the synthesis of a particular protein by recognizing a stop codon on mRNA. A codon is three ordered bases on RNA that codes for a particular amino acid or for stop or start. While Caskey was on vacation, Scolnick and a technician named Theresa Caryk discovered a peak of protein in the column indicating the factor and, more surprisingly, a second peak. There appeared to be two terminating proteins, not one as expected. Excitedly and quickly, they assayed each peak against codons thought to be stop (or termination) codons in the genetic code. One protein peak matched UAA and UAG, and the other UAA and UGA. (There are 3 positions in a triplet that can be filled with 4 different bases,

12.1. Edward Scolnick in 1980. From the *NIH Record*, vol. 32.

so there are $4 \times 4 \times 4 = 64$ triplets in the genetic code; most triplets code for a specific amino acid, but some are grammatical and code for start and stop directions for the ribosome.) This was a "terrific result,"[1] and it was written up and sent to *PNAS* in August 1968.[2] The thrill of research had Scolnick hooked, and consequently, he did not return to practice medicine.

By the end of the 1960s, research on the genetic code was winding down, and Scolnick needed a new focus. Like many others, he thought he should transition from prokaryotic biology to animal biology. He enrolled in an intensive summer course in animal virology at CSHL in 1969 to brush up on his skills.

Edward Scolnick Moves into Virology

After returning to Bethesda, Scolnick realized that he needed to move from the Heart Institute if he wanted to focus on virology. He approached George Todaro, who, along with Robert Huebner, was driving tumor virology at the NIH. Todaro welcomed him into his lab. Scolnick set to work increasing his knowledge about RNA tumor viruses whose mechanism of replication was largely unknown. He attended a lecture by Sol Spiegelman, who suggested that the viruses used an RNA-dependent RNA polymerase to replicate, but as this claim left out the role of DNA, it did not make sense to Scolnick.

Moving labs involved a demotion for Scolnick, who joined Todaro's National Cancer Institute–supported lab based in Springfield, Virginia, as a senior postdoc, not a permanent staff member as he had been at the Heart Institute. He was in his new position for about a week when David Baltimore contacted Todaro to ask for a large quantity of Rauscher leukemia virus so he could purify an enzyme that he thought would crack the puzzle of RNA virus replication. With the subsequent discovery of this new enzyme—reverse transcriptase—a big piece of the mechanism of RNA tumor virus replication was effectively solved (see chapter 8). The new enzyme could also be used to detect viral activity and thus could be a very useful tool in the search for a human RNA tumor virus, the principal goal of the Todaro lab. When Baltimore visited the lab, Scolnick met with him. Their discussion reinforced Scolnick's view that reverse transcriptase, while helping explain RNA tumor virus replication, did not reveal the oncogenic potential of these viruses.

After fruitlessly using reverse transcriptase for a year or so to try to find a human RNA virus, Scolnick decided to study oncogenesis directly. The lab had worked with Kirsten sarcoma virus (KiSV), named after the University of Chicago virologist who discovered it, Werner Kirsten. KiSV was replication defective: it could cause cancer but could not replicate unless in the presence of a second virus, known as a helper virus, called Kirsten leukemia virus (KiLV). To study KiSV, the Todaro lab often had to study a mixture of KiSV and KiLV. Using reverse transcriptase, Scolnick worked on subtractive hybridization. More specifically Scolnick used the leukemia virus as a probe to show that there were pieces of the sarcoma virus that did not hybridize with the leukemia virus probe. Perhaps there was genetic information in the sarcoma virus that was not in the leukemia virus and was responsible for the potential for KiSV to cause sarcomas.

The Key Insight: A Gene Stolen from Rats

The data from these subtractive hybridization experiments sat in Scolnick's notebooks for a couple of months because Scolnick did not have an "intellectual handle" on how to use it.[3] Then one Sunday, Scolnick was reading Ludwik Gross's *Oncogenic Viruses* in the bathroom of all places. He read passages like the following: "The virus isolated by Kirsten and his colleagues induced also typical erythroblastosis in some of the inoculated rats. Furthermore, after passage in rats, when harvested from spleens or plasma of the rat donors, the virus induced in either mice or rats erythroblastosis and multiple pleomorphic sarcomas."[4] In other words, after the virus replicated in rats, it had different oncogenic properties. "Oh my God," said Scolnick as he realized the mouse virus must have transduced (taken up) a rat gene or genes, and these genes were what he was seeing in KiSV when he subtracted the genes from KiLV. Even though it was Sunday, much to the consternation of his wife, he abruptly left for the lab to recheck his notebooks. By this stage, with help from Huebner, Scolnick had set up his own laboratory with Wade Parks in Rockville, Maryland, in a renovated cow barn.

Convinced that he was right, Scolnick spent the next six months devising experiments to prove his insight. Once he made the proper reagents, his experiments were "beautifully clear." Along with solving the puzzle of the origin of the sarcoma-causing ability of KiSV, Scolnick and his colleagues proposed that KiSV could be a model of the origins of oncogenic viruses in general. These viruses were created when a non-oncogenic virus transduced specific cellular genes from the host cell.[5] Scolnick presented his results at the important 1974 CSHL meeting on tumor viruses. Stehelin, Bishop, Varmus, and Vogt cited Scolnick in their Nobel Prize–winning work on Rous sarcoma virus published in 1976. Wallace Rowe, also a virologist at the NIH, told Scolnick about another tumor virus that Jennifer Harvey had discovered in a similar way. Scolnick obtained some virus and showed that Harvey sarcoma virus also had rat sequences.

The *ras* Oncogene

The next step for Scolnick was to try to find the oncogene itself. One approach was to find a mutant virus that was temperature sensitive for transformation, in other words, a mutant virus that would transform cells at one temperature but not at another. The mutation would be in the gene responsible for the transformation. Steve Martin had used the same approach for Rous sarcoma virus in California, and the use of temperature-sensitive mutants was a standard approach to characterizing a gene. Scolnick and his colleagues screened thousands of clones to find one that was temperature sensitive. They then looked for the protein by transplanting Kirsten transformed cells or the membrane thereof into various species and screening for antibodies that would detect a unique protein band in cells transformed by Kirstein virus but not by viruses from cats or mice. It was a brute force approach, but it finally paid off when injected rats produced antisera that reacted to a protein unique to Kirsten transformed cells. That they had the right protein could be checked using the temperature-sensitive mutant: changing the temperature would affect the reaction between the protein and the antisera. The protein turned out to be a small protein that they called p21, reflecting its molecular mass of 21 kilodaltons.[6] A similar protein could be found in Harvey sarcoma virus. A mammalian oncogene protein product was cutting-edge science, so Scolnick asked Wally Rowe, a member of the National Academy of Science, to submit his findings to *PNAS*, which Rowe was happy to do. The paper was published in 1979.

In addition to the size of the oncogene protein product, Scolnick wanted to know what it did, or what molecular biologists call the function of the protein. He assembled all the known protein cofactors and made an assay to check for any cofactor that stabilized the protein at higher temperature. His thinking was that if the protein was bound by a cofactor, it would become more stable. Knowing which cofactors were involved with the protein would give a clue to its enzymatic activity. Scolnick's group aimed to consider "every cofactor known to man"—TPN, NAD, NADH, ATP, GTP, vitamins, and so on—to discover what type of enzyme he was dealing with. In the very first experiment, they discovered that GTP protected the enzyme from thermal degradation. "It was dramatic, like black and white; it was such a beautiful experiment," Scolnick later recalled.[7] This result was unexpected because Rous Src protein was a kinase that used ATP, not GTP. Although initially called *src*, like the Rous sarcoma virus gene, the new oncogenic gene was now to be called *ras*, for rat sarcoma.[8]

The Early Career of Robert Weinberg

The next major step forward came after Scolnick was contacted by Robert Weinberg, an MIT biologist (figure 12.2). Weinberg was a student at MIT in the early 1960s, when the genetic code was being worked out, and a genetics class with Maurice Fox exposed him to the excitement of the prospect of successful reductionism, the notion that understanding the biosphere could come from working out the central dogma of the then-nascent field of

12.2. Robert Weinberg and Phil Sharp at Cold Spring Harbor Laboratory in 1974. Image courtesy of Cold Spring Harbor Laboratory Library and Archives

molecular biology: that information flows from DNA to RNA to protein. Weinberg had started as a premed student but being on call at night did not appeal to him, so he decided to pursue a research career in biology instead. He continued at MIT for graduate school, followed by a postdoc at the Weizmann Institute in Israel, where he worked on SV40, and another postdoc at the Salk Institute in California. At Salk, he joined Dulbecco's laboratory and continued working on SV40. Dulbecco took a hands-off approach to Weinberg, but Dulbecco's longtime collaborator Marguerite Vogt, whom Weinberg saw as the "heartbeat" of the lab, spent more time with him, conveying her expertise in growing viruses and transforming cells.

Weinberg returned to Boston in 1972 to join the new MIT Center for Cancer Research headed by Salvador Luria, the Nobel laureate who made his name working with bacterial viruses. Luria recruited him at the prompting of David Baltimore, even though Weinberg's research interests overlapped with those of Baltimore, who himself had already committed to the new center. Typically, substantial overlap was avoided since it would diminish the breadth of the research center, but Luria decided to go for talent in this case. Unlike other cancer centers, the MIT center was focused on only basic cancer research and therefore did not include a cancer clinic. It aimed to take advantage of the abundant federal funds flowing from Nixon's War on Cancer. Weinberg started at MIT in a corner of Baltimore's laboratory but was soon given his own laboratory, as Luria had promised. Although independent, Weinberg's laboratory remained closely associated with David Baltimore's. They were "cheek to jowl," as he put it. Baltimore had cloned reverse transcriptase, which provided a new tool to refocus Weinberg's interest in RNA.

In 1972, Weinberg switched from working on the DNA virus SV40 to the RNA virus Moloney leukemia virus (MLV). His lab began to accumulate personnel to study the new viral DNA as it is reverse transcribed from the viral RNA genome. The goal was to purify the viral DNA before it had a chance to integrate into the genome of the host cell. The first significant results showed that the MLV purified from infected cells was infectious, and David Baltimore submitted the results to *PNAS* for Weinberg's team.[9] Geoffrey (Geof) Cooper in Howard Temin's lab had previously reported infectious DNA from Rous sarcoma virus–infected cells, and Temin was similarly working on unintegrated circular DNA.[10] Further work in the Weinberg lab showed that the DNA made from MLV RNA was double stranded and could be supercoiled and still infectious, that is, when introduced as naked DNA into cells, it could drive the formation of infectious virus particles. The trick in these experiments involved this technique of introducing naked DNA into cells, termed transfection.

David Smotkin, Weinberg's PhD student, had suggested the transfection experiment.[11] He was inspired by the recent work of biologists Frank Graham and Alex van der Eb.[12] They had shown that cells would take up foreign DNA 100 times more easily, in their case adenovirus DNA or SV40 DNA, if the DNA were precipitated together with calcium phosphate crystals. Why calcium phosphate has this effect is still not fully understood. Weinberg was skeptical that it would work for MLV as it was quite different from adenovirus or SV40. Smotkin pushed forward anyway. The lab did not have any human kidney cell lines, the material the Dutch had used, so Smotkin used what they had on hand, mouse NIH 3T3 fibroblastic cells. After two weeks of incubating cells, Smotkin could see that viral DNA had indeed been transfected—the DNA infected a small percentage of the cells on his petri dishes and infectious virus was being released by the cells.

With the transfection experiment refined, Weinberg and Smotkin asked another question. The MLV DNA that Smotkin had transfected only yielded infectious virus particles, otherwise leaving the virus-producing cells essentially unchanged. They moved to look at another retrovirus, in this case one that could infect normal cells and convert them into cancer cells, specifically Scolnick's Harvey sarcoma virus. In the interim, Smotkin decided to leave bench research and pursue an MD at Yale University, but he first helped another graduate student, Mitch Goldfarb, learn how to transfect viral DNA.

Robert Weinberg's Epiphany

A massive winter storm in Boston in February 1978 provided a break from routine and time for Weinberg to think about new experiments. With three feet of snow, the roads were closed. As he walked in the snow to his laboratory, he had what he thought was a wonderful idea. Instead of transfecting Harvey sarcoma virus DNA, his lab could transfect cellular DNA from cancer cells. The work of Stehelin, Bishop, and Varmus suggested that viral oncogenes had cellular counterparts that needed to be mutated to cause cancer. Perhaps cells that had been transformed through exposure to chemical carcinogens (which were known to mutate cellular genes) also carried mutant oncogenes in their DNA. Perhaps these cellular oncogenes could be transferred via Weinberg's lab's transfection procedure to healthy normal cells and cause cancer in them. If so, then the approach could be used to screen for new cellular oncogenes. Weinberg's thinking was inspired in part by the work of Berkeley biologist Bruce Ames, who showed that various chemical carcinogens are also DNA-damaging agents and induce mutations; this led directly to the deduction that cells transformed by chemical carcinogens are likely to carry mutant genes, among them mutant oncogenes. Weinberg believed that this new set of experiments would create a new direction for his lab. As he later recalled, "It was an epiphany, a moment of great excitement. For eight weeks, I lived with the thrill of having had an original idea that was mine and mine alone."[13]

The Spandidos Affair

Unfortunately for him, the excitement Weinberg felt was short lived. It soon became apparent that Demetrios Spandidos, a Greek biologist working in Toronto, had published precisely such experiments in the widely read journal *Nature* several months earlier. Weinberg realized that he must have been peripherally aware of these other experiments, squirreled them away deep in his brain, and somehow, at one level, forgot about them. In late March, the Greek biologist gave a stunning talk at Harvard Medical School. He was onto something hot and rumors buoyed the crowd.[14] He was the protégé of Lou Siminovitch, one of the most eminent Canadian molecular biologists, and the rumors did not disappoint. Spandidos's presentation was slick, more so than an average talk by a virologist, and it was titled "Genetic Analysis of Malignancy Using Gene Transfer Mechanisms and Mutation Frequency Studies." He claimed to have conceived and enacted the same strategy that Weinberg had thought of while trudging through the snow, presenting a large volume of data that suggested he had been extraordinarily productive and had a number of collaborators. It was a totally new way of understanding the functions of mammalian genes, but it was built on traditions going back to Oswald Avery, Colin MacLeod, and Maclyn McCarty, at Rockefeller, who showed in the mid-1940s that functional DNA could be transduced or transfected from one bacterial strain to another. Spandidos claimed that he had perfected transfection and shown that cellular oncogenes from hamster tumor cells could be transfected into normal cells and turn them cancerous.

Weinberg was dejected. The elaborate experiments he planned had already been done by someone else. He had been, in the parlance of science, scooped. Making it worse, Weinberg heard rumors that Spandidos was being considered for a job at MIT in addition to Harvard. It did not hurt Spandidos's chances that the head of the MIT cancer center, Salvador Luria, was an old friend of Spandidos's mentor. Spandidos was also being courted by Cold Spring Harbor Laboratory after Joe Sambrook saw him give an impressive talk in Greece and invited him to give one in the United States where James Watson could hear the new approach and make him a job offer. Weinberg realized he had been eclipsed. His brilliant and original insight made in the aftermath of the February blizzard was hardly original and hardly brilliant.

And then something shocking happened. Several months later, just before Spandidos was to give an invited lecture at the MIT Center for Cancer Research, David Baltimore wheeled around the wall that separated their two office and informed Weinberg of news that was totally unanticipated: "You won't believe what just happened. You won't believe it. Lou Siminovitch has just thrown Spandidos out of his lab!"[15] The results presented in the Harvard talk had been submitted to *Cell*. The editor of *Cell*, Benjamin Lewin, was impressed with the paper but sent it out to be reviewed. An anonymous reviewer, who we now know was Richard Axel at Columbia, praised the paper, which purported to show that two genes needed to be transfected into a healthy cell to turn it into a cancerous cell. Axel was impressed but expressed a note of caution: the amount of work needed to generate the data in the paper would take his own lab three to four times longer to generate. Lewin planned to publish the groundbreaking paper in *Cell* anyway, but the cautionary note set off alarm bells for Siminovitch. The number of petri dishes needed to carry out the experiments detailed in the paper—perhaps as many as 10,000—exceeded the number used by the entire Siminovitch laboratory in a year. Siminovitch accused Spandidos of fraud and ejected him from the laboratory. He called Benjamin Lewin at *Cell* and had the paper retracted before it was printed. Spandidos did not admit guilt, and Siminovitch did not organize an investigation. Nonetheless, postdocs very rarely leave a laboratory under circumstances like this, and the work Spandidos did was taken to be fraudulent by many in the field.

The Transfection Assay for Oncogenes

This scandal was bittersweet for Weinberg. He had in fact not been scooped, which was good news. But the general approach was now tainted in many eyes, and he would now need to produce exceptional, ironclad data to convince his peers that transfection experiments could elucidate the genetics of cancer by revealing previously undiscovered oncogenes in the genomes of cancer cells.

Mitch Goldfarb worked to perfect the transfection assay. One challenge was that clusters of transformed cells were difficult to distinguish from other cell clusters that were not cancerous—Weinberg himself often could not tell the difference.[16] Goldfarb had to improve his observational acumen, so they developed a double-blind procedure so that Goldfarb could not tell whether he was investigating cells transfected with tumor cell DNA or control DNA. After many attempts, in the summer of 1978, Goldfarb had a clear signal using cells from a cancerous cell line from the lab of Charles Heidelberger in Wisconsin, who had made cancerous mouse cells by exposing them to a coal-tar carcinogen. And then Goldfarb dropped the transfection assay. He needed to concentrate on finishing his other PhD research so he could graduate, and he was also worried that the new restrictions on recombinant DNA technology coming from Asilomar would stop him taking the natural next step, cloning the mammalian oncogenes in bacteria.[17]

Luckily, another graduate student was looking for a new research project. Chiaho Shih had left Howard Green's MIT laboratory after creative and personality differences led to a rupture between them. Green was a leading cell biologist who, among other accomplishments, created the first widely used 3T3 cells. Consequently, Shih had some training in cell biology and cell culture. Weinberg strong-armed Shih into diving into the transfection experiments in his lab even though others there considered them a project with a high likelihood of failure. After he mastered the procedure, he started getting positive results. Weinberg was excited: "Goldfarb had been on the right track all along. I could not contain my excitement. My God, I thought, chemically transformed cells actually carry oncogenes in their DNA!"[18]

The small and exclusive Gordon Conference in late June 1978 provided the venue to announce the good news to the broader biology community. It was held at the Tilton School in New Hampshire, and recording of talks was prohibited to encourage scientists to present their newest findings before publication. Weinberg's talk, however, did not elicit the praise and appreciation he was expecting. Incredibly, their work had unified two approaches to cancer. External chemical mutagens could create oncogenes within the cell itself. The complete cause of cancer simultaneously involved both external and internal factors. Tumor viruses presumably could pick up these genes and transmit them. But, beyond a couple of pointed questions after the talk, it fell flat. Virologists, it would seem, were still burned from the Spandidos affair and were skeptical of Weinberg's claims. Nonetheless, the results were submitted to *PNAS* in August and published in November 1979.[19] As a member of the National Academy, David Baltimore submitted the paper for his colleague's group.

The Spandidos affair did not scare everyone from this line of research. Two other groups were pursuing the same objective, and the subfield developed into a three-way race. In addition to Weinberg's group, Geoffrey Cooper at Boston's Dana-Farber Cancer Center and Michael Wigler at CSHL, which Goldfarb had joined, bringing with him the transfection technology, had competing research goals. Interestingly, the competition between Weinberg and Cooper mirrored the earlier competition between their respective mentors, Baltimore and Temin.

Geoffrey Cooper

Like Weinberg, Cooper was an undergraduate at MIT in the 1960s, working in the laboratory of Maurice Fox on bacterial transformation (transfection in bacteria), presaging his work on oncogenes (figure 12.3). After graduating in 1969, he wanted to switch from work on bacteria to work on cancer, so he enrolled in an MD/PhD program at the University of Miami, which had the added benefit of deferring the Vietnam draft.[20] (Being in an MD program allowed a deferment, but the PhD program did not.) He worked on chemotherapeutic treatments for cancer, and, after learning about the discovery of reverse transcriptase in 1970, decided he wanted to pursue a postdoc under Howard Temin in Wisconsin. Cooper's advisor helped him land the coveted position by arranging for him to have lunch with Temin during a visit to Miami.[21] Once in Wisconsin, Cooper drew on his experience with bacterial transformation by building upon the work of the Czech husband-and-wife team of Miroslav Hill and Jana Hillova, who claimed to be able to transfer information from the Rous sarcoma virus by transferring DNA between chicken cells.[22] Temin was skeptical, but Cooper succeeded in developing the transfection technique, which was then used to characterize RSV proviral DNA. Temin and Cooper used the transfection assay to look for cellular transforming DNAs in chicken and quail tumors. They were unsuccessful, possibly because they used chicken embryo fibroblasts and not NIH 3T3 cells as recipients.

After two years in Wisconsin, Cooper took a job at the Dana-Farber Cancer Institute at Harvard Medical School. Temin warned him of the feudal nature of biological science at Harvard.[23] There were many warring factions there, which is one reason Temin had declined an offer from Harvard himself.

After some time in Boston, Cooper switched from chicken cells to mammalian cells. He got some NIH 3T3 cells from Weinberg.[24] By December 31, 1979, the paper summarizing the results was written and sent to *Nature*. "Transforming Activity of DNA of Chemically Transformed and Normal Cells" was published in April 1980, five months after Weinberg's paper. Like Weinberg's group, Cooper's group thought that the data supported the hypothesis that cellular genes can cause transformation of normal cells into cancer cells when expressed at abnormal levels. Unlike Weinberg, Cooper's group thought small pieces of normal cellular DNA could also transform cells.

12.3. Geoffrey Cooper at the microscope. Image courtesy of Geoffrey Cooper

The findings that cellular genes could transform mouse fibroblasts was good news. It suggested that the same general oncogenes might be found in many species, and both Weinberg and Cooper went hunting for oncogenes in human cancers. Shih got a human bladder carcinoma cells line called EJ from a friend at the Dana-Farber Cancer Institute. DNA from these cells, too, could transform NIH 3T3 cells. Cooper and his new postdoc Ted Krontiris initiated experiments to look for transforming genes in human tumor cells and succeeded with EJ bladder carcinoma, which they also obtained from

Lan Bo Chen at Dana-Farber. So, the two groups were using the same cell line. Cooper's paper was communicated to *PNAS* by Temin and published in February 1981. Weinberg's paper was published in *Nature* the following month.[25] The field of human oncogenes was starting to heat up.

The Charmed Life of Michael Wigler

As mentioned above, a third scientist, Michael Wigler, entered the oncogene race (figure 12.4). He did not have a straight path to oncogene research. Called a child prodigy, Michael Wigler attended Princeton to study mathematics, before going to medical school at Rutgers. Finding medicine boring, he flunked out. He successfully applied for a technical assistant position in a biochemistry laboratory at Columbia University after seeing it advertised in the *New York Times*. While there, he read James Watson's *The Double Helix*: "It was an epiphany. I decided to become a molecular biologist."[26] As a graduate student at Columbia, he worked under the guidance of Richard Axel and developed a powerful transfection technique. Like Frank Graham and Alex van der Eb, Wigler used calcium phosphate to slip new DNA into NIH 3T3 cells. He spent time optimizing reaction conditions, including the most effective concentration of calcium. He also realized that he needed carrier DNA. In 1975, he made cells deficient in thymidine kinase (tk– cells). They then cut up herpes simplex virus using restriction enzymes to take advantage of the fact that some pieces of the herpes simplex virus contain a thymidine kinase. They selected for cells that took up the herpes virus thymidine kinase by growing them in media in which having a working thymidine kinase was necessary for survival and consequently had the tk+ phenotype.[27]

Extending Wigler's earlier ideas, the group saw that they could pair any gene with a selective marker like tk. Cells that took up the tk would likely take up the other gene, too. This process was called "co-transformation." Again, they published their results in *Cell*, which, along with *Nature* and *Science*, was the most prestigious journal in molecular biology. Axel suggested that they patent the process.[28] Wigler was unsure but Axel insisted, and Columbia University was awarded a number of patents starting in 1983.[29] These patents proved to be very lucrative, generating nearly $1 billion over their 20-year lifespans. Most of the proceeds went to the institution, but one-third was split among the scientists themselves.

James Watson at CSHL was thinking of hiring Spandidos before the Greek

12.4. Michael Wigler at his lab in 1985. Image courtesy of Michael Wigler

biologist was thrown out of his Toronto lab. With a well-known eye for scientific talent, Watson instead hired Wigler, who also had offers from Columbia and Harvard. MIT was also interested in him, but his job talk did not go well there given his somewhat eccentric speaking style.[30] With the recent discovery of split genes and Richard Roberts's work on restriction enzymes, Wigler judged CSHL to be the center of the molecular biology world.[31] Watson thought that Wigler could redo the Spandidos experiments, this time correctly.

Despite the scandal, Spandidos's claims did motivate Wigler to study oncogenes and malignant transformation. Before hearing of Spandidos's work, Wigler had thought the idea that there were dominant oncogenic mutations was likely false.[32] Henry Harris, in 1969, had shown that if you fused a cancer cell with a normal cell, you could stop the transformation.[33] These results made people like Wigler think that oncogenes were recessive. Wigler also heard through the grapevine that Weinberg was doing similar experiments. His own transfection work could be modified relatively easily for oncogene work. When Wigler hired Mitch Goldfarb as a newly minted PhD from Weinberg's lab at MIT, all the pieces were in place.

Using his approach, Wigler showed that a single gene could cause the transformation of 3T3 cells, reinforcing the work of Weinberg and Cooper earlier in 1981. A further question was whether the multiple cell lines that caused transformation by transfection did it the same way with the same gene, or whether different genes were in play in each case.

Cloning the Oncogene

The next goal was obvious to all three groups: identify and sequence the DNA that can transform cells. An important step toward this goal was to clone the responsible gene. In 1980, cloning a particular gene out of a genome of tens of thousands was like finding a needle in a haystack.

In addition to the laboratories of Wigler and Cooper, a fourth entered the race to identify a human oncogene: that of Mariano Barbacid, a Spanish biologist working at the NCI in Bethesda. Wigler's lab, Weinberg's lab, and Barbacid's lab appeared to have isolated a similar gene. Both the Wigler and Weinberg camps rushed to publish their findings. Weinberg's paper was hand-carried to *Cell*, Wigler sent his to *Nature*, and Barbacid published in *PNAS*.[34]

Wigler's team won the race, publishing in April 1982. All camps agreed on the small size of the gene and that it was of human origin. It would be found to be homologous to a gene expressed in HeLa cells and T24 cells, which derived from a bladder cancer patient. They did not know it at the time, but the T24 and EJ cells derived from the same patient.

The Function of the Oncogene

The next step was to further characterize the function of the protein expressed by the cellular oncogene and the difference between the cellular gene in a normal cell and a transformed cell. Some communication among the laboratories occurred at conferences and by telephone. In one call, Wigler asked Weinberg whether the MIT group had isolated an oncogene known to be one carried by a known tumor virus. There was a growing number of such genes, *src*, *mos*, *ras*, *myc*, *mil*, and so on. Weinberg said he did not know of any match. Weinberg's student Luis Parada was comparing the new human cellular oncogene with the viral oncogenes using hybridization. After some false starts, Parada got a positive from *ras* and negative result from *src*, *myc*, *mos*, and the rest. "The irony, painful as it was, became apparent only at the very end. We had spent a year searching for something we could have found in one day," Weinberg later wrote.[35]

Weinberg contacted Wigler, who, based on the earlier phone call, had decided not to pursue this line of research. It was hard news for Wigler to hear. After all, Goldfarb, who had worked on *ras* in Weinberg's lab and was an expert on the gene, was now in Wigler's lab working on different projects. Weinberg also called Scolnick to tell him the good news. Scolnick agreed to collaborate with Weinberg and send him reagents and antibodies that he had developed to work with Ras. Consequently, Scolnick appears as a coauthor with Weinberg on one of the MIT papers.[36] Although Weinberg announced this discovery first,[37] his lab was actually beaten to this result by Geoffrey Cooper's.[38] When Channing Der joined Cooper's lab as a postdoc, he undertook a systematic comparison of the human tumor–derived transforming genes and retroviral transforming genes. Publishing in *PNAS*, Cooper's showed that the *ras* found in Harvey sarcoma virus, called *h-ras*, was homologous to an oncogene gene involved human bladder cancer and that the *ras* found in Kirsten sarcoma virus, called *k-ras*, was homologous to the oncogene in a lung cancer. This was an important result because Wigler later found the *k-ras* gene to be involved in additional lung and colon cancers and subsequently found it to be the most common oncogene in human tumors. Cooper also called Edward Scolnick with his results shortly after Weinberg had done so. Scolnick had already agreed to collaborate with Weinberg, but he agreed to also supply Cooper with his reagents. Barbacid's group also identified the bladder carcinoma gene as *h-ras*. That *ras* genes were a factor in different types of cancer suggested that among the 100 or more types of cancer, there might be a simpler genetic basis.

The way ahead lay in characterizing how the *ras* gene that transformed cells was different from normal, or wild type, *ras*, which did not.[39] To do that required sequencing the DNA of both normal and transforming types of the *ras* gene. The three laboratories—Weinberg's, Wigler's, and Barbacid's—that had cloned the human oncogene now raced to find that difference. Wigler had a head start with the clone of *ras*. Weinberg's graduate students narrowed down the site of the mutation to 350 bases, which they sequenced multiple times for accuracy with the help of a collaborator at NCI, Ravi Dhar. In late August 1982, the answer that emerged was stunning. There appeared to be only a point mutation in a single base pair between the transforming *ras* and normal *ras* gene. Although he had not done the work, Weinberg's post-doctoral fellow François Dautry presented the result at Roswell Park Cancer Institute in Buffalo.[40] Weinberg wanted the lab to have priority for the discovery so decided to allow someone not associated with the work to present the findings at a talk that had been scheduled for a different topic. Once a *New York Times* reporter heard the news and interviewed Dautry, the word was out. Their work was published in *Nature*, back to back with a paper from Barbacid's group, who had found the same mutation.

Michael Wigler drew the short straw. He had found the same mutation as the other groups, and quite likely before them. However, he also found a second mutation in the promoter region and carefully conducted more experiments to determine whether the second mutation was significant. It wasn't. These additional experiments took more time, and his paper came out a couple of months after the others were published. Consequently, he did not receive the amount of credit he was due and was effectively punished for his care and rigor. Weinberg offered to write a letter to a journal pointing this out, but Wigler declined the offer,[41] thinking that the field had had enough drama with the Spandidos affair. Such is the imperfect nature of science—sometimes there is a cost to being careful.

The mutation that caused the Ras protein to transform 3T3 cells was a single amino acid substitution in the 12 position of 192 amino acid long chain. Instead of glycine, the oncogenic protein had a valine. This simplicity of this result was shocking. It suggested a type of reductionism first articulated by Linus Pauling and colleagues in 1949 when sickle cell anemia was shown to be caused by a single amino acid substitution.[42] Perhaps cancer, like sickle cell anemia, was a "molecular disease." Weinberg and others had expected the mutation to change the levels at which the Ras protein was synthesized, but instead it was clearly a mutation that changed the 3-D structure of the protein.

Further work showed that the situation was more complex. The NIH 3T3 cells first used by Weinberg and then the other groups had preexisting mutations that allowed a single mutation in *ras* to cause tumorigenic transformation. Other cells, such as primary rodent cells, were not transformed by the mutated *ras*. They required additional mutations in other oncogenes such as *myc*. Nonetheless, Ras is an important signaling protein and is found in almost all animal cells.

Wigler continued to work on *ras* genes. He hoped that the flexibility of yeast genetics would shed light on the function of the Ras protein. Scolnick had found *ras* in the well-studied yeast. Wigler found a third version of *ras*, called *n-ras*, in neuroblastoma cells. He has remained at CSHL and works on the molecular basis of cancer and autism.

Six weeks before he got the call from Weinberg, Scolnick accepted an offer from the pharmaceutical company Merck, who wanted him to head a group in molecular virology. He would be working under Maurice Hilleman, the world-class virologist who developed mumps, measles, and rubella vaccines. One idea Scolnick brought with him to Merck was to make an anti-Ras drug as a way of treating cancer. The Ras protein is thought to be mutated in approximately 30% of cancers. Unfortunately, finding a drug that inactivates mutated Ras has proven very difficult. and none has been approved that targets Ras despite Harold Varmus devoting $10 million to the goal in more recent times when he was head of NCI.[43] (Some cancer drugs target downstream targets of Ras.) At Merck, Scolnick rose to become head of research in 1985, a position he would hold for 17 years. He was an important player in the creation of the HPV vaccine Gardasil (see chapter 15) and a variety of other blockbuster drugs including Vioxx, which was famously pulled from the market for serious safety concerns. After Merck, Scolnick directed a research center at the Broad Institute. MIT named the Edward M. Scolnick Award in Neuroscience in his honor.

Geoffrey Cooper continued to study oncogenes. He moved to Boston University in 1998, where he served as chair of the Biology Department for 13 years, and then became associate dean of the Department of Natural Sciences. He writes novels about science, including one titled *The Prize*, which tells the fictional story of a competitive race for the Nobel involving fraud and treachery. He also coauthored the textbook *The Cell: A Molecular Approach*, which is in its eighth edition.

Weinberg remained at MIT. His laboratory was the subject of a fine-grained book called *Natural Obsessions*, on the internal laboratory dynamics of his lab during the race for the discovery of oncogenes. His work has been cited over 273,000 times, according to Google Scholar. He has won scores of prizes. Although there was talk about the Nobel Prize, it has so far eluded the discoverers of the Ras protein.

13

A Molecular Brake on Cancer

David Lane, Arnold Levine, and the Tumor Suppressor p53

Arnold (Arnie) Levine started working on SV40 at Princeton in 1968. He earned his PhD from the University of Pennsylvania in 1966, working on adenovirus, and then completed a two-year postdoc at Caltech. Like many other virologists at this time, he transitioned from bacterial viruses—in his case ΦX174—to animal viruses, after judging that the maturing of the field of bacteriophage biology meant that undeveloped animal virology was a richer area for a new assistant professor to investigate. Though not employed in his lab at Caltech, Levine was influenced by Max Delbrück's reductionism, his view that facts about simple systems could be extrapolated to more complicated systems. SV40 was a simple DNA virus, so Levine could bring his skills working with DNA to bear.

Knowing that he had to have tangible results in four years to get tenure and promotion at Princeton, he chose to work on SV40 replication rather than its oncogenic properties, which he thought could take a decade of work.[1] He could turn to the causal properties of SV40 after he got tenure, he thought. This strategy proved effective, as Levine and his students showed that SV40 replicates as a circle of DNA, and there was some evidence that replication was bidirectional.[2] This work on viral replication was enough to get tenure at Princeton.

As he planned, Levine then turned his attention to the question of the oncogenic properties of tumor viruses (figure 13.1). He was influenced by papers that suggested that tumors abnormally replicate early fetal development. Some evidence supported this idea. For example, some colon cancers involve the reexpression of an embryonic antigen called carcinoembryonic antigen, which is still used as a marker to detect cancer. Additionally, liver

13.1. Michael Fried, Janet Arrand, and Arnold Levine at Cold Spring Harbor Laboratory in 1979. Image courtesy of Cold Spring Harbor Laboratory Library and Archives

cancer can involve the reexpression of alpha-fetoprotein (AFP). Levine was interested in whether tumor viruses reprogram cells so they reexpress fetal proteins. Since the mature immune system has not seen these proteins, it should respond by making antibodies, he reasoned. Levine and his colleagues made tumors in hamsters using SV40 and antisera from tumor-bearing mice. They achieved promising results: mouse cells infected with SV40 produced a 54-kilodalton protein that interacted with large T and small t antigens and could be copurified with them, presumably because they bound together. The same 54-kilodalton protein could also be found in two different, uninfected murine embryonal carcinoma cell lines. In other words, Levine had identified a cellular protein that interacted with a transforming viral gene.[3]

David Lane and the Imperial Cancer Research Fund

This was a significant outcome, but Levine was beaten to his result by David Lane and an English team at the ICRF. Lane was driven to study cancer after his father died while he as a student.[4] He was trained in immunology at University College London by Av Mitchison, a nephew of the great English biologist J. B. S. Haldane[5] and an eccentric friend of James Watson.

(Watson was Mitchison's best man at his wedding.) He described Lane as the "best student I ever had." Some of Mitchison's approach to science, particularly his emphasis on original thinking, rubbed off on Lane. Lane called him "a remarkable man who had a profound effect on me."[6] Working with an American postdoc Don Silver, Lane established radioimmunoassays to investigate autoimmune reactions in mice.[7] After completing his doctoral research, Lane secured a postdoc in Lionel Crawford's ICRF laboratory in 1976. Among other abilities, Crawford was interested in Lane's skill in using radioactive iodine to label proteins.[8] At the ICRF, Lane worked with technician Alan Robbins to make specific antiserum against SV40 large T antigen. "This was my first experience of the white-hot competitive world of tumour virology."[9] Crawford thought the existing antisera for detecting T antigens were awful and hoped Lane could improve the assays.[10] It was an exciting time to be at the ICRF, with Renato Dulbecco down the hall and Harold Varmus taking a sabbatical there. "I felt we were right there at the center of things," Lane later recounted.[11]

After a few months, Lane made enough progress to publish a paper titled "An Immune Complex Assay for SV40 T antigen."[12] Just has he was ramping up his research, Crawford took a sabbatical to the United States, leaving Lane to complete the work and take care of the laboratory and its management by himself. "We made what we thought was a terrifically good specific reagent to large 'T' antigen. We were really happy with it because we have done all kinds of clever things to make right. But when we used it, instead of—as we hoped—just bringing down one protein [large T antigen], it always brought down this other protein with a molecular weight of 53 kilodaltons, as well."[13] Some in the laboratory told him that it was a contaminant or perhaps a degradation product of large T antigen. Lane resisted this interpretation, influenced by Joe Sambrook's view that large T antigen must interact with the host cell machinery. When Crawford returned from sabbatical, he repeated the experiment and got the same results. The two submitted a paper to *Nature* in November 1978, which was rejected at first, but a revised version was published in mid-March 1979.[14] One author described Lane in virtue of his work at the ICRF as a "scientific superhero."[15] By the time the paper came out, Lane and his wife, who was also a scientist, had moved to Cold Spring Harbor Laboratory in Long Island, New York. "It was the best thing that happened to me," he later said.[16] There Lane created monoclonal antibodies for large T and, using immunofluorescence to "see" the antibodies, found the large T in

13.2. David Lane and Keith Willison at the bench at Cold Spring Harbor Laboratory in 1979. Image courtesy of Cold Spring Harbor Laboratory Library and Archives

the nuclei of SV40 transformed cells (figure 13.2). Monoclonal antibodies bind one epitope and are derived from clones of a single white cell. At the end of 1979, Joseph Sambrook and James Watson offered Lane a position at CSHL, but he decided to honor an earlier commitment and return to his position at Imperial College London.

Two other groups also discovered a 53-kilodalton protein involved in cancer in 1979. Lloyd Old (1933–2011), at Memorial Sloan Kettering in New York City, who was involved in the feline leukemia virus story, used modified antibodies to look for proteins expressed in cancerous cells and not in normal cells. He found a candidate, which he called p53 because it was measured to be 53 kilodaltons in mass. (Later, more accurate measurements showed the protein was smaller, but the name stuck.) Since his team found it in tumors that were not caused by viruses, they concluded that it was either a cellular gene or a gene from an endogenous tumor virus.[17] The results were significant enough for Old to submit to *PNAS* in February 1979.

At the same time, Pierre and Evelyne May's laboratory in Villejuif, France, was investigating SV40 and polyoma virus. Pierre May's doctoral student Michel Kress was investigating whether SV40 had a protein homologous to polyoma's middle T protein. It was known that SV40 and polyoma were very closely related viruses. Kress thought he found what he was looking for when he identified a 53-kilodalton protein, and there was general excitement in the research facility. When further work showed the protein was from the cellular host, it was unclear what to make the result. Nonetheless, it was published in the *Journal of Virology*. Unfortunately for Kress, he moved to a new facility and onto other topics, and consequently, he is largely forgotten in the p53 story.

There was not a lot of communication among these four groups, so it took a year before everyone realized they were working on the same protein. Despite this, the next steps were obvious to everyone involved, just as they were for the *ras* oncogene: first, clone the p53 gene, second, sequence it, and third, find its biological function.

The Race to Clone the p53 Gene

One of the first researchers to clone p53 was Moshe Oren, who began the process in Arnold Levine's lab at Princeton but completed it at the Weizmann Institute in Israel. It took him two and a half years.[18] He succeeded by making a DNA copy (cDNA) of all the mRNAs from a cell line—what is called a cDNA library—and selecting for the clone that contained cDNA for p53. The state of the technology meant that this type of screening approach was still labor intensive in the early 1980s. After more than 1,800 attempts, he succeeded, wrote up his results, and had James Darnell send them to *PNAS* in October 1982.

Arnold Levine, who had moved to Stony Brook University on Long Island near Cold Spring Harbor Laboratory, was also trying to clone p53. He teamed up with Genentech, a biotechnology firm in San Francisco with expertise in cloning. Others, too, were in the race: Russian scientist Peter Chumakov had in fact cloned p53 before Oren, but because he published in Russian in *Proceedings of the USSR National Academy of Sciences*, his work went largely unnoticed in the West.[19] By 1984, a number of labs had clones of p53, and they began to experiment with them to try to determine its function.

Oren used p53 combined with the oncogene *ras* to see whether it had a similar effect as the oncogene *myc*.[20] Oncogenes *ras* and *myc* together could transform many types of cells. (By itself *ras* could transform NIH 3T3 cells, but not many other cell lines.) Overall, p53 seemed to have the same effect as *myc*, although it was slightly less potent. Oren and others who did similar experiments in 1984 concluded that p53 was another oncogene. Other groups reported immortalizing cells with p53.[21] Lionel Crawford and colleagues showed that many different tumors had elevated levels of p53.[22] The protein had all the hallmarks of an oncogene.

Back in the United States, Arnold Levine was frustrated. He had a p53 clone like the other groups, but he could not replicate the oncogene-like results. His clone did not transform cells. "It was a depressing time," he recalled.[23] For a few years, Levine and his postdocs and students tried to make his p53 clone transform, but to no avail. Nonetheless, in the mid-1980s, oncogenes became increasingly important in the study of cancer, and the field moved from an "enemy from the outside" view of cancer to an "enemy from the inside" view.[24] For many biologists, cellular oncogenes replaced tumor viruses as the proximate cause of cancer. This reorientation of the field was little consolation to Levine, who could not replicate the successes of the competing labs. The reason for his lack of results started to come into focus once his clone and others were sequenced. It turned out that Levine had the wild-type p53 gene, while other groups had various mutant versions. The mutants had at least one amino acid difference relative to the wild type. Initially, the small differences were dismissed as sequencing errors, but it appeared that a single amino acid change could turn a normal wild-type p53 protein into a transforming gene.

While the different clones of p53 explained some of the data, the function of p53 in the cell was still unclear. One approach was to use the different clones together in an experiment. Levine used both Oren's clone that transformed and his clone that did not. The transformation was stopped. His clone was a transformation stopper, or tumor suppressor. He, his postdoc Cathy Finlay, and his student Philip Hinds sent the paper to *Cell* to be published in 1989.[25] "I was a little bit nervous about that [paper] because we were saying that everything in the field was wrong. Here are a dozen papers saying that it was an oncogene and they were all wrong."[26] While the paper was in press, Levine attended a meeting at Cold Spring Harbor Laboratory to present his results.[27] He anticipated significant pushback from the community. But he was pleasantly surprised when, prior to delivering his talk, he listened to a cancer researcher from Johns Hopkins Medical School, Bert Vogelstein, make the case that p53 is a tumor suppressor, just as he was going to. Levine relaxed and gave his talk taking on the field, knowing there was additional evidence that he was on the right track. Levine and Vogelstein decided to collaborate since they had complementary approaches to understanding the function of p53. The two talks created a buzz at the conference.[28]

Bert Vogelstein, p53, and Human Cancer

Vogelstein approached p53 from the point of view of human cancer (figure 13.3). He, his PhD student Suzanne Baker, and colleagues analyzed the connection between human colon cancer and p53.[29] Small regions of chromosome 17 are missing in 75% of colorectal cancers. Vogelstein's team wanted to identify the missing gene—a potential tumor suppressor related to this type of cancer. In the beginning, they believed the existing literature and were trying to show that p53 was not the important missing gene even though it was known to be on chromosome 17.[30] However, they discovered the opposite. In many cases the p53 gene was deleted in both copies of chromosome 17. In cases where a copy of the gene still existed in the tumor, Baker et al. used DNA sequencing to show that it was mutated. As important was the observation that the previously published sequence of human p53 thought to be a normal wild type was in fact a mutated version, an oversight that had initially concealed the function of the p53 protein. Baker's findings were significant and were submitted to *Science* in early February 1989. Later work in the Vogelstein lab showed that wild-type p53 protein could bind specific human DNA sequences, but mutated forms could not.[31]

Following the Cold Spring Harbor Laboratory meeting, the Levine and Vogelstein labs swapped clones and repeated experiments. Vogelstein and Levine drew inspiration from a famous tumor suppressor gene called retinoblastoma, or *Rb*. This gene was discovered by researchers studying eye cancer. Cancer of the retina primarily afflicts children, and treatment often requires removal of the eye. A genetic model proposed by Alfred Knudson in 1971 suggested that two genetic mutations were needed to cause retinoblastoma tumors. Subsequent work showed that the two mutations corresponded to mutations or deletions in each of the *Rb* genes on the two copies of chromosome 13 in each cell.[32] Both copies of the *Rb* gene have to be mutated or lost to cause cancer. One mutation or a loss of one copy of the gene, greatly increase the chances of cancer developing. This explains the disproportionate levels of childhood retinoblastoma tumors in certain families.

In the 1980s, Weinberg's laboratory was involved with cloning the *Rb*

13.3. Bert Vogelstein at Cold Spring Harbor Laboratory in 1994. Image courtesy of Cold Spring Harbor Laboratory Library and Archives

gene, which would be one of the keys to understanding the genetics of cancer. Work by Ed Harlow and colleagues showed that *Rb* was connected to tumor virology. Harlow was a friend of David Lane and trained in immunology at the Imperial Cancer Research Fund, where he obtained his PhD under Lionel Crawford in 1982. He was recruited to Cold Spring Harbor Laboratory in 1983 by James Watson, in part for his skill in making monoclonal antibodies. Harlow and Lane cowrote an influential handbook called *Antibodies: A Laboratory Manual* that was published by Cold Spring Harbor Laboratory Press in 1988. At the time he was writing the manual, Harlow and his lab were investigating an early region of the adenovirus genome that encoded for E1A proteins implicated in the transformation potential of adenovirus. They found a protein-protein linkage between E1A and a 105-kilodalton protein. Harlow was in London in 1987 when he read a paper from Wen-Hwa Lee's laboratory on Rb that said, among other things, that the protein was about 110 kilodaltons. Perhaps the cellular protein interacting with E1A was Rb, he thought. Back at CSHL, his graduate students Karen Buchkovich and Peter Whyte had the same thought and began the relevant experiments to check before he returned. Harlow contacted Weinberg to obtain some of the necessary reagents to work with Rb, and they soon found that their hunch was correct. They published the positive results in *Nature* in July 1988.[33] These findings uncovered the first known physical linkage between an oncogene and what Harlow and colleagues called an "anti-oncogene" but is now better known as a "tumor suppressor." They also showed why tumor suppressors were important to understand transformation and how they could be approached from a molecular perspective. The idea was that the viral E1A protein bound to the tumor suppressor p53 to inactivate its tumor-suppressing function.

Building on the insight about p53, Vogelstein took information from a variety of clinical and biological sources and constructed an influential genetic multihit theory of cancer that could be seen as an articulation of earlier somatic mutation theories.[34] Transforming normal epithelial cells into colon cancer, he argued, took at least six consecutive genetic events, including a mutation in the *ras* gene and the loss of p53 function, which Vogelstein hypothesized turned a late adenoma into a carcinoma. He called it a rudimentary model, but it would have the virtue of focusing cancer research on the various steps of carcinogenesis and of generalizing to other forms of cancer.

The realization that p53 was not a classical oncogene but a tumor suppressor involved in human cancers accelerated research into this protein. By 2020, nearly 100,000 papers were published on the role of p53 in a wide variety of cell types and species. It is arguably the most important protein in basic cancer research and the "most intensively studied molecule in biomedical research."[35] In 1993, *Science* magazine deemed it "Molecule of the Year."[36] The editor-in-chief Daniel Koshland Jr. wrote, "Some molecules are good guys, some are bad guys, and some are bad because they fail in their functions,"[37] noting that p53 was a member of the third group. *Science* selected p53 because of its effects on the cell cycle, DNA repair, and cancer. More than 50% of human cancers involve a loss of function due to a mutation in p53.

Vogelstein, Lane, and Levine teamed up to summarize the state of the field in 2000. They used a common analogy: p53 functions as a brake in the cell cycle. Traditional oncogenes function as the accelerator. Just as a functioning brake can slow down and stop a car with the accelerator stuck down, properly functioning p53 controls the replication of cells even in some cases when oncogenes are active. If p53 fails to function properly, uncontrolled growth—cancer—can ensue. The trio of researchers considered six different ways p53 can be inactivated, including by a virus, by a deletion in the gene, or by being mislocated to the cytoplasm. They asked how to make sense of all the activating signals and downstream regulators uncovered in the vast literature of the previous decades. Their solution was to draw an analogy: "The cell like the internet appears to be a 'scale free network': a small subset of proteins are highly connected (linked) and control the activity of a large number of other proteins, whereas most proteins interact with only a few others."[38] The idea is that p53 is one of the highly connected nodes in the cell. Thus, to simplify somewhat, inactivating it can likely bring down a whole host of cell signaling pathways and create a cancer cell.

The role of p53 in the modern understanding of cancer is one of the most important fruits of basic research into tumor virology. Temin's serious joke about studying chickens to eventually understand human cancer was fulfilled. Focusing first on researching oncogenes and their cellular homologs, and second on which parts of the cellular machinery interact with the oncogenes were fruitful approaches to understanding the fundamental mechanisms of cancer in the late twentieth century.

Sir David Lane was elected a fellow of the Royal Society in 1996 and knighted in 2000. He also branched out into industry by founding a biotechnology company called Cyclacel Limited and cofounding FogPharma. Based in Dundee, Scotland, and New Jersey and founded with more than $100 million of private equity, Cycacel company focuses on developing drugs to treat cancer, including drugs that target p53 pathways. The tagline "translating cancer biology into medicines" appears prominently on its website. Three drugs are currently in various stages of development. Cyclacel is listed on the NASDAQ with a market capitalization of $53 million in 2021. FogPharma, still in the startup phase, aims to develop mini-proteins that can penetrate the cell membrane to target proteins like Ras and p53. Lane himself is now chief scientist of Singapore's Agency for Science, Technology and Research, where his laboratory continues to work on p53 and cancer biology.

Arnold Levine returned to Princeton in 1984 and became the chair of the Department of Molecular Biology. In 1998 he left to become president of Rockefeller University. He left that position following a scandal associated with a relationship with a female graduate student.[39] David Baltimore, who had also filled the position of president of Rockefeller, credited Levine with reinvigorating the research-focused university.[40] Levine returned to New Jersey and in 2002 was made a professor in the Institute of Advanced Study at Princeton, famous for being the final academic home of Albert Einstein.

Bert Vogelstein has remained at Johns Hopkins Medical School. In 2021, webometrics ranked him the seventh–most cited living scholar in terms of his Google h-index of 274 and total citation count of 433,410.[41] While the webometrics ranking is not complete—only those with Google Scholar profiles are represented—it does illustrate that Vogelstein is arguably the most influential living cancer researcher. A small protein discovered by looking at a relatively obscure DNA tumor virus has reshaped biologists' understanding of cancer at the molecular level.

Tumor virology influence extends beyond cancer biology, however, as we will see in the next chapter.

14

Unplanned Practical Payoffs

Robert Gallo, Luc Montagnier, Françoise Barré-Sinoussi, HTLV, and HIV

Robert (Bob) Gallo was raised in Waterbury, Connecticut, the grandson of immigrants from northern Italy. His father, a hardworking self-taught metallurgist and his uncle, a zoologist, served as role models. The "dominant impression of his youth" was the treatment and eventual death of his younger sister from leukemia—she was diagnosed at age 5. Although treated in Boston by Sidney Farber with antimetabolite chemotherapy, Judy Gallo died in 1948 when Robert was 11.[1]

The loss of his sister helped push Gallo into science and away from religion. *Microbe Hunters*, by Paul de Kruif, and *Arrowsmith*, by Sinclair Lewis, two books that valorized scientific medicine, also had an impact on the young Gallo. Before he began college, he knew he wanted to pursue a career in biomedical research, possibly focusing on blood cells and diseases like leukemia. He studied at Providence College, majoring in biology, and then Thomas Jefferson School of Medicine in Philadelphia. Following advice from his mentor Alan Erslev, Gallo left for a medical internship at the University of Chicago before heading to the National Institutes of Health.

Like other physicians interested in research, such as Bishop and Varmus, Gallo joined the Public Health Officer program at the NIH. Competition was fierce since it counted in lieu of a military obligation in Vietnam. Gallo joined in July 1965 and by 1966 began working with Sy Perry on the growth characteristics of leukemia cells at the NIH and with Sid Pestka of the Marshall Nirenberg laboratory on tRNA at the fast-growing National Cancer Institute. In 1972, Perry formed a new department—Human Tumor Cell Biology—and appointed Gallo to head it. Gallo created his first laboratory that attracted an international "hodgepodge" of clinical and basic research people.[2] A motivating idea for the lab was to investigate the causes of diseases using the new tools of molecular biology, including molecular hybridization.[3]

Gallo's friendship with Robert (Bob) Ting at the NCI helped to shape the direction of his lab in the 1970s.[4] Ting had studied with Luria at MIT and Dulbecco at Caltech, and consequently, he was well versed in tumor virology. RNA tumor viruses seemed especially interesting to Gallo as they could cause leukemia and solid tumors in a variety of species. Ting introduced Gallo to Huebner, who with Todaro in the late 1960s hypothesized that the cause of cancer was tumor viruses not integrated into the genomes of the host cells.[5] Huebner's emphasis on endogenous viruses downplayed exogenous viruses, but nonetheless, because of promising results by Jarrett in cats, Gallo decided he would look for exogenous (transmitted from host to host) human tumor viruses. This was a risky approach as everyone who looked for these viruses previously had failed. "It was for virologists a blind trail, a graveyard of experiments."[6]

What made Gallo's decision to hunt for human cancer viruses easier was Temin and Baltimore's discovery of reverse transcriptase in May 1970. As described in chapter 8, this discovery legitimated Temin's speculation that an RNA tumor virus makes a DNA version of its genome that can insert itself into the DNA genome of the host cell. If reverse transcriptase was coded for only in viral genetic material, it could be used in an assay to look for tumor viruses. Gallo's lab developed the test for reverse transcriptase and began to refine it to prevent false positives for cellular DNA polymerases.

Refining and extending this approach to virus hunting would consume the Gallo lab for the next decade. Positive results in other species spurred the group on. Carl Olson and Janice Miller and a group at Wisconsin and others at the University of Pennsylvania reported a leukemia-causing retrovirus in cows.[7] Other results beginning with an Oregon study showed that a retrovirus isolated in woolly monkeys could infect other primates, raising the possibility that humans could be infected by other simian viruses.

Then, in 1972, Gallo's group appeared to have a big break. His postdocs M. Sarngadharan and Marvin Reitz detected reverse transcriptase activity in cells from a patient with leukemia. The lab was elated: finally, positive results. It was the cherry on the top to have their findings published in *Nature*.[8] Unfortunately and surprisingly to the team, the paper had little effect. Many scientists were skeptical of the report since there had been other false reports of human tumor viruses. Some called them "human *rumour* viruses"[9]

To convince the skeptics, Gallo and his team would need additional evidence, ideally, to isolate the virus from a cell culture. However, cell biologists at the time could not culture the white blood cells needed. Gallo's group took up the challenge, following clues from Leo Sachs that the key to cell culture was finding the appropriate growth factor, a protein that induces cells to grow. This task was especially difficult because the various subtypes of white blood cells were only just being worked out, and there are potentially as many types of leukemia as there are subtypes of white blood cells. The group looked at blood from many types of leukemia, and in 1973 they found a factor that would keep myeloid leukemic white blood cells growing continuously.[10] Then they got lucky. For reasons that they did not understand, one of the myeloid cell lines became "immortalized" in the sense that it continued to grow without the need for growth factor. This line, called HL-60, came to be widely used in the field. Another important myeloid cell line was HL-23. It required growth factor to grow, but it tested positive for reverse transcriptase activity. Gallo's goal of demonstrating a human tumor virus seemed within reach.

Then disaster struck. Somehow the freezer storing the embryo cells that produced the growth factor was left unplugged. Without refrigeration, they died. Without the growth factor, HL-23 died also. Gallo was devastated, but there was nothing he could do but to start the search again.

After many unsuccessful tries, Robin Weiss and Natalie Teich, who were visiting Gallo's lab from London, achieved positive results with a growth factor. But their success was apparently too good to be true. When Gallo presented the results at one of the annual meetings of the Special Virus Cancer Program in Hershey, Pennsylvania, other researchers who had examined Gallo's cells announced that they were in fact contaminated, perhaps with two or even three viruses. The problem could have been caused by viral contamination in the lab, in which a previously identified retrovirus might have infected the samples in the lab itself. In any case, it was no new virus discovery. Their claims that Gallo had made a mistake, asserted in a public forum in front of Gallo's colleagues, were painful. After the talk, Ludwik Gross put his arm around Gallo and consoled him, reminding him that he and Temin were both ridiculed for their respective ideas about tumor viruses, but neither had given up.[11]

Chastened but undeterred, Gallo's group continued to look for growth factors for human myeloid cells. Doris Morgan, who was a trained clinician from Anderson Hospital and Tumor Institute (later MD Anderson) began her postdoc in Gallo's lab around this time. He suggested she try a different approach using PHA (phytohemagglutinin), a plant extract that could induce human lymphocytes to divide a couple of times in culture. Morgan mixed human blood and bone samples from healthy donors in a nutrient broth, added PHA, and sampled the nutrient broth for its potential to induce long-term growth of myeloid cells.[12] Surprisingly, the broth contained a growth factor, but the cells it induced to grow were not the expected myeloid cells; instead, they were T-cells. Epstein had discovered that Epstein-Barr virus immortalized B-cells. Morgan and Gallo had discovered was something similar in T-cells, for which no growth factor was previously known, and testing showed no presence of EBV. They named it "T-cell growth factor," and after growing cultures for nine months, published the results in *Science* in 1976.[13] This growth factor was later renamed Interleukin-2 (IL-2). Having IL-2 to culture T-cells was a major step forward, but the team at the Gallo lab tempered their enthusiasm after the disastrous contamination problem they had experienced earlier.[14]

The Discovery of HTLV

With the ability to culture T-cells, Gallo continued to look for human RNA tumor viruses, or retroviruses as they began to be called in 1975, by looking for novel reverse transcriptase (RT) activity. After more than a year of searching, Bernard Poiesz, a clinical postdoc in Gallo's lab, detected RT activity in a T-cell line, which was induced to grow with their growth factor and sourced from a patient with lymphoma of the T4 lymphocyte. To rule out an already known retrovirus, they tested for antibodies against known reverse transcriptases.[15] There was no reaction, so it was likely that the RT and thus the virus were new. But Gallo was still a little paranoid and wanted to rule out the possibility that the virus had come from a cow since the nutrient broth derived from bovine sources. He called his friend Arsene Burney, an expert in bovine viruses, for help. It did not matter that it was Christmas Day. Burney sent additional reagents to do the necessary tests, which ruled out known bovine viruses.

The group named the new virus human T-cell leukemia virus, following standard naming conventions. Later, some, including Gallo himself, thought "lymphotropic" was more accurate than "leukemia," and both variants of the name can be found in the medical literature. They hoped the name change would bring order to the field. It did not. Some of their material was presented in 1979, and in mid-1980 Gallo wanted to send a foundational paper to *PNAS* and supporting papers to other journals. The paper sent to *PNAS* was accepted, as *PNAS* papers normally are, since they had to be submitted by a member of the National Academy of Science (in this case, Henry Kaplan of Stanford University).[16] However, the paper sent to the *Journal of Virology* was rejected outright with no opportunity to revise and resubmit. The editor, Robert Wagner, wrote to Gallo, "Enclosed are the comments of two reviewers, both of whom expressed grave doubts about the evidence that your protein is really analogous to a retroviral structural protein. I completely agree with reviewer #1 that there is little point in perpetuating this controversy about the 'presumed viral nature of this material.'"[17]

This letter expressed the general resistance of the field to human cancer retroviruses. Not surprisingly, Gallo did not take no for an answer and fought back, and after six months the paper was in fact accepted at the *Journal of Virology*. It probably helped that the other papers were published in the meantime, and thus, considered collectively, the evidence for HTLV was much stronger by then.

In the next phase of the research, Gallo learned about some epidemiological work in Japan that showed clusters of T-cell leukemia in certain regions of Japan, including Okinawa. This disease, studied by Kiyoshi Takatsuki as early as 1976, was called adult T-cell leukemia, or ATL, and the presentation of symptoms in Japan matched the cases from the Americas.[18] Gallo began to collaborate with Yohei Ito from Kyoto University to investigate the connections between the Japanese cases and HTLV. By late 1981, 100% of eight examined Japanese cases of ATL tested positive for antibodies for HTLV. Ito organized a conference in March 1981, where Yorio Hinuma of Kyoto University announced that he, too, had found a retrovirus associated with ATL. He called it ATLV, but it was likely the same virus Gallo and Ito had found or a very close relative. Later it was decided to call the virus HTLV and the disease ATL. The virus appeared to be transmitted vertically, through mother's milk or perhaps in utero. It could also be transmitted via sexual intercourse.

Guy de Thé, discussed in chapter 5 for his work on Epstein-Barr virus, also began to study HTLV around this time. He discovered that the virus could be found on some Caribbean islands while remaining absent from others. It was also endemic to some parts of sub-Saharan Africa. It appeared from his research that the virus emerged in Africa and had been transported to the Caribbean during the slave trade. Additional comparative virology showed that HTLV was closely related to viruses in monkeys and chimps. Perhaps it had jumped from one species to another in the evolutionary past in a process called zoonosis.

Gallo's hunt for human retroviruses bore additional fruit. In the spring 1981, at a conference in Venice, Italy, he learned from UCLA biologist David Golde about a cell line from a rare leukemia of a T-cell subtype called hairy-cell leukemia. It had this name because hair-like fibers extrude from the cancerous cells. Golde's cell line was derived from T-cells, and he called it MO, using the initials of the patient who had the rare leukemia. Gallo suggested to Golde that this rare form of leukemia might be caused by a new type of human retrovirus. Few at the conference were convinced, but a few months later, Golde sent Gallo fluid from his MO line to test for a virus. (He would not send Gallo the MO cell line itself for commercial reasons.) A retrovirus was successfully isolated. It turned out to share more than half of its sequence with HTLV. A renaming was in order: HTLV became HTLV-1, and the new virus became HTLV-2. A relatively rare virus, HTLV-2 is primarily blood-borne and transmitted person to person by IV drug users sharing needles.

The French Connection: Luc Montagnier

Luc Montagnier developed an interest in the medical sciences early. In a brief autobiography he mentions that a serious bicycle accident in his childhood shaped his future path and impressed on him the importance of medicine.[19] He enrolled in a physics, chemistry, and biology program at the Université de Poitiers when he was 17. This was a pathway to the study of medicine. He was a good student and studied anatomy under Jean Foucault, father of the famous French philosopher Michel Foucault. After he graduated from Poitiers in 1953, he moved to Paris to complete a medical degree. Interested in pure biology as well as medicine, he attended lectures at the Sorbonne. However, he considered French biology still disadvantaged relative to the United States because of his country's isolation during World War II.[20] Many advances in biology had not been incorporated into the French curriculum.

After completing his studies, Montagnier took a job as an assistant in cellular biology at the Institut Curie; named after the famous French physicists who studied radium at the turn of the twentieth century, it focused on studying and treating cancer. Montagnier became interested in foot-and-mouth disease virus. He tried to grow cultures of mouse cells but had no success. To keep abreast of the new thinking in biology, he attended Jacques Monod and François Jacob's molecular biology club at the Institute of Physico-Chemical Biology in Paris.

In 1957, Heinz Fraenkel-Conrat at Berkeley and Alfred Gierer and Gerhard Schramm at Tübingen showed that new tobacco mosaic viruses could be produced by infecting leaves with merely TMV RNA.[21] These spectacular experiments showed that the genetic information of a virus was contained in its nucleic acid and not its protein. Montagnier was impressed by these experiments and committed himself to virology on account of them. He collaborated with Jean Leclerc to purify RNA from foot-and-mouth disease virus. However, his position at the Institut Curie did not give him access to the equipment he needed to pursue this line of work. Also, his relationship with his supervisor was deteriorating, so Montagnier left his position. This decision did not go over well. His advisor would not let him defend his medical thesis, but Montagnier found another professor who would.

Luckily, Montagnier obtained a position at the Centre national de la recherche scientifique and was awarded a scholarship to take part in a three-year exchange with the British Medical Research Council. He left France for the laboratory of Kingsley Sanders on the edge of London. There Montagnier worked on the structure of viral RNA.[22] He would win a bronze medal from CNRS for this work. Three years after Montagnier joined his group, Sanders's success was internationally recognized, and he was recruited by the Memorial Sloan Kettering Cancer Center in New York. Members of Sanders's group needed to find new positions.

The best place to do animal virology was with Dulbecco in his laboratory in California, thought Montagnier. Unfortunately for Montagnier and his wife, sunny California was not going to be their next home; Dulbecco was on sabbatical with Michael Stoker and Ian Macpherson in Glasgow, Scotland. Montagnier judged the person to be more important than the place, so the Montagniers instead drove north to Glasgow, a place that "had all the characteristics of a nineteenth-century industrial city in decline."[23] Montagnier worked with Macpherson to develop a new way to grow polyoma virus in BHK cells suspended in agar.[24] Among other uses, this technique allowed for the demonstration that naked polyoma DNA transformed cells just as the whole virus did.[25]

By 1964, his scholarship from CRNS expired, so Montagnier returned to

France and the Institut Curie. He attempted to adapt his polyoma virus agar measurement technique to Rous sarcoma virus and human cells but had limited success. In work that competed with Peter Duesberg, Vogt, and Bill Robinson (see chapter 9), he and his collaborators showed that the carcinogenic and infective properties of polyoma virus were separable capacities.[26] The different capacities could be inactivated by different levels of radiation. Given the differences, they concluded that the cancer-causing capacity of polyoma was located in only one-fifth of the viral genome. Montagnier thought about replication of the RNA virus RSV. Unlike Howard Temin, who thought there must be a DNA intermediate, Montagnier liked the hypothesis that the RNA molecules replicated themselves, possibly by making double-stranded RNA. Placing little importance in an enzyme that makes DNA from RNA, he was not part of the race for the discovery of reverse transcriptase.

Nonetheless, Temin's and Baltimore's discovery changed the field of RNA tumor virology including a change in name to retrovirology. Montagnier tried to find DNA in human tumors that would transform normal cells into cancerous cells, but to no avail. He needed a change of direction to keep his research program fresh.

Montagnier decided to study antiviral defenses, in particular, a molecule called interferon that is part of an animal's natural defense against viral infections.[27] In 1972, Montagnier and his colleagues isolated interferon RNA. A natural goal was to clone the gene for interferon; that is, put the human gene into bacteria to have them make the interferon protein. Montagnier was offered money from mid-level people at a pharmaceutical company to do the necessary work, but upper management declined. In the end, it was Biogen (see chapter 11), an American-Swiss company, that commercialized interferon production.

The search for retroviruses again became a focus of the Montagnier laboratory in 1977.[28] Along with a reverse transcriptase assay, he used an anti-interferon serum to detect a virus. The idea was that interferon in humans would often overwhelm retrovirus replication. Adding the serum might make otherwise invisible viruses detectable. Montagnier obtained leukemias from the Cochin Hospital in Paris and made cell cultures. Françoise Barré-Sinoussi, a technician, looked for reverse transcriptase activity. Although they had some success in mice, no significant results came in for humans. The group tried to publish the mouse results in *Nature*, but the paper was rejected on the basis that the mouse interferon was not pure. Given others' false positives, the evidentiary bar was high for new retrovirus discoveries.

When the news of Gallo's discovery of HTLV-1 made it to France, Montagnier focused on the isolation of TCGF (T-cell growth factor) that allowed Gallo's laboratory to culture normal T-cells. Montagnier wondered whether this growth factor might supplement his work on mice. The two men agreed to collaborate. Coincidently, Barré-Sinoussi had traveled to Gallo's laboratory to try to replicate the mouse results in monkey cells. Perhaps with TCGF and interferon together, new viruses could be discovered. Montagnier claims that Barré-Sinoussi did not learn how to grow T-cells while she was in Gallo's laboratory.[29] Gallo disagrees.

Meanwhile, back in France, Montagnier began looking for human viruses in T-cells using interferon and TCGF that Gallo sent him in a bottle of culture medium. He planned to look for a breast cancer virus by looking at Parsee women. These women tended to marry only other Parsees and had a very high rate of breast cancer. The approach built upon the success of Bittner, who had discovered a virus in mice that caused breast cancer (see chapter 1). Sol Spiegelman in the United States claimed that he had found a human breast cancer virus, but it turned out to be a false alarm. Despite a promising approach, Montagnier put this avenue of research on hold when a more pressing disease emerged.

In the summer of 1981, Gallo heard reports of a cluster of pneumonia cases among young gay men in California.[30] What made these cases unusual was the presence of a parasite *Pneumocystis carinii* (PC) that was normally not disease causing in healthy adults. The parasite was typically observed only in newborns without immune systems. Meanwhile, in New York, there was a report of a cluster of Kaposi's sarcomata in the gay community. Manifested by purple lesions on the skin, this cancer did not usually afflict the young. The Centers for Disease Control, the federal agency responsible for protecting public health, took notice.[31] James Curran of the CDC raised the alarm: he presented the epidemiological findings at the NIH in late 1981. He suspected that it might be the beginning of a new disease and wanted research scientists to begin work on it. He presented more data again in early 1982. One research scientist who heard the call was Gallo.

One fact about the afflicted patients piqued Gallo's interest. Among the unusual symptoms the young male patients had was a reduction in T4 lym-

14.1. Max Essex and Robert Gallo at Harvard School of Public Health. Image courtesy of Max Essex

phocytes (also called CD4+ T-cells.) Gallo spoke with his friends Max Essex and William Haseltine, also specialists in virology (figure 14.1). They were thinking along similar lines and pointed out that some retroviruses suppress the immune system.[32] For example, feline leukemia virus, discovered by Jarrett (see chapter 4), often causes an immune deficiency as well as leukemia. HTLV-1 also could suppress the immune system. Gallo wondered whether the new disease was due to a mutated HTLV that emerged in Africa and traveled to Haiti and then the United States.

Despite some misgivings that there was plenty of work left to do on HTLV-1 and HTLV-2, in May 1982, Gallo began a project in his laboratory to search for a retrovirus that might be a cause of the new disease. He adapted methods refined in the search for HTLV: culture T-cells from affected patients with Interleukin-2 and then assay for reverse transcriptase activity. For many months, the results were inconclusive. Unlike HTLV, which could immortalize T-cells, the proposed retrovirus seemed to kill the T-cells relatively quickly, complicating their work. Additional techniques using immunofluorescence to search for HTLV antigens also did not produce clear results.

By late 1982, the CDC named the disease acquired immunodeficiency syndrome (AIDS).[33] More than 800 people died of AIDS that year.

In addition to the Gallo and Essex labs in the United States, other teams began the search for a human retrovirus. Earlier, Montagnier had suggested that blood plasma imported into France from the United States to make hepatitis vaccines should be screened for retroviruses. A French immunologist named Jacques Leibowitch raised the alarm that American plasma might carry the AIDS virus after reading an August report on what the *Medical World News* called AID.[34] In that report, Robert Gallo was quoted as saying that a "reasonable candidate" for an infectious cause of AIDS was HTLV, as it was endemic in the Caribbean and was known to be transmitted through donor blood.[35] Leibowitch considered this an "involuntary tip" from Gallo and tried to get a French retrovirologist interested, but he was rebuffed. Eventually, in response to his concern, Luc Montagnier, Jean-Claude Chermann, Barré-Sinoussi, and others began a research project to screen plasma and also to look at the white blood cells of French AIDS patients. There were very few French cases of AIDS at the time, perhaps a dozen or two.

Françoise Barré-Sinoussi

Françoise Barré-Sinoussi came from humble roots. She was attracted to science and nature at a young age. Under the impression that training as a scientist was less expensive than medical training, she took that route. After much searching, she obtained a technical job at the Institut Pasteur in the laboratory of Jean-Claude Chermann in the early 1970s. He convinced her to pursue a PhD, which she completed in 1974 with a dissertation on a molecule that inhibited the reverse transcriptase enzyme of the Friend leukemia virus (see chapter 2). A postdoc at the NIH followed. Her experience with reverse transcriptase assays would prove valuable and as a junior scientist, she had more experience studying retroviruses than most other French biologists.[36]

The Patient Known as BRU

In December 1982, Montagnier and his group were given a test tube of a lymph node biopsy in the patient's blood with a note, "persistent lymphadenopathy in a homosexual man."[37] The patient would be known as BRU, from his name, Frédéric Brugiere. Montagnier dissociated the biopsy into single cells and made a suspension of lymphocytes. He placed the suspension at 37°C, the optimal temperature for growth. On January 6, he added some Interleukin-2 and some anti-interferon serum. The culture grew well, and every three days Barré-Sinoussi analyzed a sample for reverse transcriptase activity. Mouse virus reverse transcriptase needs manganese ions to function well; reverse transcriptases from chicken viruses and HTLV need magnesium ions. The French group decided to try both options. As Barré-Sinoussi put it, "The first week of sampling did not show any reverse transcriptase activity, but in the second week I detected weak enzymatic activity, which increased significantly in a few days."[38] The weak activity was a magnesium-sensitive enzyme. It looked like Montagnier and his collaborators were on the verge of a great discovery.

Unfortunately, just at this critical juncture, the cells in culture started to die. The team tried adding new lymphocytes, which seemed to work, but only temporarily. Chermann called Gallo for advice, and Gallo suggested they use lymphocytes from umbilical blood, which had worked for HTLV-1 and HTLV-2.[39] As it turned out, Montagnier was already aware of Gallo's protocols, and they consequently had used newborn lymphocytes. Nonetheless, the new retrovirus did not immortalize the cells like HTLV did, and the French scientists had to add umbilical blood lymphocytes every few days to keep the culture and thus the virus alive.

The next step was to see whether the virus was HTLV, a virus closely related to HTLV, or something different. Montagnier reached out to Gallo to get antibodies against HTLV.[40] If these antibodies also reacted against the new virus, then it was likely not a novel retrovirus. Gallo complied, but the antibody tests were inconclusive. Some suggested a reaction indicating a close relationship between HTLV, and others, especially one directed at the HTLV core protein p24, did not produce any reaction.

Another possible approach was to use electron microscopy to see whether the new virus looked like HTLV. Charles Dauguet was the experienced electron microscopist at the Institut Pasteur. On February 4, 1983, in a sample from BRU's lymph node, he found virus particles with dense cores. They did not look like HTLV particles, which had a different core. Of course, correlation, especially in only one patient, does not imply causation. A great deal of further evidence would still be needed to show that this new virus was the cause of AIDS.

Generalizing the results from BRU proved difficult in the beginning. A second patient, MOI, had antibodies against the BRU virus and HTLV. This was either evidence that the BRU virus was HTLV or that MOI was doubly infected with two viruses: the BRU virus and HTLV. Montagnier asked for significantly more resources from French science administrators but was only given one extra room, and a small one at that.

Nevertheless, by April 1983, Montagnier thought the team had enough data to publish in the prestigious journal *Nature*. However, before the article draft was finished, he received a call from Robert Gallo, who had been working in a similar vein and thought that the French scientists and the American scientists should publish back-to-back articles in *Science*, an equally prestigious journal. This was a polite gesture on Gallo's part—he could have tried to publish before Montagnier and then claim priority for any commonality in their respective results. From a scientific point of view, it also made sense strategically because the two papers collectively would provide more evidence of their claims about the new retrovirus.

Gallo had been working on AIDS since May 1982. His laboratory had detected some reverse transcriptase activity in some samples, but none in others. He had hired the Czech biologist Mikulas (Mika) Popovic in January 1980. Popovic was skillful in creating and maintaining cell lines, working first on HTLV and then on AIDS. Results with the AIDS virus were ambiguous, and cell cultures did not immortalize as they had for HTLV-1. Between November 1982 and February 1983, they had five positives for reverse transcriptase.[41]

Montagnier quickly wrote up a paper for *Science* and sent it to Gallo, who would be selected as a reviewer for the journal. Barré-Sinoussi would be the first of 12 authors of the French paper, "Isolation of a T-lymphotropic Retrovirus from a Patient at Risk for Acquired Immune Deficiency Syndrome (AIDS)." In his haste, Montagnier did not write an abstract, a standard part of the paper. To help the French team, Gallo wrote one and dictated it over the phone to Montagnier, who agreed it was acceptable. Later, this would be a point of contention between them as the abstract expressed Gallo's view that the new virus was a member of the HTLV group of viruses, a position Montagnier would reject. The abstract ended, "From these studies it is concluded that this virus, as well as the previous HTLV isolates, belong to a general family of T-lymphotropic retroviruses that are horizontally transmitted in humans and may be involved in several pathological syndromes, including AIDS."[42] The text of the paper showed evidence both for and against the idea that the new virus was a member of the HTLV group. "Viral core proteins were not immunologically related to the p24 and p19 proteins of subgroup I of HTLV. However, serum of the patient reacted strongly with surface antigen (or antigens) present on HTLV-I infected cells. Moreover, the ionic requirements of the viral reverse transcriptase were close to those of HTLV."[43] And the French group observed, "The virus appears to be a member of the human T-cell leukemia virus (HTLV) family."[44] In Gallo's paper, which immediately preceded the French paper, the American team reported three isolates of a virus from three different patients with AIDS: two American and one French.[45] Gallo's friend Max Essex, another early advocate of the idea that AIDS was caused by a retrovirus, also published a paper in the series: his team reported that sera from 19 of 75 AIDS patients had antibodies to an HTLV protein, whereas only 2 of 366 non-AIDS control subjects did.[46]

The differences between Gallo and Montagnier were explicitly aired in June 1983.[47] The French team was now calling the virus LAV for lymphadenopathy associated virus. (BRU did not have full-blown AIDS but rather lymphadenopathy, a precursor to AIDS.) Gallo and the Americans called it HTLV-3. The name made a difference: HTLV-3 reinforced Gallo's view that the current work was an extension of his earlier work on HTLV-1 and HTLV-2. LAV reinforced the view that Montagnier had made a significantly different finding than Gallo had in his earlier work on retroviruses that attack human T-cells. After meeting at the Paris apartment of Guy de Thé, the two men debated the nature of the virus. Montagnier "presented one argument after another. Gallo would have none of them."[48]

The discourse continued in July at Gallo's house before a meeting on HTLV in Maryland. Montagnier thought his electron micrographs of LAV were evidence that LAV was more closely related to infectious equine anemia virus, a lentivirus, but he was unable to convince most of the Americans. He decided against a previously agreed collaboration with Gallo.

The 1983 CSHL Conference on HTLV and AIDS

James Watson asked Robert Gallo, Max Essex, and Ludwik Gross to organize a Cold Spring Harbor Laboratory meeting on HTLVs and AIDS in September 1983.[49] Gallo invited Montagnier. Essex wrote the preface of the resulting volume and dedicated it to Mary Lasker, the force behind the Lasker Award, for her support of the idea that cancer can be caused by viruses.[50] Serological data presented at the meeting undermined Gallo's early hypothesis that a minor variant of HTLV-1 or HTLV-2 was the cause of AIDS. Since he did not have all the details worked out, Gallo decided not to present his most recent results: 20 new detections of the novel retrovirus.[51] Essex, Gallo, and several Japanese and British virologists signed an agreement about the international nomenclature of retroviruses that caused lymphoma and leukemia, stipulating that these viruses would be named HTLV-1, HTLV-2, HTLV-3, and so on. The implication was that Gallo's nomenclature was to be preferred over the French name. Montagnier did not sign.

Montagnier presented at the end of the conference, by which point many people had already gone home. His talk, "A New Human T-lymphotropic Retrovirus," summarized the new and old research from the French group. He presented tables showing the similarities and differences of LAV-1 and HTLV-1 and EIAV. The French team's paper argued that LAV was "an entity clearly distinct" from HTLV. Despite the thinner attendance, a barrage of questions followed his talk, including some from Gallo. The two men's respective characterizations of this event are quite different. From Gallo's perspective, Montagnier did not have good answers to his questions. But Montagnier saw himself as being open minded by not ruling out a causal role for HTLV completely. "T-lymphotropic retroviruses are the primary agents of the disease [AIDS]. Among such causative agents we include LAV-related viruses, HTLV-related viruses, and any other lymphotropic retroviruses to be discovered," he said.[52] He recalled Gallo as saying that Montagnier had "punched me out," which Montagnier took to mean that the French work had demolished all Gallo's work on HTLV and AIDS, but Gallo did not remember it that way.[53] Montagnier believed that he was leaving several doors of inquiry open.

Gallo later regretted how aggressive he was with Montagnier in the question session, although such a robust back-and-forth was not uncommon at CSHL meetings attended by the heavy hitters in the field. He was dissatisfied with Montagnier's response to his numerous criticisms of the presentation. For example, Gallo was not convinced by the electron microscopy. And the French reverse transcriptase assays were not as tightly quality controlled as Gallo's had been for the work on HTLV he presented in 1980. Did Montagnier really believe that LAV was involved in the earlier stages of AIDS but not in later? Montagnier had presented data that there were antibodies to LAV in 60% of patients with lymphadenopathy, but only 20% with AIDS. How would Montagnier explain these data? Gallo wanted to know (figure 14.2).

14.2. Robert Gallo and Luc Montagnier. Image courtesy of Cold Spring Harbor Laboratory Library and Archives

Gallo thought it was obvious that only one agent caused the disease. He pushed Montagnier on this point, and their personalities clashed. Gallo was extroverted and liked the rough-and-tumble of debate. Montagnier was quiet and reserved. He saw himself facing a "wall of indifference and bad faith."[54] The indifference was not just at Cold Spring Harbor but also at home. The French, too, did not give his work as much credit as Montagnier thought it deserved. Some of the indifference was due to AIDS having the stigma of a gay disease, something the majority of heterosexual people did not have to worry about.[55] Later he sent a paper to *Nature* expanding on his ideas, but it was rejected. He then tried to publish it in *PNAS*, but the second reviewer never responded and the paper languished.

The Cause of AIDS

By 1984, Gallo and his collaborators had assembled a large amount of data supporting the idea that the novel retrovirus was the cause of AIDS. In fact, they had enough for four *Science* papers. This was an unprecedented event—to have four back-to-back papers in one of the most prestigious scientific journals in the world. Gallo's collaborator Mikulas Popovic described the creation of a cell system for the detection of HTLV-3. When cells of the system are infected by HTLV-3 they produce a large amount of the virus and turn into giant multiply nucleated cells. In the second paper, Gallo's team used Western blots to detect surface antigens of HTLV-3 and concluded that they were distinguishable from but "significantly related to" those of HTLV-1 and 2. The third paper considered the antigenic proteins of HTLV-3, which were also distinguishable from but related to those of HTLV-1 and 2. The fourth paper showed that 88% of patients with AIDS had antigens against HTLV-3, but only 1% of the heterosexual control group. A fifth paper published in the *Lancet* found 100% of AIDS patients tested positive for the virus.[56]

Collectively, the five papers were the strongest evidence yet that the new virus was the cause of AIDS.[57] From them, a pleasing division of credit could be drawn among the contributors. Montagnier and his French team discovered the new retrovirus, and Gallo and his American collaborators showed that it was the cause of AIDS. These papers cemented Gallo's position as the leading AIDS researcher in the world. The vast majority of the scientific community took this work to "demonstrate conclusively" that the retrovirus was the cause of AIDS.[58] There were a small number of skeptics, the most prominent being Peter Duesberg (see chapter 9), the Berkeley virologist who made his later career as a contrarian about the causes of AIDS and cancer.[59]

Rising Tensions between US and French Scientists

In April 1984, lawyers from the Department of Health and Human Services (HHS) advised Gallo to file for a patent for his blood test for the virus.[60] The French had filed for a similar patent earlier, but the Americans at HHS appeared to be unaware of this development. Margaret Heckler, the secretary of HHS, which oversees the NIH, gave a news conference praising the work of Gallo and the NCI laboratory that he directed. From her perspective this was the payoff of her decision to make it the federal government's number one health priority and devote significant resources to study AIDS 10 months earlier. She did not mention the important role of Montagnier's group, which fueled the increasingly bitter dispute over priority between the Americans and the French.[61] The closest she got was to downplay the initial identification of the virus. "The NCI work provides the proof that the cause of AIDS has been found. It does this because it goes beyond the simple identification of a particular virus."[62] Others spoke at the news conference, including Gallo. He suggested the virus should be called HTLV-3/LAV, but this did little to stifle the anger felt in France.

According to Nikolas Kontaratos, Secretary Heckler intended to say more about the French work, but her voice gave out.[63] A statement released later by the HHS acknowledged Montagnier's role, but it was too late. The reporters covering the press conference focused on only the American work. The French press reported the slight against the Institut Pasteur.

This escalation in tensions between the French and American scientists could have been avoided. Originally, Gallo and Montagnier had planned to make a joint announcement about the cause of AIDS together in June. The *Science* articles would be published in May. However, news of the findings in *Science* had leaked to the media in April, and the popular magazine *New Scientist* published an article. A day before the HHS announcement, the *New York Times* ran a one-sided article claiming that the Institut Pasteur had discovered the cause of AIDS. Secretary Heckler was rushed into a press conference without the proper planning, and the resulting statements aggravated the growing rift between the NIH and the Institut Pasteur. The French scientists felt snubbed by Heckler's press conference, and rightly so.

After the *Science* papers came out, Gallo's colleague M. G. Sarngadharan, "Sarang," brought the cell line to France to compare HTLV-3B with LAV. Genetic comparisons—first a restriction map and later a DNA sequence—showed that the two viruses were essentially identical. Given that retroviruses mutate quickly, the match was not a coincidence, but rather pointed to a common origin. But where? The patient BRU had not recently been in the United States, so it was likely that in either Gallo's or Montagnier's lab a mix-up of some sort had occurred. Virologists call such mistakes "contamination events," and they are not uncommon in virology labs. Montagnier had given his LAV strain to others before Gallo, so it was likely that the error occurred in Gallo's lab. The story would become more complicated once the isolates could be sequenced. The LAV virus that Gallo's lab had studied was not LAV-BRU but rather LAV-LAI, another French isolate. This virus—LAV-LAI—had also contaminated cultures in Robin Weiss's lab as well as at the Institut Pasteur, and researchers in both locations who had thought they were studying LAV-BRU were actually studying LAV-LAI. One possible cause of the contamination was the common use of the fume hood when handling potentially dangerous cell cultures. If material from one cell culture contacted another, the more virulent virus could be transferred and out-replicate the less virulent one.[64] Much of the hands-on work was done at the same location in the laboratory, under fume hoods, and over time the chance of an accidental contamination event increases.

In September 1983, Gallo's senior scientist Popovic had used an unusual technique to establish an immortal cell infected with the retrovirus.[65] Instead of keeping the various isolates separate, he had pooled many samples together with hopes that the number of viruses and the synergy among the different viruses and cells present would produce a useful, virus-producing cell line. He learned this approach from his fellow Czech virologist Jan Svoboda.[66] He also tried to create a permanent cell line infected with LAV. The French had not been able to create one. Working with most known leukemic T4 cell lines was out of the question as they were already infected with HTLV and so would complicate any results, but the cell lines known as HUT 78 and Ti7.4 were possibilities. Popovic was successful and created a cell line that was called H9.[67] Although this was kept secret in the beginning, Popovic intentionally or unintentionally included LAV in the pool. Unfortunately for Gallo and Popovic, it was LAV that the successful cell line produced. At the time, the exact origins of the virus seemed less important as they were in a hurry to create enough virus to accelerate research against the disease that was killing an increasing number of people.

The Legal Battle over the HIV Test

Which virus was used to create the HIV blood test would be important in the legal battle over the patent rights between the NIH and the Institut Pasteur. While the French patent application had been submitted before the American one, for reasons that appear to be internal to the patent office, the French patent application fell through the cracks, and the Americans were granted a patent on the blood test first in May 1985. (The FDA had approved it in March.) One possible reason the patent office did not realize that the two tests were so similar was that one described the virus as LAV while the other used the name HTLV-3. There was also some urgency to commercialize a test in order to protect the blood supply. Not surprisingly, the Institut Pasteur was upset by the outcome, and its lawyers formally protested the decision. If it could be shown that Gallo had used a French isolate of the AIDS virus, then his priority would be seriously questioned, and the patent could possibly be overturned. There was a lot of money at stake—tens of millions of dollars—as an AIDS blood test would be used to screen all blood for the US medical supply. There were royalties, too, that could be beneficial to the NIH and Institut Pasteur as well as the individual scientists. Gallo's test was set to be commercialized by Abbott Laboratories.[68]

The lawyers for the two sides could not resolve the standoff, and so in December 1985, the Institut Pasteur filed a lawsuit against the US government. Meanwhile, the FDA approved the French blood test in February 1986. It would be manufactured by the US company Genetic Systems.

While the legal wrangling dragged on, another dispute was brewing, what to name the virus? Harold Varmus, later the director of the NIH, led 13 other retrovirologists on an International Committee on the Taxonomy of Viruses subcommittee to decide the official name. Both Gallo and Montagnier were members, as was Jay Levy, who ran the third laboratory to isolate the virus,[69] and Max Essex. Gallo and Essex argued that the group should agree to call it HTLV-3. Montagnier wanted to call it LAV. After over a year of discussion, the group took a middle path. The virus would be called human immunodeficiency virus (HIV) and significantly different isolates would be given the designation HIV-1, HIV-2, and so on. This respected a tradition in virology to begin the name of a virus by naming its host. (Think of FeLV or HTLV, for example.) The committee struggled over whether to include AIDS in the name but decided to leave it out. Gallo and Essex disapproved of the new name and did not sign on to the decision. Eventually, however, they came around, and HIV became the universally accepted name for the cause of AIDS around the world.

While this furor over naming unfolded, many in the scientific community worried that the legal dispute was detrimental for science and the fight against the AIDS pandemic. A number of Nobel laureates wrote to President Ronald Reagan, urging him to help settle the dispute. But it was not Reagan who would ultimately step in to mediate—his administration was notoriously slow to react to the "gay" pandemic—instead, it was a well-known scientist: Jonas Salk, famous for his vaccine for poliovirus. For more than a year, Salk arbitrated between Gallo and Montagnier, guiding them toward an agreement. Salk pushed Gallo and Montagnier to write a joint history of the discovery to put the dispute over priority to rest. Such a mutually agreed history might trigger a deeper agreement, Salk hoped.[70] They fought over the text and references but eventually published the joint history in *Nature* in April 1987.[71] From this process, a deeper agreement between the scientists did come.

In society more generally, gay activists pushed for a better federal response to the pandemic. The AIDS Coalition to Unleash Power (ACT UP) was formed in March 1987 to increase funding, change legislation, and enact public policy to accelerate the response to the AIDS crisis.

Salk proposed to the HHS lawyers that the French patent should also be granted, both tests be commercialized, and the profits from the two tests be paid into a joint foundation and then distributed. The agreement was to be officially made public at a White House ceremony between Reagan and French prime minister Jacques Chirac in March. The popular press reported the agreement as assigning credit equally for discovering HIV and creating the blood test for the virus, but it was mostly an agreement about patent rights and royalties.[72]

Despite the agreement over the blood test, the controversy over credit for the discovery of the cause of AIDS intensified in November 1989 with the publication of a 55,000-word piece in the *Chicago Tribune*, in which reporter John Crewdson accused Gallo of scientific misconduct in misrepresenting LAV as HTLV-3. Crewdson would later expand his research and publish a book—*Scientific Fictions*—on the "dark side" of Gallo. In addition to sullying Gallo's reputation, the *Chicago Tribune* article led to an official investigation by the Office of Scientific Integrity (OSI). In a preliminary report in 1991, OSI found him guilty of scientific misconduct, but the final report found him responsible only for inadequate oversight of work done under his leadership. The newly created Office of Research Integrity found him guilty of research misconduct but later dropped the charges after the definition of scientific misconduct was narrowed and thus more difficult to prove.

Democratic representative John Dingell, in his capacity as chair of the Energy and Commerce Committee, which asserts congressional oversight over the NIH, also investigated Gallo. (He also investigated David Baltimore. See chapter 8.) Dingell wanted Congress to endorse a scathing report on Gallo and Popovic, but the Democrats lost power to the Republicans in the election of November 1994 and the subcommittee investigating Gallo was dissolved.

The 2008 Nobel Prize

Ultimately, these controversies were not career ending for Gallo, who decided to leave the NIH in 1996 to start the Institute for Human Virology, in partnership with several Maryland universities, which is still active and employs more than 300 people in Baltimore. During most of the 1980s, he was the most cited scientist in the world.[73]

However, the taint of lawsuits and investigations might have had some effect on Gallo's legacy in the long run. In 2008, Montagnier and Barré-Sinoussi were awarded the Nobel Prize, sharing it with Harald zur Hausen for his discovery of HPV as the cause of cervical cancer. Gallo was not given the prize in spite of his pivotal role in proving that HIV was the cause of AIDS. It is likely all the negative publicity played a role in denying him the prize. The Nobel Committee did not need to recognize the HIV and HPV discoveries in one Nobel Prize and could have included Gallo. Many defended Gallo, including more than 100 scientists (including the Kleins and Jan Svoboda) who wrote a letter to *Science* calling him an "unsung hero."[74]

Mika Popovic was cleared of the charges brought by the OSI by an HHS review board in 1993, although the controversy was hard on his career. He left the NIH, was effectively blackballed from research, and was only able to find a lower position in a laboratory in Stockholm.[75] He is currently an adjunct professor at Gallo's Institute for Human Virology in Baltimore.

Barré-Sinoussi, who was a junior scientist when she began working with Gallo and Montagnier, advanced her career. In 1988, she started her own laboratory at the Institut Pasteur and in 1992 became head of the Biology of Retroviruses Unit (figure 14.3). She continued to work on HIV/AIDS throughout her career, consulting for the WHO and the UNAIDS, who lead the global effort to end AIDS as a public health threat. She advocated for an improved response to HIV in developing countries. From 2012 to 2016 she was the president of the International AIDS Society. She won a number of awards and honorary degrees. In 2013 France made her a grand officer in the National Order of the Legion of Honor. *Time* magazine featured her on a cover of the March 16, 2020, issue devoted to women throughout history.

After winning the Nobel Prize, Luc Montagnier has become a controversial figure. He published a paper in a journal over which he had control claiming that bacterial and viral DNA immersed in water is able to emit radio waves.[76] To the surprise of the virology community, he suggested that his research could be used to treat autism by taking advantage of "memory" structures in water. The homeopathy community embraced his research as validating their claims, but mainstream science has largely shunned it. In the last decade of his life, he was a professor at Shanghai Jiao Tong University in China, where he ran a research institute. He died February 8, 2022.

14.3. Françoise Barré-Sinoussi. Image courtesy of the Institut Pastuer

Despite all the drama over credit and patents, it is remarkable how quickly a test for HIV was developed. The development of the first effective drug against AIDS, AZT, which works by inhibiting reverse transcriptase, also happened rather quickly despite political barriers.[77] No doubt there was much pressure to do something to fight a deadly pandemic, complicated by sexual politics of the time, but without the basic science of retroviruses, neither testing nor treatment would have been possible. The speed of the response can thus be attributed to earlier work on retroviruses, including Temin and Baltimore's discovery of reverse transcriptase and Gallo's discovery of HTLV. The massive funding of cancer virus research paid off by enabling researchers to understand so much about retrovirology before the discovery of HIV. Thousands, if not millions, of lives were saved.

15

Planned Practical Payoffs

Harald zur Hausen, Jian Zhou, Ian Frazer, Douglas Lowy, John Schiller, HPV, and the Cervical Cancer Vaccine

Harald zur Hausen was born in 1936 and raised in northern Ruhr, Germany.[1] His parents were interested in science and technology. In the 1930s, his father, an agronomist by training, was an amateur balloonist at a time when the great German airships *Graf Zeppelin* and *Hindenburg* were making regular transatlantic trips. Zur Hausen began primary school in 1942, but the school was closed later that year because of Allied bombing of the Gelsenkirchen area. He was sent to stay with his aunt on a farm near Münster. After the war, the zur Hausens tried to resume their life in Gelsenkirchen, but it was difficult to do so for a number of reasons, including the damage their house had sustained from a bombing raid. Returning to school, Harald had to work hard to catch up with the curriculum. He loved investigating nature—at about 8 years old he already thought of himself as a scientist studying birds and animals. His best friend at the time remembered zur Hausen memorizing the Latin names for European birds.[2] His first publication was a description of a Siberian hawk attacking a duck in a German hunting journal. He was 12. He raised squirrels and watched snakes. His brother called him "an exceptional nature lover."[3] He also read about medical researchers like Pasteur and was impressed by a biography of Robert Koch, the German doctor who found the causes of tuberculosis and anthrax after whom Koch's postulates are named. Sinclair Lewis's *Arrowsmith*, a fictional story about a doctor using bacteriophages to cure bacterial infections, also influenced him.[4]

In the final years of *Gymnasium* (a German advanced high school aimed at university preparation), zur Hausen decided to study medicine and become a researcher. He enrolled in biology and medicine at the University of

Bonn in 1955. Biology instruction there was a disappointment, though, as it did not cover the new molecular biology. Interested in infections, zur Hausen left Bonn for the University of Hamburg, where there was a center for the study of maritime and tropical diseases. The medical thesis topic selected for him at Hamburg did not interest zur Hausen, so he dropped it and moved again, this time to the University of Düsseldorf, which also offered training in tropical diseases.

At Düsseldorf, zur Hausen had very little guidance, which bothered him greatly.[5] There he had some "crazy ideas that did not work out" partly because of lack of supervision. Nonetheless, it was a fruitful time as zur Hausen was "fresh and young" and full of ideas. He was drawn to virology and despite being inexperienced was asked to work in a poliovirus diagnostic laboratory. Mundane diagnostics did not interest zur Hausen; he was more interested in whether a virus could modify the genome, an idea that occupied his thoughts after taking a course in phage genetics and learning about lysogeny. He wondered whether cancer cells could pick up foreign DNA much like bacteriophage, which pick up toxin genes. His thesis topic on the antiseptic properties of floor waxes, which was given to him, displayed the gap between his interests and his mentors'.

Harald zur Hausen Trains in Philadelphia

By 1965, he was desperate to find a new place to work, a place closer to the action in molecular biology. Zur Hausen responded to a letter advertising a research position at the Henles' laboratory in Philadelphia (see chapter 5). The husband-and-wife team spoke German, which was a bonus as he spoke little English. Werner and Gertrude Henle arranged to meet with him in Heidelberg. They told him about the new tumor virus they received from Anthony Epstein. They wanted zur Hausen to join them in Philadelphia and work on it.

It was too good an opportunity to turn down. The Henles gave zur Hausen the scientific mentorship that had been missing in Germany. The three often discussed what came to be known as Epstein-Barr virus, and they did not always see eye to eye. Werner Henle suspected some patients had continuous rounds of infections of Epstein-Barr virus, which interferon production kept in check, whereas zur Hausen thought it more likely that the EBV genome persisted in infected cells. Which view was correct was a major question for zur Hausen.

Wanting to broaden his laboratory skills, zur Hausen worked on adenovirus and helped develop an immunofluorescence test for EBV. Eventually, using radioactive probes, zur Hausen, George Klein, the Henles, and others demonstrated that the EBV genome remained unintegrated in the Raji line of Burkitt lymphoma cells.[6]

While in Philadelphia, zur Hausen contemplated another cancer-virus connection. He had read a 1967 article in *Bacteriological Review* that described cases of warts, possibly caused by papillomaviruses, that turned malignant.[7] In one case from 1940, a papilloma from the eyelid of a 71-year-old man was ground up and injected into the eyelids of monkeys, and after 60 days a tumor grew at the injection site. He did not take up the idea immediately, but he did start collecting similar anecdotal cases in the literature.

In 1969, zur Hausen was offered a senior scientist position in Würzburg by the director of the local Institute of Virology, Eberhard Wecker. By this time, he was increasingly skeptical of claims that cervical cancer was caused by herpes simplex virus, as many virologists suspected at the time.[8] (That sexually active women had a higher risk of cervical cancer was one reason virologists suspected herpesvirus.) Once in Würzburg, zur Hausen hired his childhood friend Heinrich Schulte-Holthausen, now a chemist, as his laboratory assistant. Together, they tried to confirm his idea that Epstein-Barr virus maintained its DNA in the host cell. In situ hybridization would be the tool to investigate. EBV could not be cultured directly, but by culturing cancer cells, they were able to purify EBV DNA and make a probe for the virus. The viral DNA had a different density and so could be separated from host DNA in a centrifuge. Using the probe, they showed that biopsies that did not produce virus nonetheless contained EBV DNA. It was an important result, which they published in *Nature* in 1970.[9]

Early Research on HPV and Cervical Cancer

In 1972, zur Hausen received an attractive offer to set up a center for clinical virology at the University of Erlangen-Nuremberg, and he took it. He moved his entire group. It was an opportunity to begin working on a possible HPV–cervical cancer link. Just based on epidemiology, zur Hausen and others knew that there was some link between cervical cancer and sexual activity. He suspected that the virus normally caused nonmalignant warts, but when it was present in the cervix, it might be more malignant. He contacted dermatologists, who supplied his group with a large number of warts.

He ground up foot warts with a mortar and sea sand, purified the virus DNA, and made complementary RNA probes. One of the first results from his approach was that the virus that causes foot warts is different from the virus that causes genital warts since the DNA of virus-positive papillomas (genital warts) did not hybridize with the cRNA probes. Zur Hausen took the data to mean that there was a larger family of papilloma viruses with a number of genetically distinct strains. Schulte-Holthausen and zur Hausen also made probes for herpesvirus and tested more than 100 cervical cancer samples for herpes, coming up with negative results.

An opportunity for zur Hausen and his team to present their results and speculation occurred in December 1972, when the American Cancer Society organized its Conference on Herpes Virus and Cervical Cancer in beautiful Key Biscayne off the coast of Miami, Florida.[10] A number of heavyweight biologists were present, including Francis Crick, Sol Spiegelman, George Klein, Werner Henle, and Peter Wildy. The group converged on the Key Biscayne Hotel with hopes significantly increasing understanding of the cause of cervical cancer. Many were convinced it was herpes simplex virus 2. Such a viral cause would explain the fact that female sex workers suffered more frequently from cervical cancer than nuns did.

Zur Hausen told the group of his and Schulte-Holthausen's fruitless search for herpesvirus DNA in cervical cancer samples. He explained that he used the same methods that had produced positive results for EBV in nasopharynx (throat) tumors, so it was not likely the approach itself. Rather, if cervical cancer were contagious, another virus might be implicated. He suggested HPV was worth investigating as the possible culprit. He mentioned the anecdotal reports of malignancies due to warts, and all the virologists in the room would have been aware that Shope papilloma virus causes cancer-like growths in rabbits (see chapter 1).

Negative Reaction to zur Hausen's Work on HPV

The response from the virologists was the complete opposite of what zur Hausen had hoped. Nobody was convinced. Admittedly, in effect zur Hausen was telling many in the room the work they had been doing for years was misguided, so perhaps it was not surprising that zur Hausen's presentation fell flat. There was silence in the room until University of Chicago herpes virologist Bernard Roizman responded.[11] He suggested that zur Hausen's techniques were not sensitive enough to detect herpes DNA. He then went on to report that he had actually detected a fragment of herpes DNA in a cervical tumor. Zur Hausen asked him for a specimen so he could also study it. Roizman said he would send zur Hausen some, but it never arrived. George Klein in his published written summary of the meeting devoted a couple of paragraphs to Roizman's results. He mentioned neither zur Hausen's negative results nor his positive suggestion that HPV might be the culprit.[12]

While despondent about the talk—he later called it a low point in his career—zur Hausen also saw that there was one upside: since he had convinced no one that HPV was the more important virus for studying cervical cancer, zur Hausen would not face much competition from other laboratories. Nobody else was jumping into the race. It also helped that the papers resulting from the symposium took a long time to make it into print. An editor suggested that zur Hausen split his remarks into two papers: the results critical of herpes were published in 1974, and the suggestion that HPV should be the focus was published in 1976. There still was some competition, however. The French laboratory of Gerard Orth was conducting research along similar lines, although more biological than medical, motivating the Germans to work quickly.[13]

Isolating viral DNA from a cervical tumor was exceedingly difficult. It was harder than isolating virus material from foot warts. One reason was that tumors removed from German women were commonly quite small. Zur Hausen decided to try to get cervical cancer biopsy samples and genital warts from Africa, where tumors were detected were and removed later and so generally larger. He and his old friend from the Henle laboratory, Volker Diehl, who had experience collecting Burkitt lymphoma samples in Kenya, spent three months obtaining samples in Southern Africa. They were shipped back to Germany and frozen for future work. Unfortunately, it was difficult to find a permanent reliable contact in Africa to ship him samples, so a longer-term research relationship between Germany and Africa did not develop.[14]

Evidence Based on HPV DNA

In 1977, zur Hausen was offered the chair of the Institute of Hygiene at the University of Freiburg, and he moved his team once again. He assigned Lutz Gissmann the task of purifying HPV from genital warts—despite their low concentration of virus, he could purify viral DNA by first purifying all the DNA present in the wart and then separating the denser viral DNA using an ultracentrifuge. Once purified, the viral DNA could be cloned into phage lambda, a well-studied bacterial virus, and amplified. Soon they had a number of different viral DNAs from foot warts that they began numbering and making into radioactive-labeled HPV DNA probes. In 1979, they succeeded in purifying an HPV sequence from a genital wart. They called it HPV 6. Since the probe DNA was made with phosphorus 32, which is radioactive, its binding to other single-stranded DNA fragments can be detected on photographic film. On some DNA blots, zur Hausen could see very faint bands suggesting some hybridization between viral DNA and biopsies, but disappointingly, HPV 6 did not appear to be present in cervical cancer cells. Better techniques were needed. Harald zur Hausen's team adopted the techniques and approach of the American virologist Peter Howley, who was studying bovine papillomaviruses using hybridization. Howley and his colleagues at the National Cancer Institute used conditions of low stringency hybridization to detect sequences that were close to the probe but not necessarily identical at every base. Using low stringencies, researchers could discover related strains of HPV.

The less stringent conditions helped the team find a number of reactivities between probes made from viral DNA in genital warts and cervical tumors. HPV 6 seemed promising but did not match the sequences in cervical biopsies. Zur Hausen directed Matthias Dürst, a PhD student, to investigate a rare laryngeal wart-like tumor of the airway. Dürst used an HPV 6 probe to investigate the DNA of the tumor and found a new HPV sequence that was later called HPV 11. In November 1982, Dürst and Gissmann were excited to learn that the HPV 11 sequence reacted with 11 of 18 cervical cancer biopsies that they tested. They rushed to tell zur Hausen, who exclaimed, "This is what we are looking for." They broke out some cognac to celebrate.[15] These 11 biopsies contained a novel HPV that was named HPV 16, which was common in German cervical cancer samples but less so in samples from Kenya and Brazil. HPV 16 was detected in roughly half of cervical cancers but in benign tumors at a much lower rate, suggesting it was related to malignant cervical cancer.

The zur Hausen team rushed to get their results into print, asking Gertrude Henle, who was a member of the National Academy of Sciences, to submit a paper to *Proceedings of the National Academy of Sciences*. The paper was completed in March 1983. The conclusion was tempered but strong: "The regular presence of HPV DNA in genital cancer biopsy samples does not per se prove an etiological involvement of these virus infections, although the apparent cancer specificity of HPV 16 is suggestive of such a role."[16] At last zur Hausen saw significant payoff from his search for a viral cause of cervical cancer.

The good news continued when an HPV DNA probe from an African sample called HPV 18 matched just over 20% of samples. It was looking like the majority of cervical cancer might have a viral cause. Gissmann and Dürst presented their findings in Orenas, Sweden, at the Second International Papillomavirus Conference. Like zur Hausen in 1972, they expected a positive response from fellow virologists, but instead received hostile questions suggesting that they had made an experimental error. Gissmann put it this way: "The whole show went down the drain."[17]

One reason for the reluctance to accept the idea that HPV caused cervical cancer stemmed from the volume of research over the previous decade aimed at showing that herpes simplex type 2 virus was the cause of cervical cancer. The virus that causes Lucké's tumors in frogs, the virus that causes Marek's disease in chickens, and Epstein-Barr virus are all from the herpesvirus group, so it seemed natural that there might be an oncogenic herpes that caused cervical cancer in humans. Additionally, there was epidemiological evidence for a correlation between herpes infections and cervical cancer.[18] Despite significant funding from the NIH and other granting agencies, no virus could be consistently found in cervical tumors, and virologists played with ad hoc "hit-and-run" hypotheses to explain the data: that herpesvirus could initiate cancer but was not needed to sustain a tumor's growth.[19] It was this background that was used to assess zur Hausen's work, which made it seem heretical to many scientists at the time.

Gissmann and zur Hausen's disappointment at the hostility directed at their presentation was alleviated somewhat when Peter Howley reported he had replicated their results. At the same time, additional epidemiological data suggested that herpesvirus was not the cause of cervical cancer.[20] Spurred by these emerging pieces of news, other researchers began to investigate the oncogenic properties of HPV at this time. Zur Hausen sent the cloned virus genomes, which could be propagated as noninfectious replicons in bacteria, to almost anyone who asked for them. A number of researchers violated the transfer agreement that required written permission from zur Hausen before sharing the probes with additional people—they gave samples to third parties—which frustrated zur Hausen but did not dissuade him from his open, sharing attitude.

Meanwhile, zur Hausen went to several German and Swiss pharmaceutical companies suggesting that HPV would be a good candidate for a vaccine. Executives at every company he approached said that they were not interested, except one. The Behring company in Marburg began sponsoring a program in 1983. Behring had to get approval from its parent company, which undertook a market analysis of the demand for such a vaccine and concluded that it was not profitable because most children are already infected. The program was shut down in 1985, to the frustration of zur Hausen and his colleagues. In the same year, zur Hausen and colleagues published a paper in *Nature* showing the HeLa cell line contained HPV-18 sequences implying that HPV had caused Henrietta Lacks's cervical cancer.[21]

The Scottish Australian Ian Frazer

Another researcher, working on the other side of the world, would pick up the torch. Ian Frazer was born in 1953 in Glasgow, Scotland, to scientifically trained parents. He was raised in Aberdeen and as a teenager was as drawn to science as his parents were.[22] He had a natural aptitude for physics, receiving the top grade for the country, but decided to follow in his father's footsteps and study medicine at the University of Edinburgh.

In 1975, Frazer completed a bachelor of science degree. He traveled to Australia on a working visa and interned at the Walter and Eliza Hall Institute of Medical Research in Melbourne, working on a project involving high blood pressure and vasectomies. He impressed Ian Mackay, who was head of clinical research. Nonetheless, Frazer returned to Scotland at the end of his internship. His girlfriend, Caroline Nicoll, was waiting for him, and they married in July 1976. His adult life was coming together as he graduated with his bachelor of medicine and bachelor of surgery, the equivalent of an MD in the United States, the following year and began to work as a resident physician in Eastern General Hospital in Edinburgh.

After a couple of years working in hospitals, he was considering beginning a DPhil in immunology at Cambridge University, when a telegram arrived from Australia.[23] Ian Mackay was offering him a job in Melbourne. Knowing the institute's excellence in immunology, he accepted the position of senior research officer in Mackay's laboratory, although he and his pregnant wife planned their sojourn as a temporary dislocation. They packed up their life and bravely moved to the other side of the world.

Frazer's new project was to work on chronic liver disease. He also wondered why some patients recovered from hepatitis B virus infections and others did not. Frazer worked hard as his family expanded, spending some time exploring Australia whenever possible. He remained interested in immunology and in 1981 was intrigued to hear from visiting Nobel laureate Baruj Benacerraf about a mysterious new disease that destroyed the immune systems of gay men in the United States. It would later be called AIDS. He wondered whether this disease was related to his HBV puzzle. He started studying HIV and HPV. An early success showed that HPV infections were involved in anal cancer in gay men.[24] As the AIDS epidemic grew, many medical researchers devoted themselves to the study of HIV, but, although he published on the new virus, his central interest was HPV.

With the retirement of his mentor Ian Mackay imminent, Frazer moved from Melbourne to Brisbane to become a senior lecturer at the University of Queensland and the leader of his own laboratory. Queensland was a step down from the Walter and Eliza Hall Institute of Medical Research in quality of medical research, but Frazer did not care what other people thought and was determined to plow his own path. His independence pervaded many aspects of his life. He had no hesitation in criticizing the Australian government's slow response to the AIDS crisis. His outspokenness led to a tense telephone call from the minister of health, who threatened to have him fired.[25]

At Queensland, Frazer worked to expand his small laboratory, aiming to create an expertise in HPV biology. Aware of zur Hausen's work on the link between HPV and cervical cancer, he thought that his research should focus on how the virus could become oncogenic. He obtained oncogenic HPV DNA from Lutz Gissmann.[26] What was the mechanism of transformation, he wondered? He hired Robert (Bob) Tindle, who had a PhD from the Imperial Cancer Research Fund and was an expert on making monoclonal antibodies. His first students joined the lab, and Frazer cultivated a laid-back, collaborative approach to science. He encouraged morning tea and casual Fridays and hosted barbeques at his house. The laboratory secured external funding from the Lions Medical Research Foundation. A major focus was the immune response of patients infected with HPV. One challenge was that he could not grow HPV in an experimental model animal, like mice. It was possible to make transgenic mice that could express HPV proteins, but Frazer did not have the skills to do this, so he arranged for a sabbatical at Cambridge University in 1989.

In Cambridge, he was a temporary member of Margaret Stanley's laboratory. Next door was Lionel Crawford's ICRF lab, which Frazer would often visit to borrow equipment like pipettes. It was in Crawford's laboratory that Frazer met the Chinese husband-and-wife scientists, Jian Zhou and Xiao-Yi Sun. The scientific relationship between Frazer and Zhou would shape the trajectory of the remainder of his career. Like Frazer, Zhou was interested in HPV. Although conversing in English was a little tortured, the two scientists had similar interests and were both extremely hardworking and passionate about understanding HPV. Frazer could see that Zhou's recombinant DNA approach might be superior to creating transgenic mice.

Zhou had succeeded in the Chinese scientific establishment despite difficult odds. When he was young, his parents were forced to relocate to the country as part of the Cultural Revolution, and Zhou and his sister had to fend for themselves in Hangzhou. After he finished high school, Zhou himself was forced to work in the country and then in factories. His ability to build working radios and television sets from spare parts in the factory caught his superiors' attention. Eventually, he was allowed to apply for tertiary study and was admitted into the prestigious Wenzhou Medical College in 1977. Although romantic relationships between medical students were banned at the time, Zhou and Sun fell in love. Zhou worked incredibly hard, even on his wedding day. He gained a master's degree and then a PhD on HPV at Henan Medical University. After a postdoc at Beijing Medical University, he was offered a fellowship in Crawford's laboratory, the first Chinese researcher Crawford accepted at the ICRF. Since Sun's work, too, was of high quality, Crawford offered her a 10-week position in the lab to assist Zhou. Her contributions are perhaps underappreciated, as she allowed Zhou to take all the limelight.

The couple's position was precarious, though. They had left their young son Zixi in China, and now the Chinese authorities would not issue him a passport to join them in the UK. Sun had left just as the 1989 Tiananmen Square protests were in full bloom, and she had to walk 20 kilometers with a suitcase on a bicycle to the airport since martial law had shut down transportation systems.[27] She had planned only a short trip to the UK to visit her husband. But now the Chinese funding for Zhou's work was in jeopardy. The political situation in China was inflamed by the Tiananmen Square protests, and consequently there was pressure to bring Chinese researchers home

15.1. Xiao-Yi Sun, Jian Zhou, and Lionel Crawford celebrating the planned move to Australia. Image courtesy of Lionel Crawford

from foreign countries. But they did not want to return to China. Crawford tried to help: he applied to extend their visas and also for funding from the ICRF, but the visa application at the Home Office was delayed, and the ICRF would not award the 29-year-old Zhou a research fellowship because he was too old for a short-term appointment.[28] (He was older than normal because the Cultural Revolution had disrupted his career plans.) Seeing that he could get two birds with one stone—help his friends and also gain highly competent staff—Frazer offered them positions in his lab in Brisbane, which they were glad to accept (figure 15.1).

Once in Australia, they were able to have their three-year-old son and his grandmother travel to Australia, reuniting the family. Frazer, Zhou, and Sun formed a harmonious scientific team, all working hard to make their mark. Zhou published eleven papers in the next four years, but a significant immunological tool against HPV infections had not materialized. That HPV could not be cultured in the lab was a major stumbling block. Unlike many other viruses that could be grown in cell culture, HPV needed to be isolated from infected humans.

The Creation of HPV-Like Particles

Frazer and Zhou tried a different approach. They would clone the genes responsible for the viral capsid, purify the capsid proteins without any other viral product, and then see whether artificial capsids could be created. There was some precedent for "empty" viral shells. Some viruses create them naturally, and they were the basis for the hepatitis B vaccine (see chapter 7). Two types of proteins make up the HPV viral capsid, L1 and L2. Building on successes Zhou had in the Crawford lab, Frazer and Zhou published results on a modified vaccinia virus that expressed L1 and L2 from HPV 16.[29] Once they had created the recombinant virus-like particles, they used sucrose density gradients to purify them. Initially they had used a different translational start site for the L1 protein, but after comparing the genomic sequences of different strains of HPV they decided to use the start site common to the other strains. This small modification allowed the production of L1. Once the viruslike particles (VLPs) were purified, they could be identified and characterized using an electron microscope in Deborah Stenzel's laboratory. The "viruslike" particles were 35–40 nanometers in diameter and roughly spherical, just like they predicted. They were not true HPV virion shells, however, as those have a diameter of 55–60 nanometers. Instead, the L1 had assembled into something slightly smaller and sloppier.[30] There clearly was more to learn about L1 and L2 self-assembly. Excited nonetheless, Zhou drove back to the laboratory to show Frazer. He said that the team had made a VLP, but Frazer replied, "no Jian, we've got a vaccine!"[31] Of course, Zhou knew this already as he had been thinking about a vaccine since his days in the Crawford lab.[32]

Frazer was more entrepreneurial than the typical scientist and knew that they needed a provisional patent to protect their discovery. On a general level, the next steps were obvious. First, publish the results as soon as possible to establish priority, and second, see whether the VLPs generate an immune response in mice, which would indicate that they were a good candidate for a vaccine.

They wrote up the paper, titled "Expression of Vaccinia Recombinant HPV 16 L1 and L2 ORF Proteins in Epithelial Cells Is Sufficient for Assembly of HPV Virion-Like Particles" and submitted it to *Nature*. Unfortunately, the reviewers did not appreciate the novelty and importance of their work, and it was rejected. This response was not completely unexpected as *Nature* has a high rejection rate. One possible reason for it was the absence of immunogenicity data, which are important for a prophylactic viral vaccine candidate. In other words, Frazer and Zhou did not have data to show that the particles produce high levels of antibodies, which in turn can prevent virus infection. Whatever the reason, the paper was then submitted to the specialist journal *Virology*, which accepted it. They thanked Lutz Gissmann for the HPV 16 sample. The end of the abstract summed up the goal of the Australian team: "The production of HPV-like particles using recombinant vaccinia virus should be useful for biochemical studies and could provide a safe source of material for the development of a vaccine."

The paperwork for a provisional patent was assembled quickly, and in July 1991, Frazer and Zhou left for Seattle to present their findings at the 10th International Papillomavirus Conference. For fun, Frazer began the talk with an image of an elephant infected with huge genital warts. The rest of the talk went well and "stunned" the audience, but many remained "skeptical" of their results.[33] Some of the skepticism evaporated once zur Hausen, who was in the audience, endorsed their research as a major step forward. Others were more interested in getting a copy of the elephant warts slide.[34] And a few in the audience, such as Douglas Lowy at the NCI, were influenced enough to follow Frazer and Zhou into making HPV VLPs.[35] Lowy, who trained at NYU and Stanford University had directed a laboratory at the NCI since 1975. In the late 1970s and early 1980s, he collaborated with Edward Scolnick (see chapter 12) to decipher the molecular biology of *ras* and oncogenes and then moved to study papillomaviruses.

Back in Australia, Frazer began to think about a corporate partner to help develop the vaccine. The Commonwealth Serum Laboratory (CSL) approached him with a partnership proposal, which appealed to him because he had a prior relationship with the lab. CSL had a long history, beginning in 1916, of distributing vaccines for polio, influenza, and other infectious diseases. He agreed to the proposal: CSL would supply Frazer with funding to expand his research group, and he would offer CSL any intellectual property that emerged from his research. The University of Queensland would also share in royalties from any patents. As the promise of using viruslike particles to make a vaccine became more widely appreciated, large pharmaceutical companies approached CSL with various proposals.

In 1995, CSL licensed Frazer and Zhou's work to the large American pharmaceutical company Merck & Co. The driving force behind Merck's decision to work with them was Edward Scolnick, who had investigated tumor viruses before joining the company in 1982 (see chapter 12). Scolnick had some reservations, as a herpes vaccine remained elusive despite promising results in animals. But he thought that since herpes entered the body through two different types of cells and HPV had only one (epithelial cells) to block, an HPV vaccine held more promise. Merck also had a "superb" team of HPV researchers and experience making a hepatitis B vaccine.[36] The Merck team began an HPV vaccine program in early 1991.[37]

The American Connection: John Schiller and Doug Lowy

Schiller, a PhD in microbiology from the University of Washington at Seattle, had begun at the NCI as a postdoctoral fellow in Doug Lowy's lab in 1983. Initially, they focused on the identification and characterization of the papillomavirus genes that could induce cancer-like characteristics in cultured cells. However, the arrival of Reinhard Kirnbauer as a postdoc in 1991 induced them to undertake a new initiative, the development of prophylactic vaccines. This initiative rapidly led to a 1992 publication of a paper in *PNAS* demonstrating that the L1 could, in the absence of L2, assemble into viruslike particles. As in authentic virions, five L1 molecules formed rings called pentamers, 72 of which then assembled into an icosahedrally symmetrical 555–560 nanometer sphere.[38] Importantly, their VLPs generated high levels of antibodies that could prevent cultured cell infection by authentic animal papillomavirus virions. By 1995, animal challenge studies conducted in rabbits, beagles, and cows by the NCI and other groups reinforced the promise of an effective human vaccine.

The Merck team worked to clone the structural gene L1 from HPV16 and HPV18 into yeast. In other words, the yeast *S. cerevisiae* was genetically modified to produce viral protein, which could then be purified. The L1 protein self-assembles into VLPs within yeast cells. In an independent commercial initiative, MedImmune, which later sold its interest to GlaxoSmithKline (GSK), produced L1 VLPs in insect cell cultures, using the production system originally developed by Lowy and Schiller. A battery of biochemical processes was used to purify the VLPs. A great deal of effort was invested in establishing the VLPs were pure and stable and production could be scaled up to make a vaccine.

Frazer and Zhou continued to work on HPV biology. Frazer organized the

15th International Papilloma Conference on the Gold Coast of Australia in 1996. To make sure the event was a success, he took out a second mortgage on his home, but he was reimbursed once the 500 participants paid their registration fees. It was judged one of the most successful HPV meetings ever in terms of scientific content and location.[39] Acting as organizers of this event cemented Frazer's and Zhou's importance in the field of papillomavirus biology.

Merck Clinical Trials of an HPV Vaccine

In 1997, Merck began a phase 1 human clinical trial for an HPV 11 vaccine. Phase 1 vaccine trials test only whether the drug is safe and immunogenic and are not concerned with efficacy. Merck randomized 140 women to either a placebo or vaccine wing of the trial. No significant adverse reactions were reported, and it was viewed as "highly favorable." A successful phase 1 study for HPV 16 was also completed. To test whether the vaccine could prevent an infection in humans, an efficacy study using the HPV 16 vaccine was planned. More than 2,000 women were enrolled, and the results were astounding. On the placebo arm, there were 3.8 HPV 16 infections per 100 women per year.[40] On the vaccine arm there were zero infections! At the same time, the World Health Organization reanalyzed old tumor biopsies and concluded that 99.7% of cervical cancer was due to HPV.[41]

With these positive results, Merck ramped up its HPV vaccine work and started phase 2 clinical trials on a vaccine that included HPV 6, 11, 16, and 18. (HPV 6 and 11 cause genital warts and not cancer.) The best dose of each monovalent HPV was refined, and the efficacy arm had 30 months of follow-up. Subjects came from Brazil, Europe, and the United States: 277 with the vaccine and 275 on the placebo.[42] The combined incidence of persistent infection or disease with HPV 6, 11, 16, or 18 fell by 90%. In 2001, a large-scale phase 3 trial was begun to test efficacy on cervical cancer and vaginal precancerous lesions. It would enroll 16,000 patients. Concurrently, GSK successfully performed similar multicentric phase 1–3 clinical trials of a bivalent vaccine containing HPV16 and HPV18 VLPs.[43]

Lawsuits over the HPV Vaccine Patent

During this period of exciting vaccine development, lawsuits hung over the ownership of the relevant patents. In 2004, Frazer was extensively questioned by American lawyers trying to establish who made the key discoveries. Georgetown University filed a provisional patent in June 1992, the NIH in September, and the University of Rochester in March 1993. The Georgetown group had shown that L1 expressed in a mammalian system could be bound by a monoclonal antibody that could prevent infection of cultured cells.[44] The NCI group had shown that bovine papillomavirus (BPV) L1 or HPV 16 L1 by itself self-assembles into a VLP when expressed in insect cells, and, as noted above, that the BPV VLPs induced remarkably high levels of antibodies that could prevent authentic virus infection of cultured cells.[45] The Rochester team had done similar work with HPV 11.[46] In 1997, rather than just grant the patent to one applicant, the US Patent and Trademark Office began an interference hearing, which allowed lawyers from the different groups to interview the scientists in question. The NIH lawyers argued that Frazer had not created a VLP because Zhou and Frazer used a prototype HPV16 L1 protein sequence and not the wild-type L1. The prototype L1 would not have been useful for a vaccine, because the NIH group later demonstrated that assemblies of that protein did not induce antibodies that prevented HPV16 infection. Initially, the three US groups argued against the Australian team.[47] Merck sided with the Australians. GSK wanted Rochester or Georgetown to win because it had exclusive rights to their applications. The NCI had issued nonexclusive licenses to both companies. It was turning into a battle of giants.

Unfortunately, Zhou, who had done the benchwork and would best have been able to respond to the charge that he had not used a wild-type HPV, was not there to answer the lawyers' questions. Tragically, he died in 1999 of septic shock. He was only 42. Frazer and the Australian team tried to find answers in emails stored on a mothballed computer system. They were able to retrieve the email, but it did not help. In a preliminary ruling in August 2004, the Patent Office ruled against the University of Queensland and Frazer.

Before the final ruling, Merck, GSK, and the NCI made an agreement to cross-license the technology. Both Merck and GlaxoSmithKline had each invested many millions in clinical trials, and they both wanted to minimize their risk. Under the agreement announced in February 2005, CSL would receive royalties at various research milestones.[48] It was an international agreement among companies from three continents.

In September 2005, the verdict came out. The Australians lost. The CSL lawyers were upset.[49] Because of the cross-licensing agreement, CSL and the University of Queensland would receive royalties, but they would have been much higher if the Australians had won. The case had hinged on whether Zhou used wild-type L1 protein, samples of which were deposited with the American Type Culture Collection in 1992, but lawyers had advised the Australians not to test the samples in the event that they proved that it was not. Now there was nothing to lose, so CSL lawyer John Cox decided to have the samples tested.[50] To the good fortune of the Australians, the samples were in fact wild type. The CSL legal team debated appealing the patent decision as it was not clear whether the new finding would be admissible since appeals generally have to address a new issue not decided at the lower-level court. Ultimately, an appeal was lodged in September 2006. The FDA had approved the Merck vaccine Gardasil for four strains of HPV for use in the United States in June of that year.

In November, the Australian prime minister and minister of health announced that Gardasil would be offered free to Australian girls aged 12–18 and also to older women through their general practitioners. In total, 1 million Australian women would be covered. Frazer gave the first dose to an Australian woman in a widely reported event.

In the meantime, Ian Frazer was named "Australian of the Year" by the Australian government (figure 15.2). Macfarlane Burnet, the great Australian virologist, had won the award in 1960, as had other great Australian scientists, but it was open to anyone. Paul Hogan, the actor who played Crocodile Dundee, had won it, as had four different captains of the Australian cricket team over the years. The 2006 award gave Frazer a lot of exposure to the Australian public. He would carry the torch into the 2006 Commonwealth Games, an event the rivals the Olympic Games in countries of the former British Empire.

Resolution of the Legal Challenges

On August 20, 2007, the Australians got some good news. The appeals court for the federal circuit ruled in their favor. The demonstration that a wild-type L1 sequence was deposited with the American Type Culture Collection was enough to prove Frazer and Zhou had priority, and so the patent was awarded to the University of Queensland. Circuit Judge Pauline Newman wrote, "We conclude that Frazer was entitled to the priority date [1991] of the Australian patent application. Since the Australian filing date antedates any date alleged by Schlegel [of Georgetown], priority must be awarded to Frazer. . . . We conclude that the Board erred in denying Frazer's

15.2. Ian Frazer receiving the 2006 Australian of the Year Award from Prime Minister John Howard. Image courtesy of National Australia Day Council

entitlement to the date of the Australian patent application. The Australian application contained complete details of the method."[51] Luckily for the Australians, the NIH also abandoned all further patent prosecution. NIH decision makers, including Lowy and Schiller, believed that the NIH's focus on public health was best served by removing all ambiguity regarding the freedom, from an intellectual property perspective, of both pharmaceutical companies to commercialize their vaccines.[52] The US patent (number 7,476,389) would last 17 years. Consequently, hundreds of millions of dollars flowed to CSL and the University of Queensland from Merck and GlaxoSmithKline as girls and women in more than 100 countries took the vaccine. Zhou's estate and Frazer also received a slice of the revenue. It would make Frazer a wealthy man.

New revenue did not change Frazer's focus, however. He continued his cancer research in Queensland. He worked on a therapeutic vaccine for HPV for people already infected with the virus. He helped get the vaccine distributed in Vanuatu and Bhutan and other countries with minimal cervical cancer screening programs. He also lobbied the government for more spending on science. Spearheading a new building and center dedicated to translational research at the University of Queensland displayed Frazer's talent for the administrative side of science. After the new center was built, he was picked to head it. Awards started to stack up, including the William B. Coley Award with Harald zur Hausen in 2006 and the Australian Medical Association Gold Medal in 2009. He was made a fellow of the Royal Society in 2011.

Despite his premature death, Jian Zhou also achieved fame, especially in China and Australia. He had set up an exchange program between the University of Queensland and Wenzhou Medical School, the latter of which eventually erected a bronze statue of him.[53] At the time of his death, he was celebrated as a great Australian, and Prime Minister Kevin Rudd provided a tribute. In 2013, Zhou's family used money from the patent to create a foundation to fund young Chinese scientists seeking to study abroad. Xiao-Yi Sun still lives in Australia and works as a trustee of the Jian Zhou Foundation.

Douglas Lowy and John Schiller continued to work at the NCI. Lowy rose to the position of acting director of the NCI in 2015. They initiated the only publicly funded efficacy trial of an HPV vaccine, developed key assays for evaluating the immune response to the vaccine, and provided insights into how the antibodies generated by the vaccine prevent infection. They received significant recognition of their role in developing the vaccine with several awards including the National Medal of Technology and Innovation from President Obama in 2014 and the 2017 Lasker Award (figure 15.3).[54] The Lasker Foundation recognized their "technological advances" contributing to the HPV vaccine including the strategy of working out many of the details of vaccine development using the bovine form of the virus. Somewhat surprisingly Frazer was not included. If a Nobel Prize is awarded for the HPV vaccine, the Nobel Committee will face the difficult issue of ranking scientific importance and credit to the different groups that each provided necessary steps in the vaccine development. As the 2017 Lasker Award shows, the issue of scientific credit is not necessarily resolved by the legal resolution of a patent priority dispute.

In 2008 Harald zur Hausen won the Nobel Prize for discovering the link between HPV infections and cancer (figure 15.4).

Between 2006 and 2018, more that 200 million doses of Gardasil were distributed in the United States.[55] The most current version includes a vaccine against nine strains of HPV. There has been some political controversy about the vaccine, though. In 2007, as governor of Texas Rick Perry, who later ran for president, issued an executive order requiring girls to get the vaccine,

15.3. Douglas Lowy (*left*) and John Schiller (*right*) receive the National Medal of Technology and Innovation from President Barack Obama in 2014. Image courtesy of AP Photo/Evan Vucci

but Texas legislators reversed that decision. Many Republicans and especially libertarians dislike government forced vaccinations. (It was later suggested that Governor Perry had financial ties to Merck.) Others motivated by religious reasons believe the vaccine encourages more premarital sex, although studies clearly show that it does not.[56] The anti-vaccine movement also demonizes it to some extent, but the number of verified life-threatening adverse reactions remains low despite anecdotal accounts.[57] By January 2020, 147 people had been compensated for injury by National Vaccine Injury Compensation Program among 132 million doses given.[58] The WHO, FDA, CDC, and European Medicines Agency all judge the vaccine to be worth the risk. There are contrarians, including to the dismay of many scientists, Luc Montagnier of chapter 14, who wrote a preface to an anti–HPV vaccine book in 2018.[59] Merck was also criticized for marketing the vaccine only to young women and ignoring men-to-women and men-to-men transmission.[60]

In the United States alone, 19,000 women get cancer from HPV infections per year. For men, the figure is 8,000, mostly oropharyngeal (a form of throat, tongue, and tonsil cancer) cancer due to oral sex.[61] By 2017, 60% of

15.4. Harald zur Hausen at the Nobel Prize ceremony in Sweden. Image courtesy of the Nobel Foundation. © Nobel Media. Photo: Ulla Montan.

teenage boys and girls were receiving the vaccine in the United States. Globally 47 million women had received a full course of the vaccine by 2014. The rate is higher in developed countries and lower in parts of Africa and Asia. Looking at who had been vaccinated from 2006 to 2014, researchers estimate that 445,000 cases of cervical cancer and 184,000 deaths have been averted.[62] As more young people receive the vaccine every year, those numbers are predicted to continue to climb. China licensed the vaccine in 2017. Ian Frazer's biographer called him a man who saved millions. We should also include zur Hausen, Zhou, Lowy, and Schiller, among others. More distally the scientists of the earlier chapters, who developed the field of tumor virology so that successful vaccine research was possible, deserve some credit also. As the twenty-first century progresses and more girls, boys, men, and women get vaccinated, millions of people will be saved from cancer, and a century of foundational basic research in tumor virology will have paid off immensely.

Conclusion

Patterns in a Century of Research

These overlapping series of episodes in the history of tumor virology illustrate several themes in the growth of science and technology and help us derive more general conclusions extending beyond isolated case studies.[1] Given a century of research, there are numerous patterns and themes that one might consider, but I will sketch only the most prominent. Some themes can be found more generally in science—that sexism or military conflicts affect the progress of research—whereas other themes are more specific to tumor virology.

The Power of Tumor Virology

The most prominent theme is the centrality of the study of tumor viruses to the history of molecular biology.[2] Biologists studying tumor viruses made a number of highly significant discoveries, among the most important being oncogenes and RNA splicing. The former opened up an understanding both of the molecular basis of the cell cycle as well as the molecular basis of cancer; the latter revealed a new process of gene regulation found in all eukaryotic organisms. The unmutated versions of oncogenes, proto-oncogenes, gave biologists experimental access to important signaling pathways found in animal cells. Indeed, tyrosine kinases such as Src and Ras GTPases are crucial links in fundamental cellular control pathways, and subsequent work showed how the pathways extend causally upstream and downstream.[3]

Relatedly, biologists have made a vast amount of progress in uncovering the causal relationship between viruses and cancer. At the beginning of the twentieth century, nobody understood the internal composition of viruses, and the proximate causes of cancer were mostly mysterious. To understand the relationship between viruses and cancer, researchers had to also understand more about viruses and the nature of cancer itself. It took until the mid-twentieth century for the modern conception of viruses to come into focus: that they are cellular parasites that consist of nucleic acid (either RNA or DNA) and a protective protein coat. The discovery that the transforming part of the virus was located in the nucleic acid was a significant advance because it was evidence that the proximate cause of cancer could be genetic. The field of tumor virology grew tremendously in the 1960s and 1970s in terms of personnel and new experimental techniques, and the reductive power of molecular biology allowed the transforming potential of tumor viruses to be more precisely located, first to a part of the viral genome, then to a particular gene for some tumor viruses, and finally to a mutation in a gene.[4] The reductionist program, however, was not complete.[5] With some tumor viruses, such as hepatitis B virus or Epstein-Barr virus, their cancer-causing potential has a more complicated etiology, and the simple oncogene view does not emerge from the data. That seven Nobel Prizes were awarded directly or indirectly for work in tumor virology illustrates the impact of the field on biomedical understanding.

Many theoretical and practical scientific advances made by molecular tumor virologists took decades to emerge. Long-term sustained attention to basic research eventually paid off with cancer vaccines and an accelerated response to the AIDS pandemic. Achieving these advances took decades of financial and research pressure brought to bear on questions of fundamental biology: How do tumor viruses replicate? What do different parts of the viral genome do? Can reliable assays be constructed to run repeatable quantifiable experiments? Peyton Rous saw that new techniques were needed to build on his early insights, so he put tumor virus research aside. True believers like Ludwik Gross pushed unfashionable ideas; his underlying faith that practically all cancer was caused by viruses drove him to discover polyoma virus, which provided a model experimental organism that many laboratories could investigate. Dulbecco's and Temin's introductions of quantitative assays allowed tumor virology to replicate the successes of quantitative bacteriophage biology. Tumor virology also benefited from the new experimental techniques of molecular biology. At the same time, tumor virologists helped refine several new laboratory techniques in the 1970s and 1980s, as seen in the popular handbook, *Molecular Cloning: A Laboratory Manual*, published by Cold Spring Harbor Laboratory Press. That significant advances can take many decades of sustained work and the development of efficient experimental techniques is a lesson relevant to funding agencies like the NSF that require success in much shorter timeframes at the potential expense of larger, longer-term projects.

On Planned and Unplanned Consequences of Tumor Virology

Basic research can have unexpected practical and theoretical consequences. This mantra is well illustrated by the history of tumor virology. Gallo's search for a human retrovirus that causes cancer led his laboratory to develop cell culture techniques that shortened the development of a blood test for HIV by months, if not years. A good case can be made that the many millions of dollars spent by the Special Virus Leukemia Program of the National Cancer Institute were not wasted, even though the researchers they funded did not find the major human tumor virus they were looking for. Money was distributed to excellent experimentalists freeing their laboratories to pursue research agendas they deemed productive. The development of interleukin 2 and the refinement of reverse transcriptase into a research tool were both advances driven by tumor virology that had much wider benefits. Jumps in theoretical understanding also occurred. The discovery of RNA splicing by way of an unexpected phenomenon uncovered by investigating the transcription patterns of viruses has wide theoretical scope. Splicing is found throughout eukaryotic life, including in humans, where it regulates gene expression. Perhaps because of the constraint that only a limited number of genes can fit inside a viral capsid and the long history of coevolution of animal viruses and animal cells, the reductionism inherent within tumor virology was more likely to yield important jumps in understanding by locating and prioritizing important genes and biological mechanisms.

Taking a narrower focus, the favored explanatory hypotheses championed by tumor virologists in the 1970s have come to dominate much of contemporary basic cancer research. That cancer is caused by mutated genes producing malfunctioning proteins whose normal versions are involved in the normal cell cycle is the ascendant view of cancer today. Given this view, the search for oncogenes and their effects and regulation has become the dominant research program.[6] And these theoretical insights have moved into medical practice, with oncologists increasingly utilizing results from basic research.[7] As Harold Varmus observed, clinical oncology and the molecular basis of cancer were investigated by different communities of medical personnel who did not significantly interact in the twenty years between 1970 and 1990. From 1990 until the present, however, their interactions have deepened.[8] In the clinic, a specific course of treatment is becoming routinely guided by biopsying human cancer patients and sequencing the DNA from the tumor cells to see which oncogenes are mutated. Similarly, pharmaceutical companies actively use knowledge of oncogenes to narrow the search for new cancer drugs. That being said, some have argued that progress in treating cancer has been much slower than various cancer researchers predicted and that the biomedical community has a long way to go before basic biological understanding leads to highly effective, widespread treatments for cancer beyond vaccines.[9]

Model Systems in Tumor Virology

While the title of this book brings to mind viruses as the object of research, a better description of the primary research object of twentieth-century tumor virology is the system involving *both* viruses *and* host cells. Biologists work to refine a system that coordinates the causal interactions between viruses and cells. Indeed, a complete history of tumor virology, or animal virology more generally, cannot be disentangled from a history of cell culture.[10] Green and Todaro's creation of the 3T3 cell lines in the 1960s made mammalian cells much easier to use. Likewise, Michael Stoker and Jan Svoboda contributed useful cell lines. Gallo's discovery of interleukin 2 in the 1970s drove the study of T-cell viruses. Developments in cell culture were important in opening new avenues of research. The ability of Epstein and Barr to culture white blood cells was vital to the growth of EBV research. Even in the dispute between Montagnier and Gallo about who discovered the AIDS virus, the importance of the creation of cell lines is evident. Gallo and his coworkers deserve more credit on this point. All this work formed the backbone for further investigations that used cell lines as a tool to manage viral stocks.

The importance of model systems in twentieth-century biology and medicine is also apparent.[11] Rous sarcoma virus and various cultured animal cells were the most important model systems for RNA tumor virology. SV40, polyoma virus, and adenovirus and their associated cell lines played a similar role in DNA tumor virology.[12] A focus on a limited number of viruses for basic research, which marked the earlier phage group, continued when animal virology ascended and bacterial virology declined. Such specialization created synergy among research groups from around the world. Communities of scientists emerged around the study of a particular virus, or virus family, whether it was EBV, HPV, HTLV, or HIV.

Sexism and Recognition in Twentieth-Century Biology

The history of tumor virology also reveals the challenges of being a woman biologist in the twentieth century.[13] Sarah Stewart struggled for recognition in tumor virology, eventually finding a place in a national laboratory, but not before she was first assigned to gynecology. Marguerite Vogt did not seek the spotlight, but it is nevertheless significant that Renato Dulbecco received the lion's share of the accolades for work that they did together. That she believed that she needed to choose between being a pure scientist and being a woman striving for fame is itself a symptom of the sexism of the time. Dulbecco did not have to choose between science and fame; he could have both. The fact that Vogt worked long hours in the laboratory while he presented their results and won awards is a stark reminder of the challenges that women scientists faced that the men simply did not. The omission of Louise Chow from the 1993 Nobel Prize awarded for RNA splicing was probably in large part due, as her former mentor Norman Davison suggested, to her to being viewed as a quiet, Asian woman, and not advocating for herself as strongly as Roberts and the other men did. What Davidson omits is the likely possibility that, had she advocated for herself, a combination of racism and sexism would have made her self-advocacy less likely to pay off than it did for Caucasian men, all other things being equal. Yvonne Barr dropped out of research science—it did not seem possible for her to join the "boys club" of research and raise a family. Even Xiao-Yi Sun received only a small fraction of the recognition that her husband Zhou did.

The competition for credit and lack of agreement about who should be most rewarded for major discoveries is a recurrent theme. That biologists vie for credit is not surprising to biologists or historians of science, even though the public is rarely aware of the disputes behind the scenes. When a discovery is Nobel Prize–worthy, then otherwise private jostling is often pushed into the public sphere. Stehelin's objections to being left out of the 1989 Nobel Prize for Medicine and Physiology are a case in point, as is the dispute between Gallo and Montagnier over the discovery of HIV. When more than three people have contributed to a discovery, inevitably someone is left out of the Nobel Prize. In some of these cases, writing the history of science is especially difficult as the actors do not agree on their relative contributions, and even the chronology of events can be contested. In extreme cases, there are no extant documents to settle the dispute. Neither are disputes necessarily extinguished when the Karolinska Institutet awards a Nobel Prize or a court of law rules on a patent dispute as in the case of the HPV vaccine. It is thus tempting to speculate that the Nobel Prize does more harm than good for science—it certainly can create long-lasting fractures in scientific communities.

The Influence of War

It is also noteworthy how military conflict has shaped the history of tumor virology. Scholars have studied how war accelerates research into new areas. Think of the atomic bomb and atomic physics, for example.[14] However, my point here has more to do with how war modifies the social graph of scientists. World War II changed Gross's career, forcing him to emigrate to the United States and serve in the US military. Dulbecco was motivated to do good in the face of German atrocities. His collaborator Vogt also fled the Nazis, as did George and Eva Klein, who settled in Stockholm, and Werner and Gertrude Henle, who ended up in Philadelphia. Stoker found his way to virology in the Royal Army Medical Corps. Shope was given virological assignments by the US Navy. The boost that the Allies got from German scientists emigrating west had a flipside: the slow start that molecular biology had in Germany after World War II.[15]

The next generation of virologists were similarly shaped by the Vietnam War. The US Public Health Service Commissioned Corps—the so-called Yellow Berets—provided a way for talented physicians like Michael Bishop, Harold Varmus, Edward Scolnick, Robert Gallo, and Harvey Alter to contribute to their nation while avoiding service in Vietnam.[16] One highly visible leader of the federal response to the AIDS pandemic and the COVID-19 pandemic caused by SARS-CoV-2, Anthony Fauci, was also a Yellow Beret.[17] Although two years of service at the NIH was enough to avoid the draft, many Yellow Berets devoted themselves to research for much longer. Without the war, many talented physicians might have made a career outside of basic research. Given how competitive the US Public Health Service program was—roughly 100 physicians accepted from a potential pool of about 15,000—this program turned some of America's best young physicians into productive research scientists. Nine Nobel Laureates in the period 1985–2007, including Varmus and Bishop, were former Yellow Berets.[18]

Institutional Factors in the History of Tumor Virology

While this volume has concentrated on the biographies of scientists, the cases described here also illuminate the larger institutional structures that support the development of science. Let me focus here on three institutional factors. First, a small number of research institutions played a disproportionate role in this history. In the early days, the Rockefeller Institute for Medical Research supported Peyton Rous's and Richard Shope's research by providing the laboratories and funding that kept tumor virology alive. Bittner would not have been able to conduct his research without the standardized mouse lines developed and maintained by Jackson Laboratory in Maine.[19] With the molecularization of tumor virology in the 1960s and 1970s, the Cold Spring Harbor Laboratory in the United States and the Imperial Cancer Research Fund in the United Kingdom played important institutional roles in the research itself, including training biologists in the latest techniques. The CSHL provided a location for many important meetings that disseminated the latest results and could be used to establish the priority of findings. During the 1960s and 1970s, the tumor virus community was split into the RNA tumor virus community and the DNA tumor virus community. Both had regular meetings at CSHL. To be a player in the field meant regular pilgrimages to that biological mecca. The canon of the field was crystallized by the publication of CSHL Press volumes such as *Readings in Tumor Virology*, edited by Harold Varmus and Arnold Levine, a retrospective reprinting of landmark papers.

A second institutional factor that becomes apparent from this history is the importance of certain journals in establishing priority for discoveries. In the early days, the journal that Peyton Rous later edited, the *Journal of Experimental Medicine*, was a prestigious place to publish results. Later in the twentieth century, *Nature*, *Science*, and then *Cell* assumed that role for biomedicine. Before the mid-1990s, given the fast turnaround time and the near guarantee of a paper's acceptance after submission by a member of the academy, the *Proceedings of the National Academy of Sciences* was a popular place to establish priority for biologists in a close race with other labs. Professional connections and friendships with academy members were an advantage in promoting one's research agenda during this time. It was not enough to be a capable bench scientist. Being connected in the right way was instrumental to become a leader in the field.

The US federal government, especially the National Cancer Institute, also played an important role in driving the field forward. Throughout the century, government scientists, such as Ludwik Gross, Sarah Stewart, Robert Huebner, Edward Scolnick, Robert Gallo, and Harvey Alter, were often the first to make new discoveries. As mentioned above, the US Public Health Service Commissioned Corps also played an important role in training talented physicians who wanted to make the switch to research science. Even though it did not find a widespread human tumor virus, the NCI Special Virus Leukemia (or Cancer) Program funded many of the most important laboratories in the 1960s and 1970s. Here, Huebner played an important administrative role: having a virologist with significant administrative power and large funds to distribute made a difference. His willingness to fund work in the veterinary sciences allowed studies on nonhuman tumor viruses to influence human medicine. Perhaps this is not too surprising, as the discovery of Gross murine leukemia virus in mice and feline leukemia virus in cats were the very reasons he and others hoped and expected to find an equivalent human leukemia-causing virus. And as I argued in chapter 14, the NCI funded Gallo's work on human T-cell leukemia virus, which ultimately greatly hastened the response to the AIDS pandemic. It is remarkable how great a role the US government played in the progress of tumor virology for decades. The government's desire to "win the war" against human cancer meant research funding was channeled into basic virological research.

Continuities and Discontinuities among the Case Studies

Looking at the history of molecular biology from the 1940s to the 1970s, it is natural to see continuity between earlier work on bacterial viruses and later work on animal viruses. Many of the central players saw the cutting edge of research in molecular biology move from prokaryotic bacterial systems to eukaryotic animal cells; they themselves made the same shift. Renato Dulbecco is a pivotal figure here. By inventing a way of precisely measuring infectivity of animal viruses modeled on the plaque assay for bacteriophages, he allowed the "phage school" approach to bacterial genetics to be imported into animal virology.[20] One significant difference was the need for techniques to make immortal cell lines, which would play the functional role that bacterial cells did in earlier work. Compared to the creation of immortal cell lines derived from animals, host bacteria like *E. coli* are relatively easy to grow; animal virology benefited from earlier work on such simpler systems (phage and bacteria). The reductionistic trend to examine smaller and smaller parts of systems contrasts with the transition to examining more complex systems (animal viruses and eukaryotic cells).

There are significant historical connections among people, laboratories, and approaches among the various episodes. Many researchers appear in multiple episodes. In some cases, an entire laboratory approach was replicated in a new place.[21] In an important sense, Watson's creation of a center for tumor virology at CSHL represented a continuation of Dulbecco's West Coast group, and both Watson and Dulbecco inherited an approach to virology explicated by their mentor Luria. Other figures like Harald zur Hausen worked with Epstein-Barr virus but also made extensive contributions with HPV. Additionally, meetings such as the Gordon Conferences and CSHL meetings brought the community together, allowing major players to interact. Conferences fostered the development of scientific communities that in turn allowed for new ideas, both published and unpublished, to be quickly propagated.

Interestingly, the hepatitis B virus story is less connected to the others. The reasons for the relative isolation of HBV research include it being driven more by medical and immunological concerns than by the study of basic mechanisms of viral reproduction. HBV does not contain oncogenes as RSV and many other tumor viruses do, and the causal relationship between viral infection and carcinogenesis is more complicated. Relatedly, HBV research did not generate comparable new basic techniques that could be exported to other areas of biology.

Finally, having a connection to the larger virology community helps keep researchers aware of the latest techniques and approaches. Jan Svoboda, isolated behind the Iron Curtain in Czechoslovakia, had only partial access to the RNA tumor virus community in the West. This isolation limited the impact and reach of his research. Had he worked in the West and had access to the newest tools of molecular biology, he might have been an even more significant figure.

In addition to the continuities in the histories, there are discontinuities too. Our understanding of the nature of viruses changed radically over the century. At the beginning of the twentieth century, a virus was considered akin to a living poison, but by mid-century, a modern conception came into focus: a virus was a packaged piece of DNA or RNA encoding genes that allowed the virus to replicate itself inside a host cell.[22] Many virologists did not think viruses were alive. Despite this radical change, the innovations of early virologists often still proved useful to later virologists. Rous's work in purifying RSV, for example, was necessary for later study of the virus—even if Rous and everyone else at the time had little idea about their internal structure and composition. Likewise, the dominant conception of cancer changed significantly. First seen as being caused by an external irritant, it is now largely viewed as a genetic disease caused by mutated control genes. Here the conceptual change was more consilient.[23] Put simply, newer theories of cancer could incorporate significant aspects of both the external-cause and internal-cause views. Since mutations can be caused by mutagenic chemicals and other factors in the environment, cancer incidence is also related to the environmental context.

One can also find a variety of philosophies of science or conceptions of the scientific method guiding the various scientists in different episodes. Peyton Rous and Baruch Blumberg, for example, emphasized the careful observation of nature. Rous had no conception of the internal structure of the virus and in early years was reluctant to even use the word "virus" to describe his "agent." Blumberg's project on the geographical distribution of blood proteins was fieldwork as much as it was bench research. Robert Tjian, who investigated SV40 T antigen, explicitly approached science in a Popperian fashion: aiming to make falsifiable hypotheses. Across a century of research one can see a wide variety of methodological approaches and theoretical commitments. Ludwik Gross was motivated by the view that all cancer was virally caused; others were less optimistic, and the contemporary view is that only about 20% of cancer is caused by viruses.

What is remarkable—and perhaps one of the central strengths of science itself—is how the output from various researchers differing in theoretical and methodological orientation can still supply information needed for other researchers to drive additional investigations forward. *Scientific advancement happens despite, or in some cases, because of, methodological pluralism.*[24] The pluralism need not lead to metaphysically disunified science as asserted by the philosopher of science John Dupré or incommensurable paradigms as Thomas Kuhn famously suggested.[25] Rather diversity of approach and method can strengthen the entire field.[26] Some scholars call a pluralist, heterogeneous approach to scientific change "bricolage" and have observed it in the transfer of insights or methods between different disciplines. But here we see a form of bricolage within the same field over long periods of time.[27] Perhaps it is more common than we think.

The End of the Beginning: A New Era of Virology

More than a century after Peyton Rous's research on the sarcoma agent, a new phase in the history of virology is underway. What is called "viral oncolysis," or "virotherapy," is emerging as a new focus of study and hope in the struggle against human cancer.[28] The goal is to reengineer viruses to selectively target and kill cancer cells. Since viruses can selectively attach to certain proteins on the surface of cells, in theory, they could be engineered to attack cells that have distinctive surface proteins, as many tumor cells do. A popular virus to reengineer is adenovirus, but biologists are also experimenting with herpesvirus, poliovirus, and reovirus, among others. One major challenge is that the human immune system attacks engineered viruses before they have enough time to work. There have been clinical trials on adenovirus ONYX-015, an engineered virus with an inactive E1b gene, which when active is necessary to infect normal cells but is not necessary in tumor cells with a damaged p53 gene.[29] The hope is that it will be able to replicate only in tumor cells. While the engineered virus did gain approval for therapeutic use in China, it has not been approved in the United States. A genetically modified herpes virus was approved by the FDA in 2015 (drug name: talimogene laherparepvec) to treat inoperable melanoma. Additional work is needed to demonstrate the power of such an approach to fight the war on cancer. More than 50 clinical trials are now active on a variety of oncolytic viruses.[30] The better biologists understand the mechanism of oncogenesis, the more likely we will be able to domesticate viruses to effectively fight cancer cells. Virology is entering what might be called a new agricultural era, involving extensive genetic viro-engineering conducted in a manner that is more international, more collaborative, less sexist, and with new, additional institutional structures supporting the research. While there is still much work to do, the age of pure virus hunting is over.

Acknowledgments

I thank Bob Olby, who fostered my interest in the history of biology when I was a history and philosophy of science graduate student at the University of Pittsburgh, and Roger Hendrix, who introduced me to the joys of virology in his bacteriophage laboratory. Jim Watson provided the impetus for me to write this book in 2012. I thank him for convincing me that the history of tumor virology was an important constitutive strand in the development of molecular biology and contemporary medicine that is currently underappreciated by historians of science and researchers in biomedicine, and for his expert advice in how to approach this vast topic.

I thank Mila Pollack and the Cold Spring Harbor Laboratory Library and Genentech Center for History of Molecular Biology, who awarded me the Sydney Brenner Fellowship that supported my travels around the United States to interview American virologists and visit archives. I also thank Doron Weber and the Alfred P. Sloan Foundation, who supported my writing in the summers of 2016 and 2017 and allowed me to travel to Europe to interview European virologists. Many people generously shared information: Jim Watson, Arnie Levine, Geof Cooper, Max Essex, Bob Weinberg, Phil Sharp, Richard Roberts, Ed Scolnick, Bob Gallo, Ian Frazer, David Livingston, Ed Harlow, Augusta Gross, Michael Bishop, Peter Duesberg, David Baltimore, Peter Vogt, Hamilton Smith, Paul Berg, Bob Pollack, Steve Martin, Harry Rubin, Nancy Hopkins, Joe Sambrook, Mary Jane Gething, Tony Epstein, Carel Mulder, Howard Green, Wally Gilbert, Jim Darnell, Harold Varmus, Os Jarrett, Lionel Crawford, Robin Weiss, Robert Tjian, Tom Maniatis, Jennifer Stoker, Bill Hardy, George Klein, Betsy Loague (née Bittner), Jan Svoboda, Dominique Stehelin, Harald zur Hausen, Louise Chow, Tom Broker, Richard Gelinas, Claire Moore, David Rekosh, Joe Weber, Jim Pipas, John McLauchlan, Kirsten Balding, Graham Hatfull, and William Haseltine.

I thank a number of institutions that allowed me access to their archives: the American Philosophical Society, home of the Peyton Rous Papers and the Baruch Blumberg Papers; UC San Francisco, home of the Varmus Papers; UC Berkeley, home of the Wendell Stanley and the Harry Rubin Papers; the

University of Wisconsin Archives, home of the Howard Temin Papers; UC San Diego, home of the Marguerite Vogt Papers; the National Library of Medicine, home of the Ludwik Gross and the Werner Henle Papers; the Rockefeller Archives, home of the Richard Shope Papers; Icahn School of Medicine at Mount Sinai, home of the Charlotte Friend Papers; and CSHL Library and Archives, home of the James Watson and Sydney Brenner Papers.

Jennifer McBryan, Sondra Schlesinger, Karen-Beth Scholthof, Angela Creager, John Horgan, Jim McClelland, Kathleen Orange, and three anonymous reviewers provided numerous editorial and noneditorial improvements on the entire draft of the book. Matt Kaufmann, Aleksandra Petelski, Alex Gann, Jim Watson, Jan Witkowski, Joe Sambrook, Robert Weinberg, Nathaniel Comfort, and Robert Olby provided help on early ideas and drafts. Olivia Schreiber, Grant Vallance, John Wilkins, Dawn Digrius, Os Jarrett, David Baltimore, Phil Sharp, Richard Gelinas, Bob Gallo, Lionel Crawford, Geof Cooper, Ton van Helvoort, Xiao-Yi Sun, Bob Weinberg, Sir Anthony Epstein, Ramareddy Guntaka, Robert Tjian, Nathaniel Comfort, Joan Brugge, Bert Vogelstein, Harvey Alter, Harold Varmus, Ian Frazer, Robin Scheffler, Debbie Spector, John Schiller, Inina Kachelmeier, Neeraja Sankaran, Steve Martin, Harald zur Hausen, Art Levinson, Warren Levinson, Jay Levy, Ryan Dahn, Edward Welch Morgan, Chris Smeenk, Stacey Welch, Sam Muka, Brad Fidler, Michael Steinmann, Bill Summers, and Rachel Ankeny provided comments on early versions of various chapters.

A draft of chapter 10 on tumor virology at CSHL was presented at the 2017 meeting of the International Society for History, Philosophy, and Social Studies of Biology in Sao Paulo, Brazil. I thank Michel Morange, Jim Griesemer, and Marsha Richmond for comments and company in Brazil. An earlier version of chapter 2 was published in *Studies in History and Philosophy of Biology and Biomedical Sciences*. I thank Clare Clark and Stephanie Satalino from CSHL Archives for help with numerous images. David Goodsell painted the cover artwork. Stevens Institute of Technology awarded me a sabbatical in the spring of 2019 to write, and for that I thank Dean Kelland Thomas and Provost Christophe Pierre. Matt McAdam from Johns Hopkins University Press supported this project from its initial stages. Carrie Watterson copyedited the manuscript. Adriahna Conway and Kim Johnson were much help during production. Naturally, all remaining errors are mine.

Interviews and Archival Sources

Interviews Conducted by the Author

Watson, Jim. Cold Spring Harbor Laboratory, NY, May 18, 2012, August 1, 2013, August 19, 2014.
Levine, Arnold. Princeton University, July 28, 2012.
Bishop, Michael. University of California, San Francisco, January 8, 2013.
Baltimore, David. California Institute of Technology, January 10, 2013.
Vogt, Peter. Scripps Research Institute, January 11, 2013.
Martin, Steve. By telephone, February 4, 2013.
Rubin, Harry. By telephone, March 1, 2013.
Hopkins, Nancy. By telephone, June 21, 2013.
Sambrook, Joseph. By telephone, June 21, 2013.
Scolnick, Edward. By telephone, July 3, 2013.
Darnell, James. Rockefeller University, July 3, 2013.
Livingston, David. By telephone, July 10, 2013.
Berg, Paul. by telephone, July 15, 2013.
Coffin, John. By telephone, July 30, 2013.
Wigler, Michael. Cold Spring Harbor Laboratory, August 1, 2013.
Duesberg, Peter. By telephone, August 1, 2013.
Epstein, Anthony. By telephone, August 6, 2013.
Cooper, Geoffrey. Boston, MA, August 12, 2013.
Sharp, Phillip. Cape Cod, MA, August 12, 2013.
Essex, Max. Boston, MA, August 13, 2013.
Harlow, Ed. Boston, MA, August 13, 2013.
Weinberg, Robert. Rindge, NH, August 14, 2013.
Roberts, Richard. Ipswich, MA, August 15, 2013.
Gilbert, Walter. By telephone, August 22, 2013.
Varmus, Harold. New York, August 30, 2013.
Gross, Augusta. New York, August 31, 2013.
Klein, George. By telephone, September 3, 2013.
Green, Howard. By telephone, October 16, 2013.
Jarrett, Os. By telephone, October 22, 2013.
Crawford, Lionel. By telephone, October 22, 2013.
Gallo, Robert. By telephone, October 22, 2013.
Hardy, Bill, Franklin Lakes, NJ, January 31, 2014.
Pollack, Robert. Columbia University, June 4, 2014.
Svoboda, Jan. By telephone, September 23, 2014; Brnik, Czech Republic, August 7, 2016.
Mulder, Carel. By telephone, December 9, 2014.

Tjian, Robert. By telephone, February 20, 2015.
Stehelin, Dominique. Lille, France, August 4–5, 2016.
zur Hausen, Harald, Heidelberg, Germany, August 6, 2016.
Smith, Hamilton, Cold Spring Harbor Laboratory, October 24, 2016.
Broker, Tom. Cold Spring Harbor Laboratory, October 24, 2017.
Moore, Claire. Cold Spring Harbor Laboratory, October 24, 2017.
Weber, Joe. By telephone, October 10, 2018.

Archival Sources

American Philosophical Society
  Peyton Rous Papers
  Baruch Blumberg Papers
Cold Spring Harbor Laboratory Library and Archives
  James Watson Papers
  Sydney Brenner Papers
Ichan School of Medicine at Mt. Sinai
  Charlotte Friend Papers
National Library of Medicine
  Ludwik Gross Papers
  Werner Henle Papers
  Edward Shorter Papers
Rockefeller Archives
  Richard Shope Papers
University of California, Berkeley
  Wendell Stanley papers
  Harry Rubin Papers
University of California, San Diego
  Marguerite Vogt Papers
University of California, San Francisco
  Harold Varmus Papers
University of Cincinnati
  Albert Sabin Papers
University of Wisconsin Archives
  Howard Temin Papers

Notes

Introduction. The Untold Story of How a Century of Tumor Virology Changed Biomedicine

1. Kevles and Geison, "The Experimental Life Sciences in the Twentieth Century." Keller, *The Century of the Gene*. Rheinberger, "What Happened to Molecular Biology?" Morange, *The Black Box of Biology*. Grote et al., "The Molecular Vista."

2. There are exceptions. For example, the work of Ton van Helvoort stands out; see his "The Start of a Cancer Research Tradition"; and "A Century of Research into the Cause of Cancer." See also Morange, "From the Regulatory Vision of Cancer to the Oncogene Paradigm, 1975–1985"; and Yi, "Cancer, Viruses, and Mass Migration." Also see the work of science writer Carl Zimmer, particularly his *Rabbits with Horns and Other Astounding Viruses*. Virologists are aware of the broad impact of retroviruses. See, for example, Peter Vogt's introduction to Coffin, Hughes, and Varmus, *Retroviruses*; or Howley and Livingston, "Small DNA Tumor Viruses." And recently, after *Cancer Virus Hunters* was largely written, a book on the history of American tumor virology was published: Scheffler, *A Contagious Cause*.

3. Tumor virologists are generally aware of the influence of their discipline. See, for example, Pipas, "DNA Tumor Viruses and Their Contributions to Molecular Biology."

4. Kevles, "Pursuing the Unpopular." See also Proctor and Schiebinger, *Agnotology*.

5. Hull, *Science as a Process*. See also Pickering, *Science as Practice and Culture*.

6. A good overview of the vast literature on sexism in science and the feminist response can be found here: https://plato.stanford.edu/entries/feminist-science/; and Tonn, "Gender." See also Creager et al., *Feminism in Twentieth-Century Science, Technology, and Medicine*; and Kourany, *The Gender of Science*.

7. On the importance of institutions in the history of virology, see Méthot, " Writing the History of Virology in the Twentieth Century."

8. Robin Scheffler's recent book *A Contagious Cause* documents this funding.

9. For a recent volume dedicated to the international nature of twentieth-century biology, see Manning and Savelli, *Global Transformations in the Life Sciences, 1945–1980*. The international nature of molecular biology has been argued more generally. See, for example, Abir-Am, "From Multidisciplinary Collaboration to Transnational Objectivity"; and Rheinberger, "Patterns of the International and the National, the Global and the Local in the History of Molecular Biology."

10. For those interested in the sociological context of the rise of oncogenes in the tradition of Latour and Woolgar's influential *Laboratory Life*, I recommend Joan Fujimura's *Crafting Science*. Robin Schaffer's *A Contagious Cause* is an excellent foray into the federal bureaucracy of the NIH and NCI and how it interacted with virologists. One interesting proposal for the rethinking the relationship between history and philosophy of biology

and biology is given by Stotz and Griffiths, "Biohumanities." For an older discussion of the relationship between internalist and externalist approaches to history of biology, see Allen, "Essay Review." For still one of the best discussions of social constructivism, see Hacking, *The Social Construction of What?* My historiographical approach is motivated by the principle that historians of science should be cautious in using speculative sociological conceptual machinery that has significantly less epistemic support than the actor categories they aim to illuminate and analyze.

11. Kevles, "Renato Dulbecco and the New Animal Virology."

12. Cairns, Stent, and Watson, *Phage and the Origins of Molecular Biology*. Mullins, "The Development of a Scientific Specialty."

13. It would later become clear that some retroviruses were not oncogenic. That all retroviruses are transforming viruses was a dogma that was overturned. See Levy, "Changing Dogmas in Retrovirology."

14. For a list of the many symposia offered at CSHL, see http://symposium.cshlp.org/site/misc/index_archive.xhtml.

15. Maniatis, Fritsch, and Sambrook, *Molecular Cloning*.

16. Scheffler, "Managing the Future."

17. Wade, "Special Virus Cancer Program."

18. For an explanation of mechanisms and their discovery in molecular biology, see Darden, *Reasoning in Biological Discoveries*; and Craver and Darden, *In Search of Mechanisms*.

19. Blasimme, Maugeri, and Germain, "What Mechanisms Can't Do."

20. In his author index, Gross lists more than 650 scientists involved in tumor virology by 1961! See also the recent monograph, Scheffler, *A Contagious Cause*.

21. For ways in which history of science can be useful to scientists, see Maienschein, Laubichler, and Loettgers, "How Can History of Science Matter to Scientists?"; and Morange, "What History Tells Us."

22. Lustig and Levine, "One Hundred Years of Virology."

23. Rybicki, "A Top Ten List for Economically Important Plant Viruses."

24. Beijerinck, "Ueber ein contagium vivum fluidum als Ursache der Fleckenkrankheit der Tabaksblatter"; Sankaran, "On the Historical significance of Beijerinck and His *Contagium Vivum Fluidum* for Modern Virology."

25. Knight, *Molecular Virology*, 10.

26. Waterson and Wilkinson, *An Introduction to the History of Virology*, 13. It is possible that the foot and mouth discovery happened before TMV. For a detailed history of history of TMV research, see Creager, *The Life of a Virus*.

27. Ellermann and Bang, "Experimentelle Leukämie bei Hühnern. II."

28. Payne and Nair, "The Long View."

29. Mukherjee, *The Emperor of All Maladies*.

30. Davis, *The Secret History of the War on Cancer*. "Understanding What Cancer Is: Ancient Times to Present; Oldest Descriptions of Cancer," American Cancer Society, 2021, www.cancer.org/cancer/cancer-basics/history-of-cancer/what-is-cancer.html.

31. Alternatively, perhaps the name alludes to the pain of cancer as being pinched by a crab. See Porter, *The Greatest Benefit to Mankind*, 575.

32. Agricola, *De Re Metallica*.

33. Langård, "Gregorius Agricola Memorial Lecture."

34. Hill, *Cautions against the Immoderate Use of Snuff.*

35. Adler, *Primary Malignant Growths of the Lungs and Bronchi.*

36. Pott, "Chirurgical Observations Relative to the Cataract, the Polypus of the Nose, the Cancer of the Scrotum, the Different Kinds of Ruptures, and the Mortification of the Toes and Feet." Brown and Thornton, "Percivall Pott (1714–1788) and Chimney Sweepers' Cancer of the Scrotum."

37. Rigoni-Stern, "Fatti statistici relativi alle malattie cancerose." Aviles, "The Little Death."

38. Singer, *Galen: Selected Works*. Olson, ed. *The History of Cancer*; Hajdu, "A Note from History: Landmarks in History of Cancer, Part 1."

39. de Kruif, *Microbe Hunters*. For later works in the same genre, see Williams, *Virus Hunters*; and Radetsky, *The Invisible Invaders*.

40. Summers, "Microbe Hunters Revisited."

41. For some reflection on the strengths and weaknesses of using biographies in the history of science, see Selya, "Primary Suspects."

42. For a thoughtful consideration of collaboration, see Maienschein, "Why Collaborate?"; and Andersen, "Collaboration, Interdisciplinarity, and the Epistemology of Contemporary Science."

43. A similar point is made by Creager, " 'Happily Ever After' for Cancer Viruses."

1. The Beginnings

1. Becsei-Kilborn, *Going Against the Grain*, 53. This chapter draws on Becsei-Kilborn's PhD dissertation on Rous and her article "Scientific Discovery and Scientific Reputation."

2. Corner, *A History of the Rockefeller Institute, 1901–1953*.

3. Rous, "An Inquiry into Some Mechanical Factors in the Production of Lymphocytosis"; Rous, "The Effect of Pilocarpine on the Output of Lymphocytes through the Thoracic Duct"; Rous, "Some Differential Counts of the Cells in the Lymph of the Dog."

4. Sankaran, *A Tale of Two Viruses*, 12.

5. Rous, "The Virus Tumors and the Tumor Problem."

6. For more on the rise of cancer research in this period, see Marcus, *Malignant Growth.*

7. Rous, "An Experimental Comparison of Transplanted Tumor and a Transplanted Normal Tissue Capable of Growth."

8. Marcus, *Malignant Growth*, 118.

9. Becsei-Kilborn, *Going Against the Grain*, 56.

10. Becsei-Kilborn, *Going Against the Grain*, 90.

11. Corner, *A History of the Rockefeller Institute.*

12. The practice of successfully transplanting tumors began with the Russian veterinarian Mstislav Novinsky's work on dogs in 1876: Novinsky, "O privivanii trakovikh novoobrazovanii." Shabad and Ponomarkov, "Mstislav Novinsky, Pioneer of Tumour Transplantation."

13. Rous, "A Transmissible Avian Neoplasm (Sarcoma of the Common Fowl)."

14. Rous, "A Sarcoma of the Fowl Transmissible by an Agent Separable from the Tumor Cells."

15. Van Helvoort, "The Start of a Cancer Research Tradition."

16. J. W. Williams, "The Development of the Ultracentrifuge and Its Contributions." In 1914 Japanese researchers also reported isolating a sarcoma-inducing virus: Fujinami and Inamoto, "Ueber Geschwülste bei japanischen Haushühnern, insbesondere über einen transplantablen Tumor."

17. Rous to Prudden, October 14, 1922, Rockefeller Archive Center, 450 Faculty/Peyton Rous R762.

18. Andrewes, "Francis Peyton Rous, 1879–1970."

19. A biblical expression. See Matthew 23:24.

20. Rous to Prudden, October 14, 1922, Rockefeller Archive Center, Record Group 450.

21. Gye, "The Aetiology of Malignant New Growths"; Barnard, "The Microscopical Examination of Filterable Viruses Associated with Malignant New Growths."

22. Sankaran, *A Tale of Two Viruses*, chapter 5.

23. As Nicolas Rasmussen discusses, in the late 1940s when scientists at the RIMR examined various tissues looking for Rous sarcoma virus, they saw particles in normal tissue and concluded that RSV was not the cause of malignancy. Rasmussen, *Picture Control.*

24. Rous, presentation of the Kober Medal to Richard Shope.

25. Rous, presentation of the Kober Medal to Richard Shope.

26. For more on the RIMR and yellow fever see, Löwy, "Epidemiology, Immunology, and Yellow Fever."

27. Lewis and Shope, "The Blood in Hog Cholera."

28. An antecedent of this theory can be found in earlier work on hog cholera by Emil Alexander de Schweinitz and Marion Dorset, who suggested that a filterable factor was needed to enhance the pathogenicity of the "hog cholera bacillus." De Schweinitz and Dorset, "Form of Hog Cholera Not Caused by the Hog-Cholera Bacillus."

29. Smith, Andrewes, Laidlaw, "A Virus Obtained from Influenza Patients." Andrewes also had a close personal relationship with Rous. See also Sankaran and van Helvoort, "Andrewes's Christmas Fairy Tale."

30. Andrewes, "Richard Edwin Shope," 357.

31. Shope, "A Filterable Virus Causing a Tumor-Like Condition in Rabbits and Its Relationship to Virus Myxomatosum."

32. Olitsky, "The Action of Glycerol on the Virus of Experimental Typhus Fever and on Proteus Bacilli."

33. Gross, *Oncogenic Viruses*, 2nd ed., 49.

34. Shope and Hurst, "Infectious Papillomatosis of Rabbits."

35. Creager and Gaudillière, "'Experimental Arrangements and Technologies of Visualization."

36. Shope, "Immunization of Rabbits to Infectious Papillomatosis."

37. Margaret Stanley, quoted in Reynolds and Tansey, *History of Cervical Cancer and the Role of the Human Papillomavirus, 1960–2000*.

38. Rous and Beard, "The Progression to Carcinoma of Virus-Induced Rabbit Papillomas (Shope)."

39. Rous to Andrewes, August 25, 1934, Rous Papers, American Philosophical Society, Philadelphia, PA, quoted in Becsei-Kilborn, *Going Against the Grain*, 172.

40. For a discussion of the reception of Rous's ideas in the 1920s and 1930s, see Neeraja Sankaran's *A Tale of Two Viruses.*

41. Murphy, "Experimental Approach to the Cancer Problem." See also Sankaran, *A Tale of Two Viruses.*

42. Rous, "Presentation of the Kober Medal to Richard Shope."

43. Kay, "W. M. Stanley's Crystallization of the Tobacco Mosaic Virus, 1930–1940."

44. W. M. Stanley to the Nobel Committee, April 13, 1959, Stanley Papers, University of California Berkeley.

45. Jackson and Little, "The Existence of Non-chromosome Influence of Mammary Tumors in Mice."

46. Things are a little more complicated with sex-linked genes on the X or Y chromosome than the simple with Mendelian autosomal genes, which is what the researchers were considering.

47. Little, "Evidence That Cancer Is Not a Simple Mendelian Recessive."

48. In addition to eugenics, C. C. Little would later take an aggressive and uncompromising approach against the epidemiological work that showed that smoking tobacco causes lung cancer. At the time of writing, the University of Maine was removing his name from an academic building. Brandt, "Inventing Conflicts of Interest." Brandt, *The Cigarette Century.*

49. Gaudilliere, "Circulating Mice and Viruses," 91. See also Bud, "Strategy in American Cancer Research after World War II."

50. Little, interview by GWG, January 19, 1950, Rockefeller Archives, RG 1.1, Series 200D, Box 144, Folder 1779, quoted in Rader, *Making Mice*, 140.

51. Bittner, "A Genetic Study of the Transplantation of Tumors Arising in Hybrid Mice."

52. For more on Bittner, see Ruddy, *Of Mice and Women.*

53. Bittner, "Some Possible Effects of Nursing on the Mammary Gland Tumor Incidence in Mice."

54. Historians and philosophers of science have written a lot about model organisms. See, for example, Creager, Lunbeck, and Wise, *Science without Laws*; Ankeny and Leonelli, "What's So Special about Model Organisms?"; Ankeny, "Fashioning Descriptive Models in Biology"; Leonelli and Ankeny, "What Makes a Model Organism?"; Endersby, *A Guinea Pig's History of Biology*; Scholthof, "Tobacco Mosaic Virus." Dietrich et al., "How to Choose Your Research Organism."

55. Bittner, "Possible Relationship of the Estrogenic Hormone's Genetic Susceptibility and Milk Influence in the Production of Mammary Cancer in Mice."

56. Strong, "Obituary John Joseph Bittner 1904–1961."

57. "Medicine: Sucklings' Cancer," *Time*, June 9, 1941.

58. Snell, "Biology of the Laboratory Mouse," 323.

59. "Cancer Virus," *Time*, March 18, 1946.

60. Bryan et al., "Extraction and Ultracentrifugation of Mammary Tumor Inciter of Mice." Even as late as 1957, Bittner was still merely claiming that the agent had the properties of a virus; see J. J. Bittner, "Studies on Mammary Cancer in Mice and Their Implications for the Human Problem."

61. I thank Ton van Helvoort for this point.

62. Andervont, "Mammary Tumors in Mice."

63. Mann, "Effect of Low Temperatures on Bittner Virus of Mouse Carcinoma."

64. For the importance of electron microscopy in virology, see van Helvoort and Sankaran, "How Seeing Became Knowing."

65. Porter and Thompson, "A Particulate Body Associated with Epithelial Cells Cultured from Mammary Carcinomas of Mice of a Milk-Factor Strain."

66. Radner, *Making Mice*, 202.

67. Clark, "'U' Tests Reveal Altered Cancer 'Virus,'" *Minneapolis Star*, November 24, 1959. This newspaper article is one of few on Bittner, who thought the press often misrepresented science and was reluctant to give interviews. Bittner, "Biological Assay and Serial Passage of the Mouse Mammary Tumour Agent."

68. Luria, *General Virology*, 328. By 1960, he was calling it a virus. See Luria, "Viruses, Cancer Cells, and Genetic Concept of Virus Infection."

69. Luria, *General Virology*, 328.

70. Waterson, *Introduction to Animal Virology*, 69.

71. Waterson, *Introduction to Animal Virology*, 70.

2. True Believers

1. Document labeled "Miss M. Culter 1941?," Ludwik Gross Papers, MSC 504, Personal and Biographical, National Library of Medicine, Bethesda, Maryland, hereafter Gross Papers.

2. Augusta Gross, interview by author, New York City, November 1, 2013.

3. "The Search for Viruses as Etiological Agents in Cancer and Leukemia," 2, unpublished manuscript, Gross Papers. The manuscript appears to an expanded version of Gross, "The Fortuitous Isolation and Identification of the Polyoma Virus." Also see Bessis, "How the Mouse Leukemia Virus Was Discovered."

4. Gross published two books popularizing science in Polish: Gross, *Ludzkość w walce o zdrowie* [Humanity fights for health]; and Gross, *Siewcy chorób i śmierci* [Sowers of death and disease].

5. Craigie, "Sarcoma 37 and Ascites Tumours."

6. Personal communication from Augusta Gross, February 16, 2014; Gross to Bittner, April 23, 1944, Gross Papers. At the Institut Pasteur, Gross spoke with Amédée Borell (1867–1936), Charles Nicolle (1866–1936), Emile Roux (1853–1933), Albert Calmette (1863–1933), Charles Oberling (1895–1960), and Jules Bordet (1870–1961) about their work, seeking to get some "tips" on the communicability of cancer. Some of his interviews were published in *Ilustrowany Kuryer Codzienny*.

7. Gross to Rous, September 3, 1935, Peyton Rous Papers, American Philosophical Society, Philadelphia, PA, hereafter Rous Papers.

8. Gross to Rous, August 11, 1936, Rous Papers.

9. Rous to Gasser, September 25, 1936, Rous Papers.

10. Translation of Besredka to dean of medicine at Yale, November 20, 1936, Gross Papers.

11. "The Search for Viruses," 6.

12. Creager and Gaudillière, "Experimental Arrangements and Technologies of Visualization." For a discussion of the role of chemicals in cancer in this period, see van Helvoort, "A Century of Research into the Cause of Cancer."

13. Personal communication with Augusta Gross, November 2018.

14. Augusta Gross, interview by author, New York City, August 31, 2013.

15. Gross to Rous, February 4, 1942, Rous Papers.

16. Rous to Warren, February 11, 1942, Rous Papers. In contrast, Rous was unwilling to write a letter of recommendation for a commission in the army, as he claimed he did not know Gross's "temperament and character." Rous to Gross, February 19, 1943.

17. Bittner, "Some Possible Effects of Nursing on the Mammary Gland Tumor Incidence in Mice"; Bittner, "Mammary Tumors in Mice in Relation to Nursing." For more on the history of Bittner's work, see Gaudillière, "The Molecularization of Cancer Etiology in the Postwar United States"; Gaudillière and Löwy, "Disciplining Cancer"; and Gaudilliere, "Circulating Mice and Viruses," 91.

18. "The Search for Viruses," 10.

19. Memo, Gross to the surgeon general, September 7, 1943, Gross Papers.

20. Gross to Bittner, April 10, 1944, Gross Papers.

21. Memo, Gross to the surgeon general, March 22, 1944, Gross Papers.

22. Memo, Gross to the surgeon general, March 22, 1944.

23. Lucké, "Carcinoma in the Leopard Frog."

24. "The Search for Viruses," 17.

25. Straus, *Rosalyn Yalow, Nobel Laureate*, 124.

26. For a fascinating history of inbred mouse lines, see Rader, *Making Mice*.

27. Rader, "The Origins of Mouse Genetics," 464

28. Daston, "The Moral Economy of Science."

29. Angevine, "Significant Events in the Life of Jacob Furth."

30. "The Search for Viruses," 16.

31. Gross, "Susceptibility of Newborn Mice of an Otherwise Apparently 'Resistant' Strain to Inoculation with Leukemia"; Gross, "Pathogenic Properties, and 'Vertical' Transmission of the Mouse Leukemia Agent"; Gross, "'Spontaneous' Leukemia Developing in G3H Mice Following Inoculation, in Infancy, with AK-Emkemic."

32. Memo from De Voe, October 8, 1952, Hauck Center for the Albert B. Sabin Archives, University of Cincinnati, Ohio, hereafter Sabin Papers.

33. Gross, "A Filterable Agent, Recovered from Ak Leukemic Extracts, Causing Salivary Gland Carcinomas in C3H Mice"; Gross, "Neck Tumors, or Leukemia, Developing in Adult C3H Mice Following Inoculation."

34. Van Helvoort, "A Century of Research into the Cause of Cancer." See also Gaudillière and Löwy, "Disciplining Cancer"; Creager and Gaudillière, "Experimental Arrangements and Technologies of Visualization"; and Galperin, "Virus, provirus et cancer."

35. "The Search for Viruses."

36. Theresa MacPhail has investigated the importance of networks in virology in much detail. See MacPhail, *The Viral Network*.

37. Gaudillière and Löwy, "Disciplining Cancer."

38. For more on the work of Joseph Beard, see Creager and Gaudillière, "'Experimental Arrangements and Technologies of Visualization."

39. Bryan to Gross, March 22, 1951, Gross Papers.

40. Furth to Gross, October 15, 1951, Gross Papers.

41. Sabin to Gross, May 13, 1953, Gross Papers.

42. Sabin to Lyon, May 13, 1953, Sabin Papers.

43. Gross, *Oncogenic Viruses*, 1st ed., 281.

44. Gross, "A Filterable Agent, Recovered from Ak Leukemic Extracts, Causing Salivary Gland Carcinomas in C3H Mice," 418; Gross, "Neck Tumors, or Leukemia, Developing in Adult C3H Mice Following Inoculation."

45. Hughes, *The Virus*. Summers, "Inventing Viruses."

46. Creager, *The Life of a Virus*, chapter 4.

47. Gross, *Oncogenic Viruses*, 284.

48. "Cornering the Killer," cover story, *Time*, July 27, 1959, 53.

49. Stewart to Gross, December 11, 1952, Sarah Stewart Papers, MSC 360, National Library of Medicine, Bethesda, Maryland, hereafter Stewart Papers.

50. Gross to Stewart, December 12, 1952, Stewart Papers.

51. Harold L. Stewart [no relation], eulogy for Sarah Stewart, May 23, 1977, Stewart Papers.

52. Sarah Stewart, oral history interview, audiocassette, 1964, Folder 3, Box 1, Stewart Papers.

53. Eddy, Bernice, interview by Edward Shorter, December 4, 1986, Shorter Papers, National Library of Medicine.

54. Stewart, oral history interview.

55. Stewart, oral history interview.

56. Stewart, oral history interview.

57. Stewart to Rhodes, October 2, 1953, Stewart Papers; Stewart, "Leukemia in Mice Produced by a Filterable Agent Present in AKR Leukemic Tissues with Notes on a Sarcoma Produced by the Same Agent."

58. Gradmann, "A Spirit of Scientific Rigour."

59. Rivers, "Viruses and Koch's Postulates," 4.

60. Stewart, oral history interview.

61. Stewart, oral history interview.

62. Stewart to Frank, May 23, 1957, Stewart Papers.

63. Memo, Stewart to Andervont and Gross, December 27, 1954, Stewart Papers.

64. Gross, "Pathogenic Properties, and 'Vertical' Transmission of the Mouse Leukemia Agent"; Gross "'Spontaneous' Leukemia Developing in G3H Mice Following Inoculation, in Infancy, with AK-Emkemic"; Gross, "Neck Tumors, or Leukemia, Developing in Adult C3H Mice Following Inoculation."

65. Dulaney to Stewart, January 3, 1955, Stewart Papers.

66. Levinthal to Stewart, June 23, 1954, Stewart Papers.

67. Bittner to Stewart, January 4, 1955, Stewart Papers.

68. Gross to Stanley, December 21, 1954, Stanley Papers.

69. Creager, *The Life of a Virus*.

70. Creager, "Wendell Stanley's Dream of a Free-Standing Biochemistry Department at the University of California, Berkeley."

71. Stewart, Eddy, and Borgese, "Neoplasms in Mice Inoculated with a Tumor Agent Carried in Tissue Culture."

72. Stewart to Schwartz, April 22, 1957, Stewart Papers.

73. Heller, "Conference Summary," 7.

74. Quoted in "Cornering the Killer," 53.

75. G. Klein, *The Atheist and the Holy City*; Kevles, "Pursuing the Unpopular," 85; see also Bishop, *How to Win the Nobel Prize*, 159. Kevles does not cite a paper of Furth's that documents these experiments but presumably he means Furth et al., "Character of Agent Inducing Leukemia in Newborn Mice." Additionally, other researchers confirmed Gross's work slightly before Furth and his associates: Woolley and Small, "Experiments on Cell-Free Transmission of Mouse Leukemia." Michael Shimkin mentions the work of NCI scientists, especially Lloyd Law, as also being important. Shimkin, "As Memory Serves."

76. I thank Sir Anthony Epstein for impressing this point on me.

77. See also Javier and Butel, "The History of Tumor Virology."

78. Graffi, Bielka, and Fey, "Leukämieerzeugung durch ein filtrierbares Agens aus malignen Tumoren." Rous viewed Graffi as "the most valuably productive worker on the oncogenic viruses of those now active in both Germanys." Rous to Hirst, August 25, 1967, Rous papers. Graffi had started the Institut für Krebsforschung at the Deutsche Akademie in 1948. See Rous to Graffi, May 14, 1968, Rous papers.

79. Gross, "Viral Etiology of Cancer and Leukemia?"

80. Moloney, "Preliminary Studies on a Mouse Lymphoid Leukemia Virus Extracted from Sarcoma 37"; Moloney, "Biological Studies on a Lymphoid Leukemia Virus Extracted from Sarcoma S. 37"; Rauscher, "A Virus-Induced Disease of Mice Characterized by Erythrocytopoiesis and Lymphoid Leukemia." For reasons of space, I have not discussed an additional facet of the history here. Developments in bacteriophage biology, particularly the study of lysogenic bacteriophage, reinforced the view that tumor virus genomes could be inserted into the genomes of the host cells. For more on this aspect of the history, see Sankaran, "When Viruses Were Not in Style." And Kostyrka and Sankaran, "From Obstacle to Lynchpin."

81. Rous, "A Sarcoma of the Fowl Transmissible by an Agent Separable from the Tumor Cells."

82. Rous to Gross, April 11, 1950, Rous Papers.

83. Rous to Friend, August 4, 1956, Charlotte Friend Papers, Icahn School of Medicine, Mount Sinai, New York.

84. Friend, "Cell-Free Transmission in Adult Swiss Mice of a Disease Having the Character of a Leukemia."

85. Furth to Stewart, October 22, 1958, Stewart Papers. Furth had not changed his mind by 1976. See Furth to Gross, April 8, 1976, Gross Papers.

86. Stewart to Furth, October 29, 1958, Stewart Papers.

87. Goldfarb to Stewart, November 5, 1958, Stewart Papers.

88. Goldfarb to Stewart, February 28, 1959, Stewart Papers.

89. Eddy, Stewart, and Berkeley, "Cytopathogenicity in Tissue Cultures by a Tumor Virus from Mice."

90. Gross to Rous, November 30, 1959, Rous Papers. Given the technology at the time, Gross was unable to separately propagate the two viruses, which would have provided definitive evidence that there were two viruses present. For the work to which Gross was referring, see Eddy, Rowe, et al., "Hemagglutination with the SE Polyoma Virus"; Eddy, Stewart, and Grubbs, "Influence of Tissue Culture Passage, Storage, Temperature and Drying on Viability of SE Polyoma Virus"; Eddy, Stewart, Young, and Milder, "Neoplasms in Hamsters induced by Mouse Tumor Agent passed in Tissue Culture"; Eddy, Stewart, Kirschstein, and Young, "Induction of Subcutaneous Nodules in Rabbits with the SE Polyoma Virus"; Eddy, Stewart, Stanton, and Marcotte, "Induction of Tumors in Rats by Tissue-Culture Preparations of SE Polyoma Virus,"; Stewart and Eddy, "Tumor Induction by SE Polyoma Virus and the Inhibition of Tumors by Specific Neutralizing Antibodies"; Stewart, Eddy, and Borgese, "Neoplasms in Mice Inoculated with a Tumor Agent Carried in Tissue Culture"; Stewart, Eddy, Gochenour, et al., "The Induction of Neoplasms with a Substance Released from Mouse Tumors by Tissue Culture."

91. Gross to Friend, February 16, 1960; and Gross to di Mayorca, February 16; Friend Papers; Di Mayorca et al., "Isolation of Infectious Deoxyribonucleic Acid from SE Polyoma-Infected Tissue Cultures."

92. Di Mayorca to Gross, February 18, 1960, Friend Papers.

93. Bernice Eddy, interview by Edward Shorter, December 4, 1986, Shorter Papers.

94. "The Search for Viruses," 40.

95. Gross, *Oncogenic Viruses*, 285.

96. Horsfall, "Oncogenic Viruses."

97. Rous to Gross, August 15, Rous Papers.

98. Eddy to Gross, September 2, 1965, Gross Papers.

99. Gross, *Oncogenic Viruses*, 362–367.

100. Gross, *Oncogenic Viruses*, 363. For more on the relationship between heredity and infection, see Gaudillière, "The Molecularization of Cancer Etiology in the Postwar United States," Gaudillière and Löwy, "Disciplining Cancer"; and especially the chapter by Creager and Gaudillière, "Experimental Arrangements and Technologies of Visualization."

101. Huebner and Todaro, "Oncogenes of RNA Tumor Viruses as Determinants of Cancer."

102. Eddy, "*Oncogenic Viruses* by Ludwik Gross."

103. Tooze, *The Molecular Biology of Tumor Viruses*.

104. Tooze, "Period Piece Revised."

105. Weiss, "Look Back at Viral Oncology."

106. Robert Weinberg, personal communication with author, April 9, 2014.

107. Scheffler, "Managing the Future"; Doogab, "Governing, Financing, and Planning Cancer Virus Research."

108. Dulbecco, "From the Molecular Biology of Oncogenic DNA Viruses to Cancer"; Gross, "The Fortuitous Isolation and Identification of the Polyoma Virus."

109. The practice of naming retroviruses after the discoverer ended in the 1960s.

Nonetheless, a much-publicized case with echoes of the polyoma virus case occurred over the naming of HIV, on which Gross was consulted, when the American Robert Gallo and the Frenchman Luc Montagnier both thought they had the right to name the virus. Gallo had earlier discovered the first human leukemia virus (HTLV I), something that Gross had hinted should exist. See chapter 15.

110. See, for example, Macpherson and Montagnier, "Agar Suspension Culture for the Selective Assay of Cells Transformed by Polyoma Virus."

111. Gaudillière and Löwy, "Disciplining Cancer"; Scheffler, "Managing the Future."

112. Yi, "Cancer, Viruses, and Mass Migration."

113. Morange, "The Discovery of Cellular Oncogenes"; van Helvoort, "A Century of Research into the Cause of Cancer."

114. An immortalized cell line is a population of cells from a multicellular organism that would normally stop proliferating but, because of mutations in the cells causing them to evade normal cessation mechanisms, can proliferate indefinitely.

115. In 2008 a polyomavirus would be discovered—Merkel cell polyomavirus—that caused skin cancer in humans. Feng et al., "Clonal Integration of a Polyomavirus in Human Merkel Cell Carcinoma." Spurgeon and Lambert, "Merkel Cell Polyomavirus."

3. The Importance of Measurement

1. Renato Dulbecco, interview by Shirley K. Cohen, September 9 and 10, 1998, Caltech Archives, Pasadena, California, http://oralhistories.library.caltech.edu/26/1/OH_Dulbecco_R.pdf.

2. Summers, "How Bacteriophage Came to Be Used by the Phage Group."

3. Dulbecco, interview by Cohen.

4. For the history of the use of radioisotopes in biology, see Creager, *Life Atomic*.

5. Dulbecco, interview by Cohen.

6. Luria and Dulbecco, "Lethal Mutations, and Inactivation of Individual Genetic Determinants in Bacteriophage." Dulbecco, "Reactivation of Ultra-Violet-Inactivated Bacteriophage by Visible Light."

7. Renato Dulbecco, "Discovery of Photoreactivation," Web of Stories, www.webofstories.com/play/renato.dulbecco/13 (retrieved January 20, 2020).

8. Kevles, "Renato Dulbecco and the New Animal Virology."

9. McKaughan, "The Influence of Niels Bohr on Max Delbrück."

10. Renato Dulbecco, "Travelling to Caltech," Web of Stories, https://www.webofstories.com/play/renato.dulbecco/15 (retrieved January 20, 2020). For an account of the rise of Caltech as a molecular biology powerhouse, see Kay, *The Molecular Vision of Life*.

11. Dulbecco to Delbrück, November 22, 1948, Delbrück Papers, Caltech Archives.

12. Delbrück to Dulbecco, January 17, 1949, Delbrück Papers, Caltech Archives.

13. In an influential early book on bacteriophages, Dulbecco is described as independently discovering photoreactivation. Adams, *Bacteriophages*, 71.

14. Dulbecco, interview by Cohen.

15. Delbrück, foreword to *Viruses 1950*.

16. Dulbecco, "The Plaque Technique and The Development of Quantitative Animal Virology."

17. Skloot, *The Immortal Life of Henrietta Lacks*.

18. Dulbecco to Delbrück, February 4, 1951, Delbrück papers, Box 6, folder 22, Caltech Archives.

19. Dulbecco to Delbrück, February 9, 1951, Caltech Archives.

20. Renato Dulbecco, "Animal Virology," Web of Stories, www.webofstories.com/play/renato.dulbecco/16 (retrieved January 30, 2020).

21. Dulbecco, "The Plaque Technique and the Development of Quantitative Animal Virology."

22. Angela Creager discusses the new assay in the context of the creation of a new Kuhnian paradigm. Creager, "Paradigms and Exemplars Meet Biomedicine."

23. Dulbecco, "Production of Plaques in Monolayer Tissue Cultures by Single Particles of an Animal Virus."

24. Dulbecco and Vogt, "Some Problems of Animal Virology as Studied by the Plaque Technique."

25. Dulbecco, "The Plaque Technique and the Development of Quantitative Animal Virology."

26. Scholthof, "Making a Virus Visible." Creager et al., "Tobacco Mosaic Virus."

27. It is probable that Dulbecco means the National Foundation for Infantile Paralysis, not the National Science Foundation. For a discussion of the role of the NFIP, see Smith, *Patenting the Sun*; and Creager, "Mobilizing Biomedicine." See also Scheffler, "Protecting Children"; and Oshinsky, *Polio*.

28. Dulbecco, interview by Cohen.

29. M. Vogt, "Inhibitory Effects of the Corpora Cardiaca and of the Corpus Allatum in Drosophila."

30. "Marguerite Vogt: A Pioneer in Tumor Virology; 1913–2007," Max Delbrück Center for Molecular Medicine, www.mdc-berlin.de/marguerite-vogt (retrieved January 30, 2020).

31. They ordered macaques from Okatie Farms in South Carolina as a source of monkey kidney cells.

32. Dulbecco and Vogt, "Plaque Formation and Isolation of Pure Lines with Poliomyelitis Viruses."

33. Dulbecco and Vogt, "Plaque Formation and Isolation of Pure Lines with Poliomyelitis Viruses."

34. Caltech, press release, December 18, 1953, Biodiv 49.23, Caltech Archives.

35. Dulbecco to Sabin, August 22, 1956, MSS 688, Box 5, Folder 49, Caltech Archives.

36. Application for grant (NFIP), May 28, 1954, Biodiv 49.23, Caltech Archives.

37. Dulbecco, "Quantitative Aspects of Virus Growth in Cultivated Animal Cells."

38. Rubin, "A Disease in Captive Egrets Caused by a Virus of the Psittacosis-Lymphogranuloma Venereum Group."

39. The Spanish biologist Duran-Reynals infected ducks with Rous sarcoma virus. Duran-Reynals, "Age Susceptibility of Ducks to the Virus of the Rous Sarcoma."

40. Harry Rubin, telephone interview by author, March 1, 2013.

41. Manaker and Groupé, "Discrete Foci of Altered Chicken Embryo Cells Associated with Rous Sarcoma Virus in Tissue Culture."

42. Rubin, interview by author.

43. Rubin, interview by author.

44. Rubin, interview by author. Also see Sankaran, " When Viruses Were Not in Style."

45. Rubin to Vogt, June 25, 1998, letter in possession of Peter Vogt.

46. Temin, "The Effects of Actinomycin D on Growth of Rous Sarcoma Virus in Vitro." Temin, "Homology between RNA from Rous Sarcoma Virus and DNA from Rous Sarcoma Virus–Infected Cells."

47. Hanafusa, Hanafusa, and Rubin, "Analysis of the Defectiveness of Rous Sarcoma Virus, II." Rubin, "The Early History of Tumor Virology."

48. Rubin interview by author.

49. David Baltimore, "Renato Dulbecco 1914–2012," http://nasonline.org/publications/biographical-memoirs/memoir-pdfs/dulbecco-renato.pdf (retrieved January 30, 2020).

50. A column of methylated albumin used to separate nucleic acids. Mandell and Hershey, "A Fractionating Column for Analysis of Nucleic Acids."

51. Dulbecco and Vogt, "Evidence for a Ring Structure of Polyoma Virus DNA."

52. "1964 Albert Lasker Basic Medical Research Award," Lasker Foundation, www.laskerfoundation.org/awards/show/dna-and-rna-tumor-viruses/ (retrieved January 30, 2020).

53. For example, see, Rubin, Yao, and Rubin, "Relation of Spontaneous Transformation in Cell Culture to Adaptive Growth and Clonal Heterogeneity."

54. Bourgeois, *Genesis of the Salk Institute*. Jacobs, *Jonas Salk*.

55. Dulbecco, Hartwell, and Vogt, "Induction of Cellular DNA Synthesis by Polyoma Virus."

56. Personal communication with Lee Hartwell, October 1, 2021.

4. Cell Lines and Cat Leukemia

1. I thank Jennifer Stoker and the Stoker family for many unpublished details about Michael Stoker's career.

2. Black and Stoker, "Plasma Iron in New-Born Babies."

3. De Chadarevian, *Designs for Life*.

4. Stoker, "Autobiography"; personal communication with the Stoker family.

5. Stoker to Watson, March 10, 1957, James D. Watson Collection, CSHL Archive.

6. Watson, letter to the author, December 16, 2015.

7. Joklik, "When Two Is Better Than One."

8. Mullins, "The Development of a Scientific Specialty." Cairns, Stent, and Watson, *Phage and the Origins of Molecular Biology*.

9. Brenner and Horne, "A Negative Staining Method for High Resolution Electron Microscopy of Viruses." Wildy et al., "The Fine Structure of Polyoma Virus."

10. Stoker to Vogt, February 19, [1960?], M. Vogt Papers, University of California, San Diego.

11. Morgan, *Early Theories of Virus Structure*.

12. Stoker to Vogt, September 9, 1960, M. Vogt Papers.

13. Stoker to Vogt, February 19, [1960?], M. Vogt Papers.

14. Stoker, "Studies on the Oncogenic Activity of the Toronto Strain of Polyoma Virus."

15. Malinin, *Surgery and Life*.

16. Carrel, "On the Permanent Life of Tissues outside of the Organism."

17. Sanford, Earle, and Likely, "The Growth in Vitro of Single Isolated Tissue Cells."

18. Landecker, *Culturing Life*.

19. For more on the development of chemically defined media, see Landecker, "It Is What It Eats."

20. Eagle, "Nutrition Needs of Mammalian Cells in Tissue Culture." Eagle, "Amino Acid Metabolism in Mammalian Cell Cultures."

21. Vogt and Dulbecco, "Virus-Cell Interaction with a Tumor-Producing Virus." Sachs and Medina, "In Vitro Transformation of Normal Cells by Polyoma Virus."

22. Stoker and Abel, "Conditions Affecting Transformation by Polyoma Virus."

23. Stoker to Vogt, January 20, 1962, M. Vogt Papers.

24. Stoker to Vogt, March 27, [1962], M. Vogt Papers.

25. Lionel Crawford, personal communication with author, September 2018.

26. Jarrett, Crawford, et al., "A Virus-Like Particle Associated with Leukemia (Lymphosarcoma)."

27. "Clue from the Cat," *Time*, September 1973, 67. Jarrett, Jarrett, et al., "Horizontal Transmission of Leukemia Virus and Leukemia in the Cat."

28. American Association of Feline Practitioners, 2006 Feline Vaccination Guidelines, Summary: Vaccination in General Practice, https://catvets.com/public/PDFs/Practice Guidelines/VaccinationGLS-summary.pdf.

29. Mowat and Chapman, "Growth of Foot-and-Mouth Disease Virus in a Fibroblastic Cell Line Derived from Hamster Kidneys."

30. Stoker to Vogt, March 27, 1962, M. Vogt Papers.

31. Reynolds and Tansey, *Foot and Mouth Disease*, 45.

32. Austoker, *A History of the Imperial Cancer Research Fund*.

5. Insights from the Field

1. M. A. Epstein, "The Origins of EBV Research Discovery and Characterization of the Virus."

2. Quoted in "Dr. Denis Burkitt Is Dead at 82; Thesis Changed Diets of Millions," *New York Times*, April 16, 1993.

3. Burkitt, quoted in M. A. Epstein, "The Origins of EBV Research Discovery and Characterization of the Virus," 2.

4. Burkitt, "A Sarcoma Involving the Jaws in African Children."

5. Clarke, "Mapping the Methodologies of Burkitt Lymphoma."

6. Burkitt, "A Children's Cancer Dependent on Climatic Factors."

7. Anthony Epstein, telephone interview by author, August 6, 2013. Among other things, Epstein showed that RSV is an RNA virus, not a DNA virus, as the nucleic acid core of a virus particle could be removed with an enzyme (RNase) that breaks down RNA. This work was presaged by R. Bather in Edinburgh. Epstein and Holt, "Observations on the Rous Virus." Bather, "The Nucleic Acid of Partially Purified Rous No. 1 Sarcoma Virus."

8. Epstein, "The Origins of EBV Research Discovery and Characterization of the Virus," 4.

9. Stewart, Lovelace, et al., "Burkitt Tumor."

10. Epstein, interview by author. Elsewhere he says it was February 24.

11. Epstein, Achong, and Barr, "Virus Particles in Cultured Lymphoblasts from Burkitt's Lymphoma."

12. Crawford, Rickinson, and Johannessen, *Cancer Virus*, 41.

13. Crawford, Rickinson, and Johannessen, *Cancer Virus*, 47.

14. Henle, Henle, and Diehl, "Relation of Burkitt's Tumor-Associated Herpes-Ytpe [*sic*] Virus to Infectious Mononucleosis."

15. Henle, Henle, and Diehl, "Relation of Burkitt's Tumor-Associated Herpes-Ytpe [*sic*] Virus to Infectious Mononucleosis."

16. Today EBV is considered a herpes virus.

17. Zur Hausen and Schulte-Holthausen, "Presence of EB Virus Nucleic Acid Homology in a "Virus-Free" Line of Burkitt Tumour Cells."

18. Crawford, Rickinson, and Johannessen, *Cancer Virus*, 62.

19. Crawford, Rickinson, and Johannessen, *Cancer Virus*, 67.

20. Thorley-Lawson et al., "The Link between Plasmodium Falciparum Malaria and Endemic Burkitt's Lymphoma—New Insight into a 50-Year-Old Enigma."

21. Rajewsky, "George Klein."

22. Crawford, Rickinson, and Johannessen, *Cancer Virus*, 75.

23. For a history of thinking about biological membranes, see Grote, *Membranes to Molecular Machines*.

24. When EBNA was discovered it was widely analogized to SV40 T antigen, a better-known protein from a tumor virus. I thank Bill Summers for this point.

25. Manolov and Manolova, "Marker Band in One Chromosome 14 from Burkitt Lymphomas."

26. Shope, Dechairo, and Miller, "Malignant Lymphoma in Cottontop Marmosets after Inoculation with Epstein-Barr Virus." Richard Shope had three sons who became virologists, including Robert.

27. Dalla-Favera et al., "Human *c-myc* Onc Gene Is Located on the Region of Chromosome 8 That Is Translocated in Burkitt Lymphoma Cells." Taub et al., "Translocation of the *c-myc* Gene into the Immunoglobulin Heavy Chain Locus in Human Burkitt Lymphoma and Murine Plasmacytoma Cells." For a work on the history of chromosome research from the1950s to the 1980s, see de Chadarevian, *Heredity under the Microscope*.

28. Kennedy, Komano, and Sugden, "Epstein-Barr Virus Provides a Survival Factor to Burkitt's Lymphomas."

29. Graham Hatfull, personal communication with author, April 14, 2017.

30. Kirsten Balding, personal communication with author, October 2019.

31. Epstein, "Epstein-Barr Virus—Is It Time to Develop a Vaccine Program?"

32. Morgan, Finerty, et al., "Prevention of Epstein-Barr (EB) Virus-Induced Lymphoma in Cottontop Tamarins by Vaccination with the EB Virus Envelope Glycoprotein gp340 Incorporated into Immune-Stimulating Complexes." Morgan, Allison, et al., "Validation of a First-Generation Epstein-Barr Virus Vaccine Preparation Suitable for Human Use."

33. Beisel et al., "Two Major Outer Envelope Glycoproteins of Epstein-Barr Virus Are Encoded by the Same Gene."

6. Persistence despite Political Challenges

1. This chapter draws on an interview with Jan Svoboda in August 2016 and Svoboda, "Foundations in Cancer Research."

2. For more on Lysenkoism, see Krementsov, "A 'Second Front' in Soviet Genetics"; Dejong-Lambert and Krementsov, "On Labels and Issues"; Krementsov, "Lysenkoism in Europe"; Lewontin and Levins, "The Problem of Lysenkoism." For a recent contextualization, see Wolfe, *Freedom's Laboratory*.

3. Hejnar, "Jan Svoboda (1934–2017)."

4. Gross in his *Oncogenic Viruses* says that the "Prague strain" of RSV originally came from Engelbreth-Holm of Copenhagen.

5. Svoboda and Hasek, "Influencing the Transplantability of the Virus of Rous Sarcoma by Immunological Approximation in Turkeys."

6. Kusin, *The Intellectual Origins of the Prague Spring*.

7. Svoboda, "Presence of Chicken Tumour Virus in the Sarcoma of the Adult Rat Inoculated after Birth with Rous Sarcoma Tissue."

8. Kisselev, Abelev, and Kisseljov, "Lev Zilber, the Personality and the Scientist"; Zilber, "On the Interaction between Tumor Viruses and Cells."

9. Svoboda, "Further Findings on the Induction of Tumors by Rous Sarcoma in Rats and on the Rous Virus-Producing Capacity of One of the Induced Tumours (XC) in Chicks."

10. Svoboda, "Further Findings on the Induction of Tumors by Rous Sarcoma in Rats and on the Rous Virus-Producing Capacity of One of the Induced Tumours (XC) in Chicks."

11. Temin to Svoboda, September 4, 1962, Howard Temin Papers, University of Wisconsin, Madison.

12. Svoboda, "Foundations in Cancer Research," 8.

13. Svoboda, Chyle, et al., "Demonstration of the Absence of Infectious Rous Virus in Rat Tumour XC."

14. Scheffler, "Following Cancer Viruses through the Laboratory, Clinic, and Society."

15. Svoboda, "Foundations in Cancer Research."

16. Rous to Svoboda, February 7, 1964, American Philosophical Society, Philadelphia, PA, hereafter Rous Papers.

17. Jan Svoboda, interview by author, Brnik, Czech Republic, August 7, 2016.

18. Rous to Svoboda, February 7, 1964, Rous Papers.

19. Okada, "The Fusion of Ehrlichs Tumor Cells Caused by HVJ Virus in Vitro."

20. Svoboda, interview by author.

21. Coffin, "Rescue of Rous Sarcoma Virus from Rous Sarcoma Virus-Transformed Mammalian Cells."

22. Svoboda, Simkovic, and Koprowski, "Report on the Workshop on Virus Induction by Cell Association."

23. Crawford and Stoker, *The Molecular Biology of Viruses*.

24. For the context in Europe that led up to the Prague Spring, see Judt, *Postwar*.

25. Weiss, "Cancer, Infection and Immunity"; Weiss, "Remembering Jan Svoboda." The uninvited guests were the Russian tanks.

26. Svoboda, "Foundations in Cancer Research," 16. "Timeo Danaos et dona ferentes" is a phrase from the *Aeneid* (II, 49), paraphrased in English as the saying, "Beware of Greeks bearing gifts."

27. Svoboda, interview by author.

28. Svoboda, Hložánek, et al., "Problems of RSV Rescue from Virogenic Mammalian Cells."

29. For a discussion about the role of isotopes in molecular biology, see Creager, *Life Atomic*.

30. Svoboda, Lhoták, et al., "Characterization of Exogenous Proviral Sequences in Hamster Tumor Cell Lines Transformed by Rous Sarcoma Virus Rescued from XC Cells."

31. Mitsialis et al., "Studies on the Structure and Organization of Avian Sarcoma Proviruses in the Rat XC Cell Line."

32. Ramareddy Guntaka, personal communication with the author, September 2018.

7. A Surprising Discovery in the Blood

1. Block et al., "A Historical Perspective on the Discovery and Elucidation of the Hepatitis B Virus."

2. This chapter draws on Blumberg's *Hepatitis B* and information provided by Harvey Alter.

3. Blumberg to Allison, December 4, 1959, Box 2, Folder 10, Baruch Blumberg Papers, American Philosophical Society, Philadelphia, PA, hereafter Blumberg Papers.

4. Blumberg to Allison, July 18, 1961, Box 2, Folder 10. Blumberg Papers.

5. The term "Yellow Beret" started as a derogatory term for people who avoided going to Vietnam War. (Presumably it was coined as a contrast with Army Ranger Green Berets who did fight.) Over time, some, including the clinical associates themselves, reclaimed the term and took it as a badge of honor. I use it here in the latter sense. For a similar approach to the use of the term, see Delisio, "Fighting for a Cure"; and Haider J. Warraich, "During the Vietnam War, These Physician-Scientists Were Called 'Yellow Berets': They Are What We Need to Fight Covid-19," *Stat*, September 2, 2020, www.statnews.com/2020/09/02/yellow-berets-physician-scientists-covid-19/.

6. Blumberg and Alter, "A New Antigen in Leukemia Sera."

7. Harvey Alter, personal communication with author, October 12, 2018.

8. Blumberg, *Hepatitis B*, 84.

9. Blumberg, "A Short History of Australia Antigen," manuscript, Box 3888, Folder 5, Blumberg Papers.

10. Blumberg and Alter, "A New Antigen in Leukemia Sera."

11. Block et al., "A Historical Perspective on the Discovery and Elucidation of the Hepatitis B Virus."

12. Blumberg et al., "A Serum Antigen (Australia Antigen) in Down's Syndrome, Leukemia, and Hepatitis."

13. Prince, "Relation of Australia and SH Antigens." See also Prince to Bayer, October 22, 1968, Box 52, Folder 3, Blumberg Papers.

14. For correspondence regarding this collaboration, see Box 52, Prince, Alfred, Blumberg Papers.

15. Prince, *My Life with Viruses, Friends & Enemies*.

16. Kuhn, *The Structure of Scientific Revolutions*.

17. Bayer, Blumberg, and Werner, "Particles associated with Australia Antigen in the Sera of Patients with Leukaemia, Down's Syndrome and Hepatitis."

18. Dane, Cameron, and Briggs, "Virus-Like Particles in Serum of Patients with Australia-Antigen-Associated Hepatitis."

19. London et al., "Serial Transmission in Rhesus Monkeys of an Agent related to Hepatitis-Associated Antigen."

20. "Microbiologist Irving Millman, 88, Helped Develop Hepatitis B Vaccine," *Washington Post*, April 25, 2012.

21. Stevens, "Health Care in the Early 1960s."

22. Alter, Holland, Purcell, Lander, et al., "Posttransfusion Hepatitis after Exclusion of Commercial and Hepatitis-B Antigen-Positive Donors."

23. Blumberg, *Hepatitis B*, 125.

24. Harold M. Schmeck Jr., "Three Urge Tests of Blood," *New York Times*, July 29, 1970.

25. Alter, Holland, Purcell, Lander, et al., "Posttransfusion Hepatitis after Exclusion of Commercial and Hepatitis-B Antigen-Positive Donors."

26. Blumberg and Millman, Vaccine against Viral Hepatitis and Process (patent).

27. For more on the life of Hilleman, see Offit, *Vaccinated*. Blumberg wrote to Hilleman to congratulate him on his Lasker Award in 1984. "This also provides the opportunity for me, on behalf of my colleagues and myself to thank you for the great job you have done in producing the hepatitis vaccine. The promise you made to us some years ago that it would all work well, has been eminently fulfilled." February 22, 1984, Box 27, Folder 17, Blumberg Papers.

28. Offit, *Vaccinated*, 136.

29. Szmuness et al., "Hepatitis B Vaccine."

30. Beasley et al., "Prevention of Perinatally Transmitted Hepatitis B Virus Infections with Hepatitis B Immune Globulin and Hepatitis B Vaccine."

31. Huzair, Farah, and Sturdy, "Biotechnology and the Transformation of Vaccine Innovation."

32. Sherlock et al., "Chronic Liver Disease and Primary Liver-Cell Cancer with Hepatitis-Associated (Australia) Antigen in Serum"; Tabor et al., "Hepatitis B Virus Infection and Primary Hepatocellular Carcinoma."

33. I thank Angela Creager for this point.

34. NASA, "Astrobiology Strategy," 2015, https://astrobiology.nasa.gov/research/astrobiology-at-nasa/astrobiology-strategy/.

35. Alter, Holland, Purcell, and Popper. "Transmissible Agent in Non-A, Non-B Hepatitis"; Choo et al., "Isolation of a cDNA Clone Derived from a Blood-Borne Non-A, Non-B Viral Hepatitis Genome."

36. Blight, Kolykhalov, and Rice, "Efficient Initiation of HCV RNA Replication in Cell Culture."

37. Alter and Houghton, "Hepatitis C Virus and Eliminating Post-transfusion Hepatitis"; Kolykhalov et al., "Transmission of Hepatitis C by Intrahepatic Inoculation with Transcribed RNA."

8. A Breakthrough and a New Tool

1. Here I am taking a stand somewhat contrary to Fisher's. She argues that Temin's importance to the history of cancer research is overblown. For reasons that will become apparent I think his willingness to push ideas that others thought were seriously misguided was nonetheless courageous. Fisher, "Not beyond Reasonable Doubt."

2. Cooper, Temin, Sugden, *The DNA Provirus*. The material on Howard Temin in this chapter draws on this volume.

3. Quoted in Cooper, Temin, and Sugden, *The DNA Provirus*, xvi.

4. Quoted in Cooper, Temin, and Sugden, *The DNA Provirus*, xviii.

5. Temin, "Cancer and Viruses."

6. Potter to Temin, February 11, 1959, Howard Temin Papers, University of Wisconsin, Madison, hereafter Temin Papers.

7. For more on the unique, supportive environment of the McArdle Laboratory that made it an attractive place for Temin, see Rusch, *Something Attempted, Something Done*.

8. Dulbecco to Temin, July 11, 1961, Temin Papers.

9. Temin to Dulbecco, July 12, 1961, Temin Papers.

10. W. R. Bryan to Temin, November 15, 1961, Temin Papers.

11. Sarah Stewart to Temin, December 7, 1961, Temin Papers.

12. Temin, "Separation of Morphological Conversion and Virus Production in Rous Sarcoma Virus Infection."

13. Lionel and Elizabeth Crawford to Temin, June 12, 1962, Temin Papers.

14. Olby, *Francis Crick*, 313. Crick's original formulation of the central dogma was open to the possibility of an information transfer from RNA to DNA, although no cases were known to exist at the time. The way Watson popularized the central dogma in his *Molecular Biology of the Gene* made an RNA to DNA transfer appear highly unlikely. For a more recent discussion of the Central Dogma, see Darden, "Flow of Information in Molecular Biological Mechanisms."

15. Temin to Vogt, August 2, [1963], Vogt Papers, University of California San Diego.

16. Temin to Vogt, August 2, [1963], Vogt Papers.

17. Temin, "The Effects of Actinomycin D on Growth of Rous Sarcoma Virus in Vitro."

18. Barry to Temin, September 30, 1963, Temin Papers.

19. Cooper and Sugden, preface to Cooper, Temin, and Sugden, *The DNA Provirus*, xiii.

20. Bishop, *How to Win the Nobel Prize*, 61. See also Marcum, "From Heresy to Dogma in Accounts of Opposition to Howard Temin's DNA Provirus Hypothesis."

21. Baltimore to Temin, May 10, 1961, Box 120, Temin Papers.

22. Crotty, *Ahead of the Curve*, 10.

23. Crotty, *Ahead of the Curve*, 21.

24. Crotty, *Ahead of the Curve*, 31.

25. David Baltimore, interview by author, Caltech, January 10, 2013.

26. Baltimore, interview by author.

27. Baltimore, interview by author.

28. Michael Jacobsen (bio), Center for Science in the Public Interest, https://cspinet.org/Michael-Jacobson (retrieved January 20, 2020).

29. Baltimore, interview by author.

30. Schaffer, Hackett, and Soergel, "Vesicular Stomatitis Virus RNA"; Huang, Baltimore, and Stampfer, "Ribonucleic Acid Synthesis of Vesicular Stomatitis Virus: III."

31. Baltimore, "Expression of Animal Virus Genomes."

32. Shatkin and Sipe, "RNA Polymerase Activity in Purified Reoviruses."

33. Baltimore, David, interview by author, Caltech, January 10, 2013.

34. Munyon, Paoletti, and Grace, "RNA Polymerase Activity in Purified Infectious Vaccinia Virus."

35. Crotty, *Ahead of the Curve*, 78.

36. Temin to Baltimore, June 17, 1970, Temin Papers.

37. Crick, "On Protein Synthesis."

38. Judson, *The Eighth Day of Creation*, 337.

39. Crick to Temin, July 17, 1970, Temin Papers.

40. Crick, "Central Dogma of Molecular Biology." See also Olby, "Francis Crick, DNA, and the Central Dogma."

41. Fisher argues that there were two versions of the central dogma at the time. In contrast to Crick's version described above, Watson's version—roughly DNA makes RNA makes protein—would be refuted by the discovery of reverse transcriptase. See Fisher, "Not beyond Reasonable Doubt." See also Olby, *Francis Crick*, 313. I tried to get Watson to comment on the two versions, but he brushed off the distinction as uninteresting.

42. Draft letters, July 23, July 24, and July 27, 1970, Temin Papers. The July 27 letter is marked sent.

43. Temin to Crick, July 27, 1970, Temin Papers.

44. Crick to Temin, August 3, 1970, Temin Papers.

45. De Chadarevian, "Sequences, Conformation, Information."

46. Later philosophers would develop this idea further. See Griffiths, "Genetic Information."

47. Crotty, *Ahead of the Curve*, 81. These results were later found to be faulty.

48. Also called the Special Virus Leukemia Program before 1968. I thank Robin Scheffler for the date.

49. Scheffler, *A Contagious Cause*.

50. Jane Brody, "Surgeon and Four Cancer Researchers win Lasker Awards," *New York Times*, November 13, 1974.

51. Cooper et al., *The DNA Prophage Hypothesis*, xxi.

9. The Molecular-Genetic Basis of Cancer

1. M. K. Klein, "The Legacy of the 'Yellow Berets.'"

2. Michael Bishop, interview by author, San Francisco, January 8, 2013.

3. Bishop, *How to Win the Nobel Prize*, 53.

4. Perlman and Pastan. "Regulation of B-Galactosidase Synthesis in Escherichia Coli by Cyclic Adenosine 3′, 5′-Monophosphate."

5. Bader, "The Requirement for DNA Synthesis in the Growth of Rous Sarcoma and Rous-Associated Viruses."

6. Varmus, *The Art and Politics of Science*, 48.

7. Steve Martin, interview by author, February 4, 2013. Martin, "Rous Sarcoma Virus."

8. Duesberg and Vogt. "Differences between the Ribonucleic Acids of Transforming and Nontransforming Avian Tumor Viruses."

9. Bernstein, McCormick, and Martin, "Transformation-Defective Mutants of Avian Sarcoma Viruses."

10. P. K. Vogt, "Oncogenes and the Revolution in Cancer Research."

11. Duesberg and Vogt, "On the Role of DNA Synthesis in Avian Tumor Virus Infection."

12. Peter Vogt, interview by author, Scripps Research Institute, January 11, 2013.

13. Toyoshima and Vogt, "Temperature Sensitive Mutants of an Avian Sarcoma Virus."

14. Vogt, interview by author.

15. Dominique Stehelin, interview by author, Lille, France, August 4–5, 2016.

16. Varmus, Guntaka, et al., "Synthesis of Viral DNA in the Cytoplasm of Duck Embryo Fibroblasts and in Enucleated Cells after Infection by Avian Sarcoma Virus"; Guntaka et al., "Ethidium Bromide Inhibits Appearance of Closed Circular Viral DNA and Integration of Virus-Specific DNA in Duck Cells Infected by Avian Sarcoma Virus."

17. Ramareddy Guntaka, personal communication with author, September 24, 2018.

18. Neiman et al., "Nucleotide Sequence Relationships of Avian RNA Tumor Viruses."

19. Guntaka, personal communication with author.

20. Stehelin, interview by author.

21. Stehelin, quoted in Bishop, *How to Win the Nobel Prize*, 162.

22. Stehelin, interview by author.

23. Stehelin, Guntaka, et al., "Purification of DNA Complementary to Nucleotide Sequences required for Neoplastic Transformation of Fibroblasts by Avian Sarcoma Viruses."

24. Guntaka to Varmus, April 1, 1976, Varmus Papers, Archives and Special Collections, University of California, San Francisco.

25. Varmus to Guntaka, April 5, [1976], Varmus Papers.

26. Huebner and Todaro, "Oncogenes of RNA Tumor Viruses as Determinants of Cancer."

27. For the early history of this technique, see Fisher, "Not Just 'a Clever Way to Detect Whether DNA Really Made RNA.'"

28. Varmus to Wilson, May 22, 1975, Varmus Papers.

29. Harold Varmus, personal communication with author, October 10, 2018.

30. Stehelin to Baltimore, March 19, 1975; Hanafusa to Stehelin, March 17, 1975, in the possession of Dominique Stehelin.

31. Stehelin, Varmus, et al., "DNA Related to the Transforming Gene (s) of Avian Sarcoma Viruses Is Present in Normal Avian DNA."

32. Guntaka, personal communication with author.

33. Neiman to Varmus, December 27, 1974, Varmus Papers.

34. Temin to Bishop, September 25, 1975, from a copy in possession of Dominique Stehelin.

35. Neiman et al., "Nucleotide Sequence Relationships of Avian RNA Tumor Viruses."

36. Stehelin, interview by author.

37. Deborah Spector, personal communication with author, October 2018.

38. Varmus, Stehelin, et al., "Distribution and Function of Defined Regions of Avian Tumor Virus Genomes in Viruses and Uninfected Cells."

39. Stehelin to Varmus, December 10, 1975, Varmus Papers.

40. Oppermann et al., "Uninfected Vertebrate Cells Contain a Protein That Is Closely Related to the Product of the Avian Sarcoma Virus Transforming Gene (*src*)."

41. Martin, "The Road to Src."

42. Sedwick, "Joan Brugge."

43. Collett and Erikson, "Protein Kinase Activity Associated with the Avian Sarcoma Virus Src Gene Product." See also Wapner, *The Philadelphia Chromosome*.

44. Art Levinson later had an illustrious career in business. He left UCSF for Genentech, eventually becoming CEO. In 2011 he became chair of the board of directors of Apple Inc.

45. Levinson et al., "Evidence That the Transforming Gene of Avian Sarcoma Virus Encodes a Protein Kinase Associated with a Phosphoprotein."

46. Art Levinson, personal communication with author, January 27, 2019.

47. Hunter and Sefton, "Transforming Gene Product of Rous Sarcoma Virus Phosphorylates Tyrosine."

48. Witte, Dasgupta, and Baltimore, "Abelson Murine Leukaemia Virus Protein Is Phosphorylated in Vitro to Form Phosphotyrosine."

49. Karolinska Institutet, press release, October 9, 1989.

50. Vogt, interview by author.

51. Gina Kolata, "Frenchman says Panel Overlooked His Contribution," *New York Times*, October 11, 1989.

52. See also Guntaka, "Antecedents of a Nobel Prize (I)."

53. Peter Gorner, "Cancer Work Earns Nobel for 2 in the US," *Chicago Tribune*, October 10, 1989.

54. Coles, "French Researcher Asks for Share."

55. English translation of the French original.

56. "Conduct Unbecoming," *Nature* 342 (1990): 328.

57. Stehelin to Graf, September 15, 1974, original in possession of Dominique Stehelin.

58. Stéhelin and Graf, "Avian Myelocytomatosis and Erythroblastosis Viruses Lack the Transforming Gene Src of Avian Sarcoma Viruses."

59. Quoted in Varmus, *The Art and Politics of Science*, 119.

10. Mecca for Tumor Virology

1. Luria, *General Virology*, 324.

2. Luria, *General Virology*, 334.

3. Luria, *General Virology*, 362.

4. Creager and Morgan, "After the Double Helix."

5. Morgan, "Early Theories of Virus Structure."

6. Watson, *Avoid Boring People*, 141.

7. Watson, *Molecular Biology of the Gene*, 442.

8. Watson to Temin, January 13, 1965, Howard Temin Papers, University of Wisconsin, Madison, hereafter Temin Papers .

9. Watson, *Molecular Biology of the Gene*, 449.

10. Watson, *Molecular Biology of the Gene*, 458.

11. Creager and Morgan, "After the Double Helix."

12. Watson, *Molecular Biology of the Gene*, 468.

13. Watson, *Father to Son*, 231. For the history of CSHL, see Witkowski, *The Road to Discovery*; and also Comfort and Glass, "Building Arcadia."

14. Cold Spring Harbor also developed a focus on eugenics in the early twentieth century. See Allen, "The Eugenics Record Office at Cold Spring Harbor, 1910–1940."

15. Watson, *Avoid Boring People*, 262.

16. Watson, *CSHL Directors Report 1968*, 4, CSHL Archives, Cold Spring Harbor, New York, hereafter CHSL Archives.

17. In an analysis of the use of Feyman diagrams in postwar physics, Kaiser argues that there was no real pedagogy at a distance. Rather, physicists learned how to use Feynman diagrams through personal contact with the leaders in the field. Something similar occurs in the 1970s in tumor virology. Scientists would learn new techniques by taking summer courses at CSHL. Kaiser, *Drawing Theories Apart*.

18. Watson, *Father to Son*, 231.

19. Interestingly, Sambrook's characterization of the humanities resonates with C. P. Snow's classic account first published in 1959: Snow, *The Two Cultures*. I thank Angela Creager for this point.

20. Fenner and Burnet, "A Short Description of the Poxvirus Group (Vaccinia and Related Viruses)."

21. Fenner and Marshall, "A Comparison of the Virulence for European Rabbits (Oryctolagus Cuniculus) of Strains of Myxoma Virus Recovered in the Field in Australia, Europe and America"; Fenner and Woodroofe, "Changes in the Virulence and Antigenic Structure of Strains of Myxoma Virus Recovered from Australian Wild Rabbits Between 1950 and 1964."

22. Fenner and Sambrook, "The Genetics of Animal Viruses."

23. Brenner to Jon Beckwith (official letter), November 3, 1971, Sambrook Correspondence, CSHL Archives.

24. Brenner to Jon Beckwith (unofficial letter), November 3, 1971, Sambrook Correspondence. Brenner's opinion of Sambrook would improve greatly over the years. By the early 1980s, it had improved to the level that Brenner nominated Sambrook to be a fellow of the Royal Society. Sambrook was elected a fellow of the Royal Society in 1984.

25. Brenner had begun to work on a project that he refused to discuss with Sambrook. It turned out to be the worm *Caenorhabditis elegans*, which Brenner would champion as a model organism for the study of animal development and which would win Brenner the 2002 Nobel Prize.

26. Sweet and Hilleman, "The Vacuolating Virus, SV 40." Rhesus monkeys had been used as a model system in polio research since the virus was discovered in 1908. D. Crawford, *The Invisible Enemy*, 105.

27. Bäumler, *Cancer*, 117.

28. Bookchin and Schumacher, *The Virus and The Vaccine*; Stark and Campbell, "Stowaways in the History of Science."

29. Heiner Westphal, interview by Victor K. McInerny, March 28, 2001, in possession of McInerny. See Fujimura, *Crafting Science*, for an argument that tumor viruses became research tools in cancer research. This point is taken up more generally by Rheinberger, "Recent Orientations and Reorientations in the Life Sciences."

30. Sambrook et al., "The Integrated State of Viral DNA in SV40-Transformed Cells."

31. Westphal, interview by McInerny.

32. Sambrook et al., "The Integrated State of Viral DNA in SV40-Transformed Cells."

33. Watson, *A Passion for DNA*, 52.

34. Sambrook, "Tumor Viruses," 90.

35. Westphal, interview by McInerny.

36. Cold Spring Harbor Tumor Virus Center, research grant application, CSHL Archives.

37. Joseph Sambrook, personal communication with author, March 2015.

38. This idea of using reversion to a normal cellular phenotype has recently been used to develop a novel approach to cancer therapy and is in clinical trials. Powers and Pollack, "Inducing Stable Reversion to Achieve Cancer Control"; "Can We Make Cancer Cells Normal Again?," Columbia University Irving Medical Center, April 6, 2017, www.cuimc.columbia.edu/News/Can-We-Make-Cancer-Cells-Normal-Again.

39. Goldman, Pollack, and Hopkins, "Preservation of Normal Behavior by Enucleated Cells in Culture."

40. Watson, *Directors Report 1971*, CSHL Archives.

41. Carel Mulder, telephone interview by author, December 9, 2014.

42. Nancy Hopkins, telephone interview by author, June 21, 2013.

43. Danna and Nathans, "Specific Cleavage of Simian Virus 40 DNA by Restriction Endonuclease of Hemophilus Influenzae."

44. Rowe et al., "Isolation of a Cytopathogenic Agent from Human Adenoids Undergoing Spontaneous Degeneration in Tissue Culture"; Hilleman and Werner, "Recovery of New Agent from Patients with Acute Respiratory Illness."

45. Pierrel, "An RNA Phage Lab."

46. Westphal, "SV40 DNA Strand Selection by *Escherichia coli* RNA Polymerase."

47. Sambrook, Sharp, and Keller, "Transcription of Simian Virus 40."

48. Sambrook, Sharp, and Keller, "Transcription of Simian Virus 40," 69.

49. Watson, foreword to *Cold Spring Harbor Symposia on Quantitative Biology*.

50. Joe Sambrook, telephone interview by author, June 21, 2013.

51. Grodzicker et al., "Physical Mapping of Temperature-Sensitive Mutations of Adenoviruses." They also published in the *Journal of Molecular Biology*.

52. L. V. Crawford, "Transforming Genes of DNA Tumor Viruses."

53. Stoker, "Neoplastic Transformation by Polyoma Virus and Its Wider Implications."

54. Black et al., "A Specific Complement-Fixing Antigen Present in SV40 Tumor and Transformed Cells."

55. Baltimore, "Tumor Viruses."

56. Watson, *Directors Report* 1989, CSHL Archives.

57. Watson, *Directors Report* 1977, CSHL Archives. On the importance of mRNA, see Gaudillière, "Molecular Biologists, Biochemists, and messenger RNA."

58. Wright, *Molecular Politics*; Emrich, "Dr. Genelove."

59. Yi, *The Recombinant University*; Berg and Mertz, "Personal Reflections on the Origins and Emergence of Recombinant DNA Technology"; Berg, "Moments of Discovery."

60. Rasmussen, "DNA Technology."

61. For more on this topic, see Creager, *Life Atomic*; and M. Armstrong, *Germ Wars*, chapter 2.

62. Hellman, Oxman, and Pollack, *Biohazards in Biological Research*, 354.

63. Berg et al., "Potential Biohazards of Recombinant DNA Molecules."

64. McElheny, *Watson and DNA*.

65. In the next few years several books considered the risks of recombinant DNA technology: Rogers, *Biohazard*; Lear, *Recombinant DNA*; Wade, *The Ultimate Experiment*; and Krimsky, *Genetic Alchemy*.

66. Summary of NIH regulations from the Paul Berg Papers Guide, US National Library of Medicine, https://profiles.nlm.nih.gov/spotlight/cd/feature/dna (retrieved January 20, 2020).

67. Watson, *Annual Report* 1976, 6, CSHL Library and Archives.

68. Watson, *Annual Report* 1976, 8.

69. Tjian in Witkowski, Gann, and Sambrook, *Life Illuminated*, 120.

70. Robert Tjian, telephone interview with author, February 20, 2015.

71. Tjian, interview with author.

72. Weston, *Blue Skies and Bench Space*, 28.

73. Mary Jane Gething on Meeting Joe Sambrook, Oral History Collection, CSHL Archives, January 16, 2003, http://library.cshl.edu/oralhistory/interview/cshl/memories/meeting-sambrook/.

74. Thummel, Tjian, and Grodzicker, "Expression of SV40 T Antigen under Control of Adenovirus Promoters."

75. Thomas Maniatis, personal communication with author, July 14, 2015.

76. Creager, "Recipes for Recombining DNA."

77. James D. Watson—Biography, Oral History Collection, CSHL Archives, http://library.cshl.edu/oralhistory/speaker/james-d-watson/ (retrieved January 20, 2020).

78. Miller, *Experiments in Molecular Genetics*.

79. Maniatis, personal communication.

80. *The Joseph Sambrook Laboratory Dedication Ceremony* (booklet), 1985, CSHL Archives.

81. Maniatis, personal communication.

82. *The Joseph Sambrook Laboratory Dedication Ceremony*.

83. Griesemer and Gerson suggest we need a broader conception of collaboration in science to include relationships like that between Sambrook and Watson. Griesemer and Gerson, "Collaboration in the Museum of Vertebrate Zoology."

84. Sydney Brenner on John Cairns and Joe Sambrook, Cold Spring Harbor Laboratory, Oral History Collection, CSHL, June 10, 2002, http://library.cshl.edu/oralhistory/interview/james-d-watson/cshl-director-and-president/john-cairns-and-joe-sambrook-cold-spring-harbor-la/.

85. Mary Jane Gething on Marriage to Joe Sambrook, Oral History Collection, CSHL, January 6, 2003, https://library.cshl.edu/oralhistory/interview/cshl/social-life/marriage-joe-sambrook/.

86. Sambrook, personal communication.

11. Control Mechanisms beyond Viruses

1. Phillip Sharp, interview by author, Cape Cod, MA, August 12, 2013.

2. Claire Moore, interview by author, Cold Spring Harbor Laboratory, October 24, 2017.

3. Phillip Sharp, personal communication with author, September 2018.

4. Sharp, personal communication.

5. David Baltimore, personal communication with author, September 2018.

6. Jim Darnell, personal communication with author, October 23, 2017.

7. Berk, "Discovery of RNA Splicing and Genes in Pieces."

8. Sharp, personal communication.

9. Joe Weber, personal communication with author, October 8, 2018.

10. Richard Roberts, interview by author, Ipswich, MA, August 15, 2013.

11. Roberts, interview by author.

12. On a related point, see Yi, "The Scientific Commons in the Marketplace."

13. McElheny, *Watson and DNA*, 189.

14. Louise Chow in Witkowski, Gann, and Sambrook, *Life Illuminated*, 105.

15. Chow, Roberts, et al., "A Map of Cytoplasmic RNA Transcripts from Lytic Adenovirus Type 2."

16. Gelinas and Roberts, "One Predominant 5′-Undecanucleotide in Adenovirus 2 Late Messenger RNAs."

17. Darnell, *RNA*, 194.

18. Louise Chow in Witkowski, Gann, and Sambrook, *Life Illuminated*, 107.

19. Roberts, interview by author.

20. McElheny, *Watson and DNA*, 208.

21. Roberts, interview by author.

22. McElheny, *Watson and DNA*, 211.

23. Phillip Sharp, comments at the CSHL meeting, "40 Years of mRNA Splicing," Cold Spring Harbor, NY, October 22–25, 2017.

24. Quoted in Witkowski, *The Road to Discovery*, 257.

25. Gilbert, "Why Genes in Pieces?"

26. Ford Doolittle, personal communication, November 2018. Doolittle, "Genes in Pieces." Gilbert, "Genes-in-Pieces Revisited."

27. Darnell, *RNA*, 197.

28. Roberts, interview by author.

29. Roberts, interview by author.

30. Louise Chow in Witkowski, Gann, and Sambrook, *Life Illuminated*, 109.

31. Anthony Flint, "Behind Nobel, a Struggle for Recognition: Some Scientists Say Colleague of Beverly Researcher Deserved a Share of Medical Prize," *Boston Globe*, November 5, 1993.

32. Also see Abir-Am, "The Women Who Discovered RNA Splicing."

33. Louise Chow in Witkowski, Gann, and Sambrook, *Life Illuminated*, 109.

34. Richard J. Roberts, banquet speech, December 10, 1993, www.nobelprize.org/prizes/medicine/1993/roberts/speech/.

35. Dick and Jones, "The Commercialization of Molecular Biology."

36. For a history of the use of databases in biology, see Strasser, *Collecting Experiments*.

12. A Second Cancer Gene

1. Interview with Edward Scolnick, National Cancer Institute Oral History Project, https://history.nih.gov/display/history/scolnick%2c+edward+1998, June 24, 1998.

2. Scolnick, Tompkins, et al., "Release Factors Differing in Specificity for Terminator Codons."

3. Ed Scolnick, telephone interview by author, July 3, 2013.

4. Gross, *Oncogenic Viruses*, 2nd ed., 584.

5. Scolnick, Rands, et al., "Studies on the Nucleic Acid Sequences of Kirsten Sarcoma Virus." See also Scolnick, Goldberg, and Parks, "A Biochemical and Genetic Analysis of Mammalian RNA-Containing Sarcoma Viruses."

6. Scolnick, Papageorge, and Shih, "Guanine Nucleotide-Binding Activity as an Assay for Src Protein of Rat-Derived Murine Sarcoma Viruses."

7. Scolnick, telephone interview by author.

8. Coffin et al., "Proposal for Naming Host Cell-Derived Inserts in Retrovirus Genomes."

9. Smotkin et al., "Infectious Viral DNA of Murine Leukemia Virus."

10. Cooper and Temin, "Infectious Rous Sarcoma Virus and Reticuloendotheliosis Virus DNAs"; Cooper and Temin, "Infectious DNA from Cells Infected with Rous Sarcoma Virus, Reticuloendotheliosis Virus or Rous-Associated Virus-0."

11. Angier, *Natural Obsessions*, 56.

12. Graham and van der Eb, "A New Technique for the Assay of Infectivity of Human Adenovirus 5 DNA"; Weinberg, "Inadvertent Cancer Research."

13. Weinberg, *Racing to the Beginning of the Road*, 142.

14. Weinberg, *Racing to the Beginning of the Road*, 131.

15. Weinberg, *Racing to the Beginning of the Road*, 145.

16. Weinberg, *Racing to the Beginning of the Road*.

17. Angier, *Natural Obsessions*, 78.

18. Weinberg, *Racing to the Beginning of the Road*, 157.

19. Shih, Shilo, Goldfarb, Dannenberg, and Weinberg, "Passage of Phenotypes of Chemically Transformed Cells via Transfection of DNA and Chromatin."

20. Geoffrey Cooper, interview by author, Boston, MA, August 12, 2013.

21. Cooper, interview by author.

22. Hill and Hillova. "Recovery of the Temperature-Sensitive Mutant of Rous Sarcoma Virus from Chicken Cells Exposed to DNA Extracted from Hamster Cells Transformed by the Mutant."

23. Cooper, interview by author.

24. Weinberg, *Racing to the Beginning of the Road*, 161.

25. Krontiris and Cooper, "Transforming Activity of Human Tumor DNAs"; Shih, Padhy, et al., "Transforming Genes of Carcinomas and Neuroblastomas Introduced into Mouse Fibroblasts."

26. Angier, *Natural Obsessions*, 299.

27. Wigler et al., "Transfer of Purified Herpes Virus Thymidine Kinase Gene to Cultured Mouse Cells."

28. US patent #4399216. For a discussion of the rise of patents in recombinant biology, see Hughes, "Making Dollars Out of DNA."

29. Colaianni and Cook-Deegan, "Columbia University's Axel Patents."

30. Weinberg, *Racing to the Beginning of the Road*, 165.

31. Michael Wigler, interview by author, Cold Spring Harbor Laboratory, August 1, 2013.

32. Wigler, interview by author.

33. Harris et al., "Suppression of Malignancy by Cell Fusion."

34. Goldfarb et al., "Isolation and Preliminary Characterization of a Human Transforming Gene from T24 Bladder Carcinoma Cells."

35. Weinberg, *Racing to the Beginning of the Road*, 185.

36. Tabin et al., "Mechanism of Activation of a Human Oncogene."

37. Parada et al., "Human EJ Bladder Carcinoma Oncogene Is Homologue of Harvey Sarcoma Virus Gene."

38. Der, Krontiris, and Cooper, "Transforming Genes of Human Bladder and Lung Carcinoma Cell Lines Are Homologous to the Ras Genes of Harvey and Kirsten Sarcoma Viruses."

39. I have used the lowercase name for the gene *ras*, but later a convention uses uppercase *RAS* for human genes.

40. Weinberg, *Racing to the Beginning of the Road*, 206.

41. Wigler, interview by author.

42. Pauling et al., "Sickle Cell Anemia, a Molecular Disease"; Strasser, "Sickle Cell Anemia, a Molecular Disease"; Hager, *Linus Pauling*. See also Feldman and Tauber, "Sickle Cell Anemia"; Strasser and Fantini, "Molecular Diseases and Diseased Molecules"; and Bechtel and Richardson, *Discovering Complexity*.

43. Richard Harris, "The 30-Year Quest to Tame the 'Wily' Cancer Gene," NPR, March 9, 2018, www.npr.org/Sections/Health-Shots/2018/03/09/589613244/The-30-Year-Quest-To-Tame-The-Wily-Cancer-Gene. Researchers are getting closer to an approved drug. See Moore et al., "RAS-Targeted Therapies."

13. A Molecular Brake on Cancer

1. Arnold Levine, interview by author, Princeton, NJ, July 28, 2012. Levine's approach to oncoviruses contrasts with Fujimura's analysis of later cancer research of the 1980s,

where scientists jumped on the oncogene "bandwagon" because of advances in our understanding of oncogenes and the relative ease of increasingly standardized DNA recombinant technologies. Fujimura, "The Molecular Biological Bandwagon in Cancer Research."

2. Jaenisch, Mayer, and Levine, "Replicating SV40 Molecules Containing Closed Circular Template DNA Strands."

3. Linzer and Levine. "Characterization of a 54K Dalton Cellular SV40 Tumor Antigen Present in SV40-Transformed Cells and Uninfected Embryonal Carcinoma Cells."

4. Zhelev, "Man of Science."

5. Hammond, "JBS Haldane, Holism, and Synthesis in Evolution."

6. Lane, "Such an Obsession."

7. Silver and Lane, "Dominant Nonresponsiveness in the Induction of Autoimmunity to Liver-Specific F Antigen"; Lane and Silver, "Isolation of a Murine Liver-Specific Alloantigen, F Antigen, and Examination of Its Immunogenic Properties by Radioimmunoassay."

8. Armstrong, *p53*, 42.

9. Lane, "Such an Obsession," 121.

10. Weston, *Blue Skies and Bench Space*, 52.

11. Lane, quoted in Armstrong, *p53*, 43.

12. Crawford and Lane, "An Immune Complex Assay for SV40 T Antigen."

13. Lane, quoted in Armstrong, *p53*, 44–45.

14. Lane and Crawford, "T Antigen Is Bound to a Host Protein in SY40-Transformed Cells."

15. Weston, *Blue Skies and Bench Space*, chapter 3.

16. Quoted in Zhelev, "Man of Science."

17. Deleo et al., "Detection of a Transformation-Related Antigen in Chemically Induced Sarcomas and Other Transformed Cells of the Mouse."

18. Armstrong, *p53*, 60.

19. Chumakov, Iotsova, and Georgiev, "Isolation of a Plasmid Clone Containing the mRNA Sequence for Mouse Nonviral T-antigen."

20. Eliyahu et al., "Participation of p53 Cellular Tumour Antigen in Transformation of Normal Embryonic Cells."

21. Jenkins, Rudge, and Currie, "Cellular Immortalization by a cDNA Clone Encoding the Transformation-Associated Phosphoprotein p53."

22. Crawford, Pim, and Lamb, "The Cellular Protein p53 in Human Tumours."

23. Levine, Arnold, interview by author, Princeton, July 28, 2012.

24. Soussi, "The History of p53."

25. Finlay, Hinds, and Levine, "The p53 Proto-oncogene Can Act as a Suppressor of Transformation."

26. Levine, interview by author.

27. Levine, Finlay, and Hinds, "p53 Is a Tumor Suppressor Gene."

28. Levine, interview by author.

29. Baker et al., "Chromosome 17 Deletions and p53 Gene Mutations in Colorectal Carcinomas."

30. Bert Vogelstein, personal communication with author, October 11, 2018.

31. Kern et al., "Identification of p53 as a Sequence-Specific DNA-Binding Protein."

32. Knudson, "Hereditary Cancer, Oncogenes, and Antioncogenes."

33. Whyte et al., "Association between an Oncogene and an Anti-oncogene."

34. I thank Angela Creager for the point about somatic mutation theory. See her *Life Atomic*, 177.

35. Lu, "Legends of p53."

36. Harris, "p53 Sweeps through Cancer Research."

37. Koshland, "Molecule of the Year."

38. Vogelstein, Lane, and Levine, "Surfing the p53 Network." For a discussion of different metaphors applied to the cell, see Reynolds, *The Third Lens*.

39. Karen W. Arenson, "Behavior Forced Rockefeller U. Head to Resign, Trustees Say, *New York Times*, February 14, 2002, www.nytimes.com/2002/02/14/nyregion/behavior-forced-rockefeller-u-head-to-resign-trustees-say.html.

40. Check, "Rockefeller Head Quits as Scandal Looms."

41. "Highly Cited Researchers (h>100) according to their Google Scholar Citations public profiles," Ranking Web of Universities, www.webometrics.info/en/node/58. To have an h-index of 274 means that Vogelstein has published 274 papers that have at least 274 citations each.

14. Unplanned Practical Payoffs

1. The biographical material on Gallo draws upon his book *Virus Hunting* and a telephone interview by the author.

2. Gallo, *Virus Hunting*, 40.

3. Giacomoni, "The Origin of DNA." Fisher, "Not Just 'a Clever Way to Detect whether DNA really made RNA.'"

4. Robert Gallo, telephone interview by author, October 22, 2013.

5. Huebner and Todaro, "Oncogenes of RNA Tumor Viruses as Determinants of Cancer."

6. Gallo, *Virus Hunting*, 68.

7. Miller et al., "Virus-Like Particles in Phytohemagglutinin-Stimulated Lymphocyte Cultures with Reference to Bovine Lymphosarcoma 2"; Miller and Olson, "Precipitating Antibody to an Internal Antigen of the C-Type Virus Associated with Bovine Lymphosarcoma."

8. Sarngadharan et al., "Reverse Transcriptase Activity of Human Acute Leukaemic Cells."

9. William Haseltine, personal communication, July 9, 2020.

10. Gallo, *Virus Hunting*, 79.

11. Throughout the 1970s, Gross would support Gallo in his search for human tumor viruses, when many others thought he was an "oddball" for pursuing what appeared to them an impossible goal.

12. Gallo, *Virus Hunting*, 90.

13. Morgan, Ruscetti, and Gallo, "Selective in Vitro Growth of T Lymphocytes from Normal Human Bone Marrows."

14. For a discussion of the use of IL-2 as a possible cancer treatment, see Löwy, *Between Bench and Bedside.*

15. Poiesz et al., "Detection and Isolation of Type C Retrovirus Particles from Fresh and Cultured Lymphocytes of a Patient with Cutaneous T-Cell Lymphoma."

16. Poiesz et al., "Detection and Isolation of Type C Retrovirus Particles from Fresh and Cultured Lymphocytes of a Patient with Cutaneous T-Cell Lymphoma."

17. Wagner to Gallo, September 15, 1980, reproduced in Kontaratos, "Dissecting a Discovery," 33.

18. Ryan, *Virolution*, 81; Uchiyama, Yodoi, Sagawa, Takatsuki, and Uchino, "Adult T-cell Leukemia."

19. Montagnier, *Virus*, 13.

20. Montagnier, *Virus*, 22. See also Gaudillière, *Inventer la biomédecine.*

21. Creager, *The Life of a Virus*, chapter 7.

22. Montagnier and Sanders, "Replicative Form of Encephalomyocarditis Virus Ribonucleic Acid."

23. Montagnier, *Virus*, 25.

24. Macpherson and Montagnier, "Agar Suspension Culture for the Selective Assay of Cells transformed by Polyoma Virus."

25. Crawford et al., "Cell Transformation by Different Forms of Polyoma Virus DNA."

26. Latarjet, Cramer, and Montagnier, "Inactivation, by UV-, X-, and Γ-Radiations, of the Infecting and Transforming Capacities of Polyoma Virus."

27. For the history of interferon, see Pieters, *Interferon*; and Rasmussen, *Gene Jockeys.*

28. Montagnier, *Virus*, 39.

29. Montagnier, *Virus*, 240n9.

30. Gallo, *Virus Hunting*, 133.

31. Gottlieb et al., "Pneumocystis pneumonia—Los Angeles."

32. Max Essex, interview by author, Boston, MA, August 13, 2013. William Haseltine, personal communication, July 8, 2020.

33. Garrett, *The Coming Plague*, 309. For a classic work on the AIDS epidemic, see Shilts, *And the Band Played On.*

34. Harden claims that Leibovitch had read Gallo's article, "Epidemic Takes Mystery Immune Deficiency beyond Gays." Presumably in *Medical World News*, in which Gallo speculated that the cause of AIDS was a retrovirus, possibly HTLV-1 or HTLV-2. Harden, *AIDS at 30*, 54. See also the letter from Jacques Leibowitch to Maître Dominique, February 28, 1992, in the Jon Cohen AIDS Research Collection at the University of Michigan, https://quod.lib.umich.edu/c/cohenaids/5571095.0541.056?rgn=main;view=fulltext.

35. "'Epidemic' Takes Mystery Immune Deficiency beyond Gays," *Medical World News*, April 16, 1982, 6–9. Gallo meant a variant or a relative of HTLV, but this subtlety did not make it into the short article. Gallo, personal communication with author, September 1, 2018. This article appears not to be indexed properly, making it difficult to find.

36. Weiss, "On Viruses, Discovery, and Recognition."

37. Montagnier, *Virus*, 48.

38. Françoise Barré-Sinoussi, "Biographical," Nobel Prize, www.nobelprize.org/prizes/medicine/2008/barre-sinoussi/auto-biography/ (retrieved January 20, 2020).

39. Gallo, *Virus Hunting*, 147.

40. Montagnier to Gallo, February 2, 1983, www.sciencefictions.net/pdfdocs/l_montagnier_to_r_gallo_02.02.83.pdf (retrieved January 20, 2020).

41. Gallo, *Virus Hunting*, 150.

42. Barre-Sinoussi et al., "Isolation of a T-Lymphotropic Retrovirus from a Patient at Risk for Acquired Immune Deficiency Syndrome (AIDS)."

43. Barre-Sinoussi et al., "Isolation of a T-Lymphotropic Retrovirus from a Patient at Risk for Acquired Immune Deficiency Syndrome (AIDS)."

44. Barre-Sinoussi et al., "Isolation of a T-Lymphotropic Retrovirus from a Patient at Risk for Acquired Immune Deficiency Syndrome (AIDS)."

45. Gallo et al., "Isolation of Human T-Cell Leukemia Virus in Acquired Immune Deficiency Syndrome (AIDS)."

46. Essex et al., "Antibodies to Cell Membrane Antigens Associated with Human T-Cell Leukemia Virus in Patients with AIDS."

47. Montagnier, *Virus*, 60.

48. Montagnier, *Virus*, 60.

49. Gallo, Essex, and Gross, *Human T-Cell Leukemia/Lymphoma Virus*.

50. Mary Lasker's husband, a wealthy businessman, had died of cancer, fueling her later philanthropy.

51. Gallo, *Virus Hunting*, 167.

52. Montagnier, Chermann, et al., "A New Human T-Lymphotropic Retrovirus."

53. Gallo, personal communication.

54. Montagnier, *Virus*, 63.

55. Auerbach et al., "Cluster of Cases of the Acquired Immune Deficiency Syndrome."

56. Goedert et al., "Determinants of Retrovirus (HTLV-III) Antibody and Immunodeficiency Conditions in Homosexual Men."

57. These five papers reinforced further evidence from the French group. Vilmer, et al. "Isolation of New Lymphotropic Retrovirus from Two Siblings with Haemophilia B, One with AIDS."

58. Reece, "The Once and Future Robert C. Gallo."

59. Bialy, *Oncogenes, Aneuploidy, and AIDS*.

60. Kontaratos, *Dissecting a Discovery*, 92.

61. It should be noted that some American scientists from the CDC worked with the French scientists as well.

62. Quoted in Kontaratos, *Dissecting a Discovery*, appendix 14.

63. Kontaratos also published an article with Spandidos (see chapter 13) on Gallo. Kontaratos, Sourvinos, and Spandidos, "Examining the Discovery of the Human Retrovirus."

64. Skalka points out that biological containment facilities were quite limited at the time in the history of HIV research. Skalka, *Discovering Retroviruses*, 123.

65. Gallo, *Virus Hunting*, 173.

66. Gallo, personal communication with author.

67. Gallo, personal communication with author.

68. "How One Test Changed HIV," Abbott, November 27, 2019, www.abbott.com/corpnewsroom/Product-And-Innovation/How-One-Test-Changed-HIV.html.

69. Levy et al., "Isolation of Lymphocytopathic Retroviruses from San Francisco Patients with AIDS." Prusiner, "Discovering the Cause of AIDS."

70. Gallo, *Virus Hunting*, 213.

71. Gallo and Montagnier, "The Chronology of AIDS Research."

72. Lawrence K. Altman, "U.S. and France End Rift on AIDS," *New York Times*, April 1, 1987, www.nytimes.com/1987/04/01/us/us-and-france-end-rift-on-aids.html.

73. Pendlebury, "Citation Superstars of NIH."

74. Abbadessa et al., "Unsung Hero Robert C. Gallo."

75. Associated Press, " Doctor Cleared of Misconduct in AIDS Study," *New York Times*, November 5, 1993, www.nytimes.com/1993/11/05/us/doctor-cleared-of-misconduct-in-aids-study.html.

76. Montagnier, Aissa, et al., "Electromagnetic Signals Are Produced by Aqueous Nanostructures Derived from Bacterial DNA Sequences."

77. For a discussion of the various impediments to early HIV and AIDS research, see Epstein, *Impure Science*. The HIV reverse transcriptase was identified as a target for drugs against AIDS as early as 1985. See Chermann, Barre-Sinoussi, and Montagnier, "Retrovirus and AIDS," 295.

15. Planned Practical Payoffs

1. The material on Harald zur Hausen draws on Claudia Cornwall's book *Catching Cancer* and an interview with him.

2. Cornwall, *Catching Cancer*, 78.

3. Quoted in Cornwall, *Catching Cancer*, 79.

4. Harald zur Hausen, interview by author, Heidelberg, Germany, August 8, 2016; Fangerau, "The Novel Arrowsmith, Paul De Kruif (1890–1971) and Jacques Loeb (1859–1924)."

5. Zur Hausen, interview by author.

6. Zur Hausen and Schulte-Holthausen, "Presence of EB Virus Nucleic Acid Homology in a 'Virus-Free' Line of Burkitt Tumour Cells"; zur Hausen et al., "EBV DNA in Biopsies of Burkitt Tumours and Anaplastic Carcinomas of the Nasopharynx."

7. Rowson and Mahy, "Human Papova (Wart) Virus."

8. McIntyre, "Finding the Viral Link."

9. Zur Hausen and Schulte-Holthausen, "Presence of EB Virus Nucleic Acid Homology in a 'Virus-Free' Line of Burkitt Tumour Cells."

10. G. Klein, "Summary of Papers Delivered at the Conference on Herpesvirus and Cervical Cancer (Key Biscayne, Florida)."

11. Cornwell, *Catching Cancer*, 68.

12. G. Klein, "Summary of Papers Delivered at the Conference on Herpesvirus and Cervical Cancer (Key Biscayne, Florida)."

13. A good case can be made that Orth and Jablonska and colleagues from Warsaw found the first oncogenic HPVs. They provided evidence that HPV 5 and 8 were linked to

skin cancer in the rare genetic disease epidermodysplasia verruciformis four years before zur Hausen reported HPV 16 in cervical cancer. Their discovery didn't have the public health implications as the discovery of HPV 16/18, but it was an important proof of concept that HPVs are associated with a human cancer. See Orth et al., "Characteristics of the Lesions and Risk of Malignant Conversion Associated with the Type of Human Papillomavirus Involved in Epidermodysplasia Verruciformis."

14. Zur Hausen, interview by author.

15. Cornwall, *Catching Cancer*, 96.

16. Dürst et al., "A Papillomavirus DNA from a Cervical Carcinoma and Its Prevalence in Cancer Biopsy Samples from Different Geographic Regions."

17. Quoted in Cornwall, *Catching Cancer*, 99.

18. Kessler, "Human Cervical Cancer as a Venereal Disease."

19. See, for example, Galloway and McDougall, "The Oncogenic Potential of Herpes Simplex Viruses." I thank Bill Summers for his advice on this topic.

20. Cornwall, *Catching Cancer*, 99.

21. Schwarz et al., "Structure and Transcription of Human Papillomavirus Sequences in Cervical Carcinoma Cells."

22. King, *Ian Frazer*. The biographical material on Ian Frazer draws on this book.

23. King, *Ian Frazer*, 36.

24. Frazer et al., "Association between Anorectal Dysplasia, Human Papillomavirus, and Human Immunodeficiency Virus Infection in Homosexual Men."

25. King, *Ian Frazer*, 56.

26. Harald zur Hausen, personal communication with author, January 22, 2019.

27. King, *Ian Frazer*, 79

28. Lionel Crawford, personal communication with author, September 2018.

29. Zhou et al., "Expression of Vaccinia Recombinant HPV 16 L1 and L2 ORF Proteins in Epithelial Cells Is Sufficient for Assembly of HPV Virion-Like Particles."

30. Lionel Crawford, quoted in Reynolds and Tansey, *History of Cervical Cancer and the Role of the Human Papillomavirus, 1960–2000*.

31. Quoted in King, *Ian Frazer*, 89.

32. Xiao-Yi Sun, personal communication with author, September 2018.

33. Scolnick, "A Vaccine to Prevent Cervical Cancer."

34. King, *Ian Frazer*, 96.

35. For a discussion of the forces beyond the NCI and HPV research, see Aviles, "Situated Practice and the Emergence of Ethical Research."

36. Scolnick, "A Vaccine to Prevent Cervical Cancer."

37. Bryan, Buckland, et al., "Prevention of Cervical Cancer."

38. Kirnbauer et al., "Papillomavirus L1 Major Capsid Protein Self-Assembles into Virus-Like Particles That Are Highly Immunogenic."

39. Matlashewski, "Human Papillomavirus Update."

40. Koutsky et al., "A Controlled Trial of a Human Papillomavirus Type 16 Vaccine."

41. Walboomers et al., "Human Papillomavirus is a Necessary Cause of Invasive Cervical Cancer Worldwide."

42. Villa et al. "Prophylactic Quadrivalent Human Papillomavirus (types 6, 11, 16, and 18) L1 Virus-Like Particle Vaccine in Young Women."

43. Descriptions of these clinical trials can be found at http://clinicaltrials.gov (retrieved January 20, 2020).

44. Ghim, Jenson, and Schlegel, "HPV-1 L1 Protein Expressed in COS Cells Displays Conformational Epitopes Found on Intact Virions."

45. Kirnbauer et al., "Papillomavirus L1 Major Capsid Protein Self-Assembles into Virus-Like Particles That Are Highly Immunogenic."

46. Rose et al., "Expression of Human Papillomavirus Type 11 L1 Protein in Insect Cells."

47. King, *Ian Frazer*, 114.

48. CSL, press release, February 3, 2005.

49. King, *Ian Frazer*, 133.

50. King, *Ian Frazer*, 135.

51. Frazer v Schlegel, 498 F.3d 1283 (2007).

52. John Schiller, personal communication with author, November 6, 2018. That ethics played a motivational role at this point in HPV vaccine development harmonizes with the work of Natalie Aviles. See Aviles, "Situated Practice and the Emergence of Ethical Research."

53. Zhao, Zhang, and Qu, "Dr. Jian Zhou."

54. "2017 Lasker~DeBakey Clinical Medical Research Award," Lasker Foundation, www.laskerfoundation.org/awards/show/hpv-vaccines-cancer-prevention/ (retrieved January 20, 2020).

55. "Questions about HPV Vaccine Safety," (FAQ), Centers for Disease Control and Prevention, July 15, 2020, www.cdc.gov/Vaccinesafety/Vaccines/Hpv/Hpv-Safety-Faqs .html.

56. For a discussion of the social and political entanglements among various viral vaccines and sexuality, see Mamo and Epstein, "The New Sexual Politics of Cancer."

57. Gee et al., "Quadrivalent HPV Vaccine Safety Review and Safety Monitoring Plans for Nine-Valent HPV Vaccine in the United States." For a thoughtful discussion of the anti-vaccine movement's reluctance to vaccinate, see Goldenberg, "Public Misunderstanding of Science?

58. www.hrsa.gov/sites/default/files/hrsa/vaccine-compensation/data/monthly -stats-january-2019.pdf (assessed July 20, 2020).

59. Holland, Rosenberg, and Iorio, *The HPV Vaccine on Trial*. For a critical review of the book, see Dorit Rubinstein Reiss, "HPV Vaccine Fear Mongering in an Anti-vax Book—a Critical Review," Skeptical Raptor, February 3, 2019, www.skepticalraptor.com /skepticalraptorblog.php/hpv-vaccine-fear-mongering-book-review/; Derek Lowe, "Luc Montagnier Is Not Losing It: Luc Montagnier Has Lost It" (blog), *Science Translational Medicine*, May 29, 2012, https://blogs.sciencemag.org/pipeline/archives/2012/05/29/luc _montagnier_is_not_losing_it_luc_montagnier_has_lost_it.

60. Gottlieb, *Not Quite a Cancer Vaccine*.

61. "New study shows HPV vaccine helping lower HPV infection rates in teen girls"

(press release), Centers for Disease Control and Prevention, June 19, 2013, www.cdc.gov/Media/Releases/2013/P0619-Hpv-Vaccinations.html.

62. Bruni et al., "Global Estimates of Human Papillomavirus Vaccination Coverage by Region and Income Level."

Conclusion

1. Philosophers of science have discussed whether historical case studies can be used to make more general points. See, for example, Pitt, "The Dilemma of Case Studies"; and Burian, "The Dilemma of Case Studies Resolved."

2. P. K. Vogt, "A Humble Chicken Virus that changed Biology and Medicine."

3. Cox and Der, "Ras History."

4. There is a large literature on the nature of reductionism. For some examples applied to molecular biology, see Fuerst, "The Role of Reductionism in the Development of Molecular Biology"; Ruse, "Reduction, Replacement, and Molecular Biology"; Sarkar, *Genetics and Reductionism*; Rosenberg, *Darwinian Reductionism*.

5. For a discussion on the limits of reductionism, see Powell and Dupré, "From Molecules to Systems."

6. See also Morange, "From the Regulatory Vision of Cancer to the Oncogene Paradigm, 1975–1985."

7. Varmus, "How Tumor Virology Evolved into Cancer Biology and Transformed Oncology." See also van Helvoort, "A Century of Research into the Cause of Cancer"; and Keating and Cambrosio, *Cancer on Trial*.

8. For a similar convergence in cell biology, see Bechtel, "Integrating Sciences by Creating New Disciplines."

9. The prominent science writer John Horgan makes this argument in "The Cancer Industry: Hype vs. Reality," *Cross-Check* (blog), *Scientific American*, February 12, 2020, https://blogs.scientificamerican.com/cross-check/the-cancer-industry-hype-vs-reality/. See also, Raza, *The First Cell*.

10. Landecker, *Culturing Life*.

11. See also R. H. Davis, *The Microbial Models of Molecular Biology*; Müller and Grossniklaus, "Model Organisms—a Historical Perspective"; Ankeny and Leonelli, *Model Organisms*.

12. Atkin, Griffin, and Dilworth, "Polyoma Virus and Simian Virus 40 as Cancer Models."

13. There is a large literature on this topic. The case of Rosalind Franklin is perhaps the most famous. For references, see Creager and Morgan, "After the Double Helix."

14. Wellerstein, "Manhattan Project."

15. Deichmann, "Emigration, Isolation and the Slow Start of Molecular Biology in Germany." See also Strasser, "Institutionalizing Molecular Biology in Post-war Europe."

16. Khot, Park, and Longstreth, "The Vietnam War and Medical Research."

17. Harden, "Emerging Paradigm, Emerging Disease," 235.

18. Khot, Park, and Longstreth, "The Vietnam War and Medical Research."

19. Rader, *Making Mice*.

20. Bill Summers suggests the use of plaque assays to quantify bacteriophages was

initially controversial but by the early 1940s was accepted by different camps of bacteriophage workers. Summers, *Felix d'Herelle and the Origins of Molecular Biology*.

21. One insightful way to analyze this replication of laboratory approach is Rachel Ankeny and Sabina Leonelli's notion of a "repertoire": a well-aligned assemblage of the skills, behaviors, and material, social, and epistemic components that a group may use to practice certain kinds of science. Watson and Sambrook replicated the repertoire found in Dulbecco's laboratory. See Ankeny and Leonelli, "Repertoires."

22. Summers, "Inventing Viruses."

23. For a nuanced discussion of the integration of cancer models, see Plutynski, "Cancer and the Goals of Integration." For the roots of the idea that science advances by becoming more consilient, see Whewell, *The Philosophy of the Inductive Sciences*. See also Wilson, *Consilience*.

24. The philosopher of science Paul Feyerabend argued for such a position in the 1970s in *Against Method*. A related point is made by Kachelmeier in "Reflections on Image and Logic." See also Lloyd, "Feyerabend, Mill, and Pluralism"; and Mitchell, "Through the Fractured Looking Glass."

25. Dupré, *The Disorder of Things*. Kuhn, *The Structure of Scientific Revolutions*.

26. Sterner, Witteveen, and Franz, "Coordinating Dissent as an Alternative to Consensus Classification."

27. Biagioli, "Scientific Revolution, Social Bricolage, and Etiquette;" Galison and Stump, *The Disunity of Science*; Gillett, "Invention through Bricolage."

28. Taylor, *Viruses and Man*, 321. See also Ledford, "Cancer-Killing Viruses Show Promise."

29. Nemunaitis et al., "Selective Replication and Oncolysis in p53 Mutant Tumors with ONYX-015." For a review of oncolytic viruses, see Lawler et al., "Oncolytic Viruses in Cancer Treatment."

30. US National Library of Medicine, https://clinicaltrials.gov/ct2/results?term=oncolytic+virus (retrieved June 19, 2020).

Bibliography

Abbadessa, Giovanni, Roberto Accolla, Fernando Aiuti, Adriana Albini, Anna Aldovini, Massimo Alfano, Guido Antonelli, et al. "Unsung Hero Robert C. Gallo." *Science* 323, no. 5911 (2009): 206–207.

Abir-Am, Pnina. "From Multidisciplinary Collaboration to Transnational Objectivity: International Space as Constitutive of Molecular Biology, 1930–1970." In *Denationalizing Science*, 153–186. Springer, 1993.

Abir-Am, Pnina. "The Women Who Discovered RNA Splicing." *American Scientist* 108, no. 5 (2020): 298–306.

Adams, Mark H. *Bacteriophages*. Interscience, 1959.

Adler, Isaac. *Primary Malignant Growths of the Lungs and Bronchi*. Longmans, Green, 1912.

Allen, Garland. "Essay Review: Genetics, Eugenics and Society: Internalists and Externalists in Contemporary History of Science." *Social Studies of Science* 6, no. 1 (1976): 105–122.

Allen, Garland E. "The Eugenics Record Office at Cold Spring Harbor, 1910–1940: An Essay in Institutional History." *Osiris* 2 (1986): 225–264.

Alter, Harvey J., Paul V. Holland, Robert H. Purcell, Jerrold J. Lander, Stephen M. Feinstone, Andrew G. Morrow, and Paul J. Schmidt. "Posttransfusion Hepatitis after Exclusion of Commercial and Hepatitis-B Antigen-Positive Donors." *Annals of Internal Medicine* 77, no. 5 (1972): 691–699.

Alter, Harvey J., Paul V. Holland, Robert H. Purcell, and Hans Popper. "Transmissible Agent in Non-A, Non-B Hepatitis." *Lancet* 311, no. 8062 (1978): 459–463.

Alter, Harvey J., and Michael Houghton. "Hepatitis C Virus and Eliminating Post-transfusion Hepatitis." *Nature Medicine* 6, no. 10 (2000): 1082–1086.

Andersen, Hanne. "Collaboration, Interdisciplinarity, and the Epistemology of Contemporary Science." *Studies in History and Philosophy of Science Part A* 56 (2016): 1–10.

Andervont, Howard. "Mammary Tumors in Mice." In *Symposium on Mammary Tumors in Mice*, by Forest Ray Moulton, 123–139. American Association for the Advancement of Science, 1945.

Andrewes, Christopher. "Francis Peyton Rous, 1879–1970." *Biographical Memoirs of Fellows of the Royal Society* (1971): 643–662.

Andrewes, Christopher. "Richard Edwin Shope." In *National Academy of Sciences Biographical Memoir*. National Academy of Sciences, Washington, DC, 1979.

Angevine, D. Murray. "Significant Events in the Life of Jacob Furth." *Cancer Research* 26, no. 3, part 1 (1966): 351–356.

Angier, Natalie. *Natural Obsessions: Striving to Unlock the Deepest Secrets of the Cancer Cell*. Houghton Mifflin Harcourt, 1999.

Ankeny, Rachel A. "Fashioning Descriptive Models in Biology: Of Worms and Wiring Diagrams." *Philosophy of Science* 67 (2000): S260–S272.

Ankeny, Rachel A., and Sabina Leonelli. "What's So Special about Model Organisms?" *Studies in History and Philosophy of Science Part A* 42, no. 2 (2011): 313–323.

Ankeny, Rachel A., and Sabina Leonelli. "Repertoires: A Post-Kuhnian Perspective on Scientific Change and Collaborative Research." *Studies in History and Philosophy of Science Part A* 60 (2016): 18–28.

Ankeny, Rachel, and Sabina Leonelli. *Model Organisms*. Cambridge University Press, 2020.

Armstrong, Melanie. *Germ Wars: The Politics of Microbes and America's Landscape of Fear*. Vol. 2. University of California Press, 2017.

Armstrong, Sue. *p53: The Gene That Cracked the Cancer Code*. Bloomsbury, 2014.

Atkin, Sarah J. L., Beverly E. Griffin, and Stephen M. Dilworth. "Polyoma Virus and Simian Virus 40 as Cancer Models: History and Perspectives." *Seminars in Cancer Biology* 19, no. 4 (2009): 211–217.

Auerbach, David M., William W. Darrow, Harold W. Jaffe, and James W. Curran. "Cluster of Cases of the Acquired Immune Deficiency Syndrome: Patients Linked by Sexual Contact." *American Journal of Medicine* 76, no. 3 (1984): 487–492.

Austoker, Joan. *A History of the Imperial Cancer Research Fund 1902–1986*. Oxford University Press, 1988.

Aviles, Natalie B. "The Little Death: Rigoni-Stern and the Problem of Sex and Cancer in 20th-Century Biomedical Research." *Social Studies of Science* 45, no. 3 (2015): 394–415.

Aviles, Natalie B. "Situated Practice and the Emergence of Ethical Research: HPV Vaccine Development and Organizational Cultures of Translation at the National Cancer Institute." *Science, Technology, & Human Values* 43, no. 5 (2018): 810–833.

Bader, John P. "The Requirement for DNA Synthesis in the Growth of Rous Sarcoma and Rous-Associated Viruses." *Virology* 26, no. 2 (1965): 253–261.

Baker, Suzanne J., Eric R. Fearon, Janice M. Nigro, A. C. Preisinger, J. M. Jessup, D. H. Ledbetter, D. F. Barker, et al. "Chromosome 17 Deletions and p53 Gene Mutations in Colorectal Carcinomas." *Science* 244, no. 4901 (1989): 217–221.

Baltimore, David. "Expression of Animal Virus Genomes." *Bacteriological Reviews* 35, no. 3 (1971): 235–241.

Baltimore, David. "Tumor Viruses: 1974." In *Cold Spring Harbor Symposia on Quantitative Biology*, vol. 34, 1187–1200. Cold Spring Harbor Laboratory Press, 1974.

Baltimore, David. "Viral RNA-Dependent DNA Polymerase: RNA-Dependent DNA Polymerase in Virions of RNA Tumour Viruses." *Nature* 226, no. 5252 (1970): 1209.

Barnard, J. E. "The Microscopical Examination of Filterable Viruses: Associated with Malignant New Growths" *Lancet* 206 (1925): 117–123.

Barre-Sinoussi, F., J. C. Chermann, F. Rey, M. T. Nugeyre, S. Chamaret, J. Gruest, C. Dauguet, et al. "Isolation of A T-Lymphotropic Retrovirus from a Patient at Risk for Acquired Immune Deficiency Syndrome (AIDS)." *Science* 220, no. 4599 (1983): 858–871.

Bather, R. "The Nucleic Acid of Partially Purified Rous No. 1 Sarcoma Virus." *British Journal of Cancer* 11, no. 4 (1957): 611.

Bäumler, Ernst, *Cancer: A Review of International Research*. Queen Anne Press, 1968.

Bayer, Manfred E., Baruch S. Blumberg, and Barbara Werner. "Particles Associated with Australia Antigen in the Sera of Patients with Leukaemia, Down's Syndrome and Hepatitis." *Nature* 218, no. 5146 (1968): 1057.

Bernstein, A., R. MacCormick, and G. S. Martin. "Transformation-Defective Mutants of Avian Sarcoma Viruses: The Genetic Relationship between Conditional and Nonconditional Mutants." *Virology* 70, no. 1 (1976): 206–209.

Beasley, R. Palmer, Chin-Yun Lee George, Cheng-Hsiung Roan, Lu-Yu Hwang, Chung-Chi Lan, Fu-Yuan Huang, and Chiung-Lin Chen. "Prevention of Perinatally Transmitted Hepatitis B Virus Infections with Hepatitis B Immune Globulin and Hepatitis B Vaccine." *Lancet* 322, no. 8359 (1983): 1099–1102.

Bechtel, William. "Integrating Sciences by Creating New Disciplines: The Case of Cell Biology." *Biology and Philosophy* 8, no. 3 (1993): 277–299.

Bechtel, William, and Robert C. Richardson. *Discovering Complexity: Decomposition and Localization as Strategies in Scientific Research*. MIT Press, 2010.

Becsei-Kilborn, Eva. "Going Against the Grain: Francis Peyton Rous (1879–1970) and the Search for the Cancer Virus." PhD diss., 2003, University of Illinois at Chicago.

Becsei-Kilborn, Eva. "Scientific Discovery and Scientific Reputation: The Reception of Peyton Rous' Discovery of the Chicken Sarcoma Virus." *Journal of the History of Biology* 43, no. 1 (2010): 111–157.

Beijerinck, Martinus Willem. "Ueber ein contagium vivum fluidum als Ursache der Fleckenkrankheit der Tabaksblatter." *Verhandelingen Der Koninklyke Akademie Van Wetten-Schappen Te Amsterdam* 5 (1898): 3–21.

Beisel, C., J. Tanner, T. Matsuo, D. Thorley-Lawson, F. Kezdy, and E. Kieff. "Two Major Outer Envelope Glycoproteins of Epstein-Barr Virus Are Encoded by the Same Gene." *Journal of Virology* 54, no. 3 (1985): 665–674.

Berk, Arnold J. "Discovery of RNA Splicing and Genes in Pieces." *Proceedings of the National Academy of Sciences* 113, no. 4 (2016): 801–805.

Berg, Paul. "Moments of Discovery." *Annual Review of Biochemistry* 77 (2008): 15–44.

Berg, Paul, David Baltimore, Herbert W. Boyer, Stanley N. Cohen, Ronald W. Davis, David S. Hogness, Daniel Nathans, et al. "Potential Biohazards of Recombinant DNA Molecules." *Science* 26 (July 1974): 303.

Berg, Paul, and Janet E. Mertz. "Personal Reflections on the Origins and Emergence of Recombinant DNA Technology." *Genetics* 184, no. 1 (2010): 9–17.

Bessis, Marcel. "How the Mouse Leukemia Virus Was Discovered: A Talk with Ludwik Gross." *Nouvelle revue francaise d'hematologie / Blood Cells* 16, no. 2 (1976): 287–304.

Biagioli, Mario. "Scientific Revolution, Social Bricolage, and Etiquette." *The Scientific Revolution in National Context*. Eds. Roy Porter and Mikulas Teich, 11–54. Cambridge University Press, 1992.

Bialy, Harvey. *Oncogenes, Aneuploidy, and AIDS: A Scientific Life & Times of Peter H. Duesberg*. North Atlantic Books, 2004.

Bishop, J. Michael. *How to Win the Nobel Prize: An Unexpected Life in Science*. Harvard University Press, 2009.

Bittner, John J. "Biological Assay and Serial Passage of the Mouse Mammary Tumour Agent in Mammary Tumours from Mothers and Their Hybrid Progeny." In *CIBA Foundation Symposium on Tumour Viruses of Murine Origin*. Eds. G. E. W. Wolstenholme and Maeve O'Connor, 56–74. Little, Brown, 1962.

Bittner, John J. "A Genetic Study of the Transplantation of Tumors Arising in Hybrid Mice." *American Journal of Cancer* 15, no. 3 (1931): 2202–2247.

Bittner, John J. "Mammary Tumors in Mice in Relation to Nursing." *American Journal of Cancer* 30, no. 3 (1937): 530–538. (Reprinted in Michael B. Shimkin, *Some Classics of Experimental Oncology: 50 Selections, 1775–1965*, 267–276, National Institutes of Health, 1980).

Bittner, John J. "Possible Relationship of the Estrogenic Hormones Genetic Susceptibility and Milk Influence in the Production of Mammary Cancer in Mice." *Cancer Research* 2, no. 10 (1942): 710–721.

Bittner, John J. "Some Possible Effects of Nursing on the Mammary Gland Tumor Incidence in Mice." *Science* 84, no. 2172 (1936): 162.

Bittner, John J. "Studies on Mammary Cancer in Mice and Their Implications for the Human Problem." *Texas Reports on Biology and Medicine* 15 (1957): 659.

Black, D. A. K., and M. G. P. Stoker. "Plasma Iron in New-Born Babies." *Nature* 157 (1946): 658.

Black, Paul H., Wallace P. Rowe, Horace C. Turner, and Robert J. Huebner. "A Specific Complement-Fixing Antigen Present in SV40 Tumor and Transformed Cells." *Proceedings of the National Academy of Sciences* 50, no. 6 (1963): 1148–1156.

Blasimme, Alessandro, Paolo Maugeri, and Pierre-Luc Germain. "What Mechanisms Can't Do: Explanatory Frameworks and the Function of the p53 Gene in Molecular Oncology." *Studies in History and Philosophy of Science Part C: Studies in History and Philosophy of Biological and Biomedical Sciences* 44, no. 3 (2013): 374–384.

Blight, Keril J., Alexander A. Kolykhalov, and Charles M. Rice. "Efficient Initiation of HCV RNA Replication in Cell Culture." *Science* 290, no. 5498 (2000): 1972–1974.

Block, Timothy M., Harvey J. Alter, W. Thomas London, and Mike Bray. "A Historical Perspective on the Discovery and Elucidation of the Hepatitis B Virus." *Antiviral Research* 131 (2016): 109–123.

Blumberg, Baruch S. *Hepatitis B: The Hunt for a Killer Virus*. Princeton University Press, 2002

Blumberg, Baruch S., and Harvey J. Alter. "A New Antigen in Leukemia Sera." *JAMA* 191, no. 7 (1965): 541–546.

Blumberg, Baruch S., Betty Jane S. Gerstley, David A. Hungerford, W. Thomas London, and Alton I. Sutnick. "A Serum Antigen (Australia Antigen) in Down's Syndrome, Leukemia, and Hepatitis." *Annals of Internal Medicine* 66, no. 5 (1967): 924–931.

Blumberg, B. S., and I. Millman. Vaccine against Viral Hepatitis and Process. Serial no. 864,788. US Patent 36 36 191, filed October 8, 1969, issued January 18, 1972.

Bookchin, Debbie, and Jim Schumacher. *The Virus and the Vaccine: The True Story of a Cancer-Causing Monkey Virus, Contaminated Polio Vaccine, and the Millions of Americans Exposed*. Macmillan, 2004.

Bourgeois, Suzanne. *Genesis of the Salk Institute: The Epic of Its Founders*. University of California Press, 2013.

Brandt, Allan M. *The Cigarette Century: The Rise, Fall, and Deadly Persistence of the Product That Defined America*. Basic Books (AZ), 2007.

Brandt, Allan M. "Inventing Conflicts of Interest: A History of Tobacco Industry Tactics." *American Journal of Public Health* 102, no. 1 (2012): 63–71.

Breitburd, Francoise, Reinhard Kirnbauer, Nancy L. Hubbert, Bernadete Nonnenmacher, Carole Trin-Dinh-Desmarquet, Gerard Orth, John T. Schiller, and Douglas R. Lowy. "Immunization with Viruslike Particles from Cottontail Rabbit Papillomavirus (CRPV) Can Protect against Experimental CRPV Infection." *Journal of Virology* 69, no. 6 (1995): 3959–3963.

Brenner, S., and R. W. Horne. "A Negative Staining Method for High Resolution Electron Microscopy of Viruses." *Biochimica et Biophysica Acta* 34 (1959): 103–110.

Brown, John R., and John L. Thornton. "Percivall Pott (1714–1788) and Chimney Sweepers' Cancer of the Scrotum." *British Journal of Industrial Medicine* 14, no. 1 (1957): 68.

Bruni, Laia, Mireia Diaz, Leslie Barrionuevo-Rosas, Rolando Herrero, Freddie Bray, F. Xavier Bosch, Silvia De Sanjosé, and Xavier Castellsagué. "Global Estimates of Human Papillomavirus Vaccination Coverage by Region and Income Level: A Pooled Analysis." *Lancet Global Health* 4, no. 7 (2016): E453–E463.

Bryan, Janine T., Barry Buckland, Jennifer Hammond, and Kathrin U. Jansen. "Prevention of Cervical Cancer: Journey to Develop the First Human Papillomavirus Virus-Like Particle Vaccine and the Next Generation Vaccine." *Current Opinion in Chemical Biology* 32 (2016): 34–47.

Bryan, W. Ray, H. Kahler, Michael B. Shimkin, and H. B. Andervont. "Extraction and Ultracentrifugation of Mammary Tumor Inciter of Mice." *Journal of the National Cancer Institute* 2, no. 5 (1942): 451–455.

Bud, Robert F. "Strategy in American Cancer Research after World War II: A Case Study." *Social Studies of Science* 8, no. 4 (1978): 425–459.

Burian, Richard M. "The Dilemma of Case Studies Resolved: The Virtues of Using Case Studies in the History and Philosophy of Science." *Perspectives on Science* 9, no. 4 (2001): 383–404.

Burkitt, Denis. "A Children's Cancer Dependent on Climatic Factors." *Nature* 194 (1962): 232–234.

Burkitt, Denis. "A Sarcoma Involving the Jaws in African Children." *British Journal of Surgery* 46, no. 197 (1958): 218–223.

Cairns, John, Gunther Stent, and James D. Watson. *Phage and the Origins of Molecular Biology*. Cold Spring Harbor Laboratory Press, 1968.

Carrel, Alexis. "On the Permanent Life of Tissues outside of the Organism." *Journal of Experimental Medicine* 15, no. 5 (1912): 516–528.

Check, Erika. "Rockefeller Head Quits as Scandal Looms." *Nature* 415 (2002): 721.

Chermann, J. C., F. Barre-Sinoussi, and L. Montagnier. "Retrovirus and AIDS." In *International Symposium: Retroviruses and Human Pathology*, 291–299. Humana Press, 1985.

Choo, Qui-Lim, George Kuo, Amy J. Weiner, Lacy R. Overby, Daniel W. Bradley, and Michael Houghton. "Isolation of a cDNA Clone Derived from a Blood-Borne Non-A, Non-B Viral Hepatitis Genome." *Science* 244, no. 4902 (1989): 359–362.

Chow, Louise T., Richard E. Gelinas, Thomas R. Broker, and Richard J. Roberts. "An Amazing Sequence Arrangement at the 5′ Ends of Adenovirus 2 Messenger RNA." *Cell* 12, no. 1 (1977): 1–8.

Chow, Louise T., James M. Roberts, James B. Lewis, and Thomas R. Broker. "A Map of Cytoplasmic RNA Transcripts from Lytic Adenovirus Type 2, Determined by Electron Microscopy of RNA: DNA Hybrids." *Cell* 11, no. 4 (1977): 819–836.

Chumakov, P. M., V. S. Iotsova, and G. P. Georgiev. "Isolation of a Plasmid Clone Containing the mRNA Sequence for Mouse Nonviral T-antigen." *Doklady Akademii nauk SSSR* 267, no. 5 (1982): 1272–1275.

Clarke, Brendan. "Mapping the Methodologies of Burkitt Lymphoma." *Studies in History and Philosophy of Science Part C: Studies in History and Philosophy of Biological and Biomedical Sciences* 48 (2014): 210–217.

Coffin, John M. "Rescue of Rous Sarcoma Virus from Rous Sarcoma Virus-Transformed Mammalian Cells." *Journal of Virology* 10, no. 1 (1972): 153–156.

Coffin, J. M., S. H. Hughes, and H. E. Varmus. *Retroviruses: 1997*. Cold Spring Harbor Laboratory Press, 1997.

Coffin, John M., Harold E. Varmus, J. Michael Bishop, Myron Essex, William D. Hardy Jr., G. Steven Martin, Naomi E. Rosenberg, et al. "Proposal for Naming Host Cell-Derived Inserts in Retrovirus Genomes." *Journal of Virology* 40, no. 3 (1981): 953–957.

Colaianni, Alessandra, and Robert Cook-Deegan. "Columbia University's Axel Patents: Technology Transfer and Implications for the Bayh-Dole Act." *Milbank Quarterly* 87, no. 3 (2009): 683–715.

Coles, Peter. "French Researcher Asks for Share" *Nature* 341 (1989): 556.

Collett, Marc S., and R. L. Erikson. "Protein Kinase Activity Associated with the Avian Sarcoma Virus *src* Gene Product." *Proceedings of the National Academy of Sciences* 75, no. 4 (1978): 2021–2024.

Comfort, Nathaniel, and H. Bentley Glass. "Building Arcadia: A History of Cold Spring Harbor Laboratory." Unpublished manuscript.

Cooper, G. M., and Howard M. Temin. "Infectious DNA From Cells Infected with Rous Sarcoma Virus, Reticuloendotheliosis Virus or Rous-Associated Virus-o." In *Cold Spring Harbor Symposia on Quantitative Biology*, vol. 39, 1027–1032. Cold Spring Harbor Laboratory Press, 1974.

Cooper, Geoffrey M., and Howard M. Temin. "Infectious Rous Sarcoma Virus and Reticuloendotheliosis Virus DNAs." *Journal of Virology* 14, no. 5 (1974): 1132–1141.

Cooper, Geoffrey M., Rayla Greenberg Temin, and Bill Sugden, eds. *The DNA Provirus: Howard Temin's Scientific Legacy*. Zondervan, 1995.

Corner, George Washington. *A History of the Rockefeller Institute, 1901–1953: Origins and Growth*. Rockefeller University Press, 1965.

Cornwall, Claudia. *Catching Cancer: The Quest for Its Viral and Bacterial Causes*. Rowman & Littlefield, 2014.

Cox, Adrienne D., and Channing J. Der. "Ras History: The Saga Continues." *Small GTPases* 1, no. 1 (2010): 2–27.

Craigie, James. "Sarcoma 37 and Ascites Tumours: Imperial Cancer Research Fund Lecture Delivered at the Royal College of Surgeons of England on 27th November, 1951." *Annals of the Royal College of Surgeons of England* 11, no. 5 (1952): 287.

Craver, Carl F., and Lindley Darden. *In Search of Mechanisms: Discoveries across the Life Sciences*. University of Chicago Press, 2013.

Crawford, Dorothy. *The Invisible Enemy: A Natural History of Viruses*. Oxford University Press, 2002.

Crawford, Dorothy H., Alan Rickinson, and Ingólfur Johannessen. *Cancer Virus: The Story of Epstein-Barr Virus*. Oxford University Press, 2014.

Crawford, L. V. "Transforming Genes of DNA Tumor Viruses." In *Cold Spring Harbor Symposia on Quantitative Biology*, vol. 44, 9–11. Cold Spring Harbor Laboratory Press, 1980.

Crawford, Lionel, Renato Dulbecco, Mike Fried, Luc Montagnier, and Michael Stoker. "Cell Transformation by Different Forms of Polyoma Virus DNA." *Proceedings of the National Academy of Sciences* 52, no. 1 (1964): 148–152.

Crawford, L. V., and D. P. Lane. "An Immune Complex Assay for SV40 T antigen." *Biochemical and Biophysical Research Communications* 74, no. 1 (1977): 323–329.

Crawford, L. V., D. C. Pim, and P. Lamb. "The Cellular Protein P53 in Human Tumours." *Molecular Biology & Medicine* 2, no. 4 (1984): 261–272.

Crawford, L. V., and M. G. P. Stoker. *The Molecular Biology of Viruses: Eighteenth Symposium of the Society for General Microbiology Held at the Imperial College, London, April 1968*. Cambridge University Press, 1968.

Creager, Angela N. H. "'Happily Ever After' for Cancer Viruses?" *Studies in History and Philosophy of Science Part C: Studies in History and Philosophy of Biological and Biomedical Sciences* 48 (2014): 260–262.

Creager, Angela N. H. *Life Atomic: A History of Radioisotopes in Science and Medicine*. University of Chicago Press, 2013.

Creager, Angela N. H. *The Life of a Virus: Tobacco Mosaic Virus as an Experimental Model, 1930–1965*. University of Chicago Press, 2002.

Creager, Angela N. H. "Mobilizing Biomedicine: Virus Research between Lay Health Organizations and the US Federal Government, 1935–1955." In *Biomedicine in the Twentieth Century: Practices, Policies, and Politics*, edited by Caroline Hannaway, 171–201. IOS Press, 2008.

Creager, Angela N. H. "Paradigms and Exemplars Meet Biomedicine." In *Kuhn's "Structure of Scientific Revolutions" at 50: Reflections on a Science Classic*, edited by Lorraine Daston and Robert Richards, 151–166. University of Chicago Press.

Creager, Angela N. H. "Recipes for Recombining DNA: A History of Molecular Cloning; A Laboratory Manual." *BJHS Themes* 5 (2020): 225–243.

Creager, Angela N. H. "Wendell Stanley's Dream of a Free-Standing Biochemistry Department at the University of California, Berkeley." *Journal of the History of Biology* 29, no. 3 (1996): 331–360.

Creager, Angela N. H., and Jean-Paul Gaudillière. "Experimental Arrangements and Technologies of Visualization: Cancer as a Viral Epidemic, 1930–1960." In *Heredity and Infection: The History of Disease Transmission*. Eds. Jean-Paul Gaudillière and Ilana Löwy, 203–241. Routledge, 2001.

Creager, Angela N. H., Elizabeth Lunbeck, Londa L. Schiebinger, and Londa Schiebinger, eds. *Feminism in Twentieth-Century Science, Technology, and Medicine*. University of Chicago Press, 2001.

Creager, Angela N. H., Elizabeth Lunbeck, and M. Norton Wise, eds. *Science without Laws: Model Systems, Cases, Exemplary Narratives*. Duke University Press, 2007.

Creager, Angela N. H., and Gregory J. Morgan. "After the Double Helix: Rosalind Franklin's Research on Tobacco Mosaic Virus." *Isis* 99, no. 2 (2008): 239–272.

Creager, Angela N. H., Karen-Beth G. Scholthof, Vitaly Citovsky, and Herman B. Scholthof. "Tobacco Mosaic Virus: Pioneering Research for A Century." *Plant Cell* 11, no. 3 (1999): 301–308.

Crick, Francis. "Central Dogma of Molecular Biology." *Nature* 227, no. 5258 (1970): 561–563.

Crick, Francis. "On Protein Synthesis." In *Symposia of the Society for Experimental Biology*, Vol. 12, no. 138–63, P. 8. 1958.

Crotty, Shane. *Ahead of the Curve: David Baltimore's Life in Science*. University of California Press, 2001

Dalla-Favera, Riccardo, Marco Bregni, Jan Erikson, David Patterson, Robert C. Gallo, and Carlo M. Croce. "Human *c-myc onc* Gene Is Located on the Region of Chromosome 8 That Is Translocated in Burkitt Lymphoma Cells." *Proceedings of the National Academy of Sciences* 79, no. 24 (1982): 7824–7827.

Dane, D. S., C. H. Cameron, and Moya Briggs. "Virus-Like Particles in Serum of Patients with Australia-Antigen-Associated Hepatitis." *Lancet* 295, no. 7649 (1970): 695–698.

Danna, Kathleen, and Daniel Nathans. "Specific Cleavage of Simian Virus 40 DNA by Restriction Endonuclease of Hemophilus Influenzae." *Proceedings of the National Academy of Sciences* 68, no. 12 (1971): 2913–2917.

Darden, Lindley. "Flow of Information in Molecular Biological Mechanisms." *Biological Theory* 1, no. 3 (2006): 280–287.

Darden, Lindley. *Reasoning in Biological Discoveries: Essays on Mechanisms, Interfield Relations, and Anomaly Resolution*. Cambridge University Press, 2006.

Darnell, James E. *RNA: Life's Indispensable Molecule*. Cold Spring Harbor Laboratory Press, 2011.

Daston, Lorraine. "The Moral Economy of Science." *Osiris* 10 (1995): 2–24.

Davis, Devra. *The Secret History of the War on Cancer*. Hachette UK, 2007.

Davis, Rowland H. *The Microbial Models of Molecular Biology: From Genes to Genomes*. Oxford University Press, 2003.

de Chadarevian, Soraya. *Designs for Life: Molecular Biology after World War II*. Cambridge University Press, 2002.

de Chadarevian, Soraya. *Heredity under the Microscope: Chromosomes and the Study of the Human Genome*. University of Chicago Press, 2020.

de Chadarevian, Soraya. "Sequences, Conformation, Information: Biochemists and Molecular Biologists in the 1950s." *Journal of the History of Biology* 29, no. 3 (1996): 361–386.

Deichmann, Ute. "Emigration, Isolation and the Slow Start of Molecular Biology in Germany." *Studies in History and Philosophy of Science Part C: Studies in History and Philosophy of Biological and Biomedical Sciences* 33, no. 3 (2002): 449–471.

Dejong-Lambert, William, and Nikolai Krementsov. "On Labels and Issues: The Lysenko Controversy and the Cold War." *Journal of the History of Biology* (2012): 1–16.

Deleo, Albert B., Gilbert Jay, Ettore Appella, Garrett C. Dubois, Lloyd W. Law, and Lloyd J. Old. "Detection of a Transformation-Related Antigen in Chemically Induced Sarcomas and Other Transformed Cells of the Mouse." *Proceedings of the National Academy of Sciences* 76, no. 5 (1979): 2420–2424.

Delisio, John Paul, Jr. "Fighting for a Cure: The Berry Plan's Impact on Civilian Medical Research." PhD diss., George Washington University, 2017.

Der, Channing J., Theodore G. Krontiris, and Geoffrey M. Cooper. "Transforming Genes of Human Bladder and Lung Carcinoma Cell Lines Are Homologous to the ras Genes of Harvey and Kirsten Sarcoma Viruses." *Proceedings of the National Academy of Sciences* 79, no. 11 (1982): 3637–3640.

de Schweinitz, Emil Alexander, and Marion Dorset. "Form of Hog Cholera Not Caused by the Hog-Cholera Bacillus." *Circular (United States Bureau of Animal Industry)*, no. 41 (1903).

Dick, Brian, and Mark Jones. "The Commercialization of Molecular Biology: Walter Gilbert and the Biogen Startup." *History and Technology* 33, no. 1 (2017): 126–151.

Dietrich, Michael R., Rachel A. Ankeny, Nathan Crowe, Sara Green, and Sabina Leonelli. "How to Choose Your Research Organism." *Studies in History and Philosophy of Science Part C: Studies in History and Philosophy of Biological and Biomedical Sciences* 80: 101227 (2020).

Dimayorca, G. A., B. E. Eddy, S. E. Stewart, W. S. Hunter, Ch Friend, and A. Bendich. "Isolation of Infectious Deoxyribonucleic Acid from SE Polyoma-Infected Tissue Cultures." *Proceedings of the National Academy of Sciences* 45, no. 12 (1959): 1805–1808.

Doogab, Yi. "Governing, Financing, and Planning Cancer Virus Research: The Emergence of Organized Science at the US National Cancer Institute in the 1950s and 1960s." *Korean Journal for History of Science* 38, no. 2 (2016): 321–349.

Doolittle, W. Ford. "Genes in Pieces: Were They Ever Together?" *Nature* 272, no. 5654 (1978): 581–582.

Duesberg, Peter H., and Peter K. Vogt. "Differences between the Ribonucleic Acids of Transforming and Nontransforming Avian Tumor Viruses." *Proceedings of the National Academy of Sciences* 67, no. 4 (1970): 1673–1680.

Duesberg, Peter H., and Peter K. Vogt. "On the Role of DNA Synthesis in Avian Tumor Virus Infection." *Proceedings of the National Academy of Sciences* 64, no. 3 (1969): 939–946.

Dulbecco, Renato. "From the Molecular Biology of Oncogenic DNA Viruses to Cancer." *Science* 192, no. 4238 (1976): 437–440.

Dulbecco, Renato. "The Plaque Technique and the Development of Quantitative Animal Virology." *Phage and the Origins of Molecular Biology.* Eds. J. Cairns, G. S. Stent, and J. D. Watson, 287–291. Cold Spring Harbor Laboratory Press, 1966.

Dulbecco, Renato. "Production of Plaques in Monolayer Tissue Cultures by Single Particles of an Animal Virus." *Proceedings of the National Academy of Sciences* 38, no. 8 (1952): 747–752.

Dulbecco, R. "Quantitative Aspects of Virus Growth in Cultivated Animal Cells." In *The Nature of Viruses*, Ciba Foundation Symposium, 147–157. John Wiley & Sons, 1957.

Dulbecco, Renato. "Reactivation of Ultra-Violet-Inactivated Bacteriophage by Visible Light." *Nature* 163, no. 4155 (1949): 949–950.

Dulbecco, Renato, Lee H. Hartwell, and Marguerite Vogt. "Induction of Cellular DNA Synthesis by Polyoma Virus." *Proceedings of the National Academy of Sciences* 53, no. 2 (1965): 403–410.

Dulbecco, Renato, and Marguerite Vogt. "Evidence for a Ring Structure of Polyoma Virus DNA." *Proceedings of the National Academy of Sciences* 50, no. 2 (1963): 236–243.

Dulbecco, Renato, and Marguerite Vogt. "Plaque Formation and Isolation of Pure Lines with Poliomyelitis Viruses." *Journal of Experimental Medicine* 99, no. 2 (1954): 167–182.

Dulbecco, Renato, and Marguerite Vogt. "Some Problems of Animal Virology as Studied by the Plaque Technique." In *Cold Spring Harbor Symposia on Quantitative Biology*, vol. 18, 273–279. Cold Spring Harbor Laboratory Press, 1953

Dupré, John. *The Disorder of Things: Metaphysical Foundations of the Disunity of Science*. Harvard University Press, 1995.

Duran-Reynals, F. "Age Susceptibility of Ducks to the Virus of the Rous Sarcoma and Variation of the Virus in the Duck." *Science* 93, no. 2421 (1941): 501–502.

Dürst, Matthias, Lutz Gissmann, Hans Ikenberg, and Harald zur Hausen. "A Papillomavirus DNA from a Cervical Carcinoma and Its Prevalence in Cancer Biopsy Samples from Different Geographic Regions." *Proceedings of the National Academy of Sciences* 80, no. 12 (1983): 3812–3815.

Delbrück, Max. Foreword to *Viruses 1950*. California Institute of Technology.

Eagle, Harry. "Amino Acid Metabolism in Mammalian Cell Cultures." *Science* 130, no. 3373 (1959): 432–437.

Eagle, Harry. "Nutrition Needs of Mammalian Cells in Tissue Culture." *Science* 122, no. 3168 (1955): 501–504.

Eddy, B. "*Oncogenic Viruses* by Ludwik Gross." *Journal of the American Medical Association* 215 (1971): 1161.

Eddy, B. E., W. P. Rowe, J. W. Hartley, S. E. Stewart, and R. J. Huebner. "Hemagglutination with the SE Polyoma Virus." *Virology* 6, no. 1 (1958): 290–291.

Eddy, Bernice E., Sarah E. Stewart, and William Berkeley. "Cytopathogenicity in Tissue Cultures by a Tumor Virus from Mice." *Proceedings of the Society for Experimental Biology and Medicine* 98, no. 4 (1958): 848–851.

Eddy, Bernice E., Sarah E. Stewart, and George E. Grubbs. "Influence of Tissue Culture Passage, Storage, Temperature and Drying on Viability of SE Polyoma Virus." *Proceedings of the Society for Experimental Biology and Medicine* 99, no. 2 (1958): 289–292.

Eddy, Bernice E., Sarah E. Stewart, Ruth L. Kirschstein, and Ralph D. Young. "Induction of Subcutaneous Nodules in Rabbits with the SE Polyoma Virus." *Nature* 183, no. 4663 (1959): 766.

Eddy, Bernice E., Sarah E. Stewart, Mearl F. Stanton, and Jerry M. Marcotte. "Induction of Tumors in Rats by Tissue-Culture Preparations of SE Polyoma Virus." *Journal of the National Cancer Institute* 22, no. 1 (1959): 161–171.

Eddy, Bernice E., Sarah E. Stewart, Ralph Young, and G. Burroughs Mider. "Neoplasms in Hamsters Induced by Mouse Tumor Agent Passed in Tissue Culture." *Journal of the National Cancer Institute* 20, no. 4 (1958): 747–761.

Eliyahu, Daniel, Avraham Raz, Peter Gruss, David Givol, and Moshe Oren. "Participation of P53 Cellular Tumour Antigen in Transformation of Normal Embryonic Cells." *Nature* 312, no. 5995 (1984): 646–649.

Ellermann, Vilhelm, and Olaf Bang. "Experimentelle Leukämie bei Hühnern. II." *Zeitschrift für Hygiene und Infektionskrankheiten* 63, no. 1 (1909): 231–272.

Emrich, John S. "Dr. Genelove: How Scientists Learned to Stop Worrying and Love Recombinant DNA." PhD diss., George Washington University, 2009.

Endersby, Jim. *A Guinea Pig's History of Biology: The Plants and Animals Who Taught Us the Facts of Life*. Random House, 2012.

Epstein, M. A., "Epstein-Barr Virus—Is It Time to Develop a Vaccine Program?" *Journal of the National Cancer Institute* 56 (1976): 697–700.

Epstein, M. Anthony. "The Origins of EBV Research Discovery and Characterization of the Virus." In *Epstein-Barr Virus*, edited by Erle S. Robertson, 1–14. Horizon Scientific Press, 2005.

Epstein, Michael Anthony, Bert G. Achong, and Yvonne M. Barr. "Virus Particles in Cultured Lymphoblasts from Burkitt's Lymphoma." *Lancet* 283, no. 7335 (1964): 702–703.

Epstein, M. A., and S. J. Holt. "Observations on the Rous Virus: Integrated Electron Microscopical and Cytochemical Studies of Fluorocarbon Purified Preparations." *British Journal of Cancer* 12, no. 3 (1958): 363–369.

Epstein, Steven. *Impure Science: AIDS, Activism, and the Politics of Knowledge*. University of California Press, 1996.

Essex, Myron, Mary Francis Mclane, Tun-Hou Lee, L. Falk, C. W. Howe, James I. Mullins, C. T. Cabradilla, and Donald P. Francis. "Antibodies to Cell Membrane Antigens Associated with Human T-Cell Leukemia Virus in Patients with AIDS." *Science* 220, no. 4599 (1983): 859–862.

Feldman, Simon D., and Alfred I. Tauber. "Sickle Cell Anemia: Reexamining the First 'Molecular Disease.'" *Bulletin of the History of Medicine* 71, no. 4 (1997): 623–650.

Fangerau, H. M. "The Novel Arrowsmith, Paul De Kruif (1890–1971) and Jacques Loeb (1859–1924): A Literary Portrait of 'Medical Science.'" *Medical Humanities* 32, no. 2 (2006): 82–87.

Feng, Huichen, Masahiro Shuda, Yuan Chang, and Patrick S. Moore. "Clonal Integration of a Polyomavirus in Human Merkel Cell Carcinoma." *Science* 319, no. 5866 (2008): 1096–1100.

Fenner, Frank, and F. M. Burnet. "A Short Description of the Poxvirus Group (Vaccinia and Related Viruses)." *Virology* 4, no. 2 (1957): 305–314.

Fenner, Frank, and I. D. Marshall. "A Comparison of the Virulence for European Rabbits (Oryctolagus Cuniculus) of Strains of Myxoma Virus Recovered in the Field in Australia, Europe and America." *Epidemiology & Infection* 55, no. 2 (1957): 149–191.

Fenner, Frank, and J. F. Sambrook. "The Genetics of Animal Viruses." *Annual Reviews in Microbiology* 18, no. 1 (1964): 47–94.

Fenner, Frank, and Gwendolyn M. Woodroofe. "Changes in the Virulence and Antigenic Structure of Strains of Myxoma Virus Recovered from Australian Wild Rabbits between 1950 and 1964." *Australian Journal of Experimental Biology and Medical Science* 43 (1965): 359–370.

Feyerabend, Paul. *Against Method*. Verso, 1993.

Finlay, Cathy A., Philip W. Hinds, and Arnold J. Levine. "The p53 Proto-oncogene Can Act as a Suppressor of Transformation." *Cell* 57, no. 7 (1989): 1083–1093.

Fisher, Susie. "Not beyond Reasonable Doubt: Howard Temin's Provirus Hypothesis Revisited." *Journal of the History of Biology* 43, no. 4 (2010): 661–696.

Fisher, Susie. "Not Just 'a Clever Way to Detect Whether DNA Really Made RNA' 1: The Invention of DNA-RNA Hybridization and Its Outcome." *Studies in History and Philosophy of Science Part C: Studies in History and Philosophy of Biological and Biomedical Sciences* 53 (2015): 40–52.

Frazer, I. H., R. M. Crapper, G. Medley, T. C. Brown, and I. R. Mackay. "Association between Anorectal Dysplasia, Human Papillomavirus, and Human Immunodeficiency Virus Infection in Homosexual Men." *Lancet* 328, no. 8508 (1986): 657–660.

Friend, Charlotte. "Cell-Free Transmission in Adult Swiss Mice of a Disease Having the Character of a Leukemia." *Journal of Experimental Medicine* 105, no. 4 (1957): 307–318.

Fuerst, John A. "The Role of Reductionism in the Development of Molecular Biology: Peripheral or Central?" *Social Studies of Science* 12, no. 2 (1982): 241–278.

Fujimura, Joan H. *Crafting Science: A Sociohistory of the Quest for the Genetics of Cancer*. Harvard University Press, 1996.

Fujimura, Joan H. "The Molecular Biological Bandwagon in Cancer Research: Where Social Worlds Meet." *Social Problems* 35, no. 3 (1988): 261–283.

Fujinami, A., and K. Inamoto. "Ueber Geschwülste bei japanischen Haushühnern, insbesondere über einen transplantablen Tumor." *Z. Krebsforsch.* 14 (1914): 94–119.

Furth, Jacob, Rita F. Buffett, Maria Banasiewicz-Rodriguez, and Arthur C. Upton. "Character of Agent Inducing Leukemia in Newborn Mice." *Proceedings of the Society for Experimental Biology and Medicine* 93, no. 2 (1956): 165–172.

Galison, Peter Louis, and David J. Stump. *The Disunity of Science: Boundaries, Contexts, and Power*. Stanford University Press, 1996.

Gallo, Robert C. *Virus Hunting: AIDS, Cancer, and the Human Retrovirus: A Story of Scientific Discovery*. Basic Books, 1993.

Gallo, Robert C., Myron Essex, and Ludwik Gross, eds. *Human T-Cell Leukemia/Lymphoma Virus: The Family of Human T-Lymphotropic Retroviruses, Their Role in Malignancies and Association with AIDS*. Cold Spring Harbor Laboratory Press, 1984.

Gallo, Robert C., Prem S. Sarin, E. P. Gelmann, Marjorie Robert-Guroff, Ersell Richardson, V. S. Kalyanaraman, Dean Mann, et al. "Isolation of Human T-Cell Leukemia Virus in Acquired Immune Deficiency Syndrome (AIDS)." *Science* 220, no. 4599 (1983): 865–867.

Galloway, Denise A., and James K. McDougall. "The Oncogenic Potential of Herpes simplex viruses: Evidence for a 'Hit-and-Run' Mechanism." *Nature* 302, no. 5903 (1983): 21–24.

Galperin, Charles. "Virus, provirus et cancer." *Revue d'histoire des sciences* (1994): 7–56.

Garrett, Laurie. *The Coming Plague: Newly Emerging Diseases in a World Out of Balance*. Macmillan, 1994.

Gaudillière, Jean-Paul. "Circulating Mice and Viruses." In *The Practices of Human Genetics*. Eds. M. Fortun and E. Mendelsohn, 89–124. Springer Netherlands, 1999.

Gaudillière, Jean-Paul. *Inventer la biomédecine: La France, l'Amérique et la production des savoirs du vivant (1945–1965)*. Éditions La Découverte, 2002.

Gaudillière, Jean-Paul. "Molecular Biologists, Biochemists, and messenger RNA: The Birth of a Scientific Network." *Journal of the History of Biology* 29, no. 3 (1996): 417–445.

Gaudillière, Jean-Paul. The Molecularization of Cancer Etiology in the Postwar United States: Instruments, Politics and Management. In *Molecularizing Biology and Medicine*. Eds. S. De Chadarevian and H. Kamminga, 139–170. Harwood, 1998.

Gaudillière, Jean-Paul, and I. Löwy. "Disciplining Cancer: Mice and the Practice of Genetic Purity." In *The Invisible Industrialist: Manufactures and the Production of Scientific Knowledge*, 209–249. Macmillan, 1998.

Gaudillière, Jean-Paul, and Ilana Löwy. *Heredity and Infection: The History of Disease Transmission*. Routledge, 2012.

Gelinas, Richard E., and Richard J. Roberts. "One Predominant 5′-Undecanucleotide in Adenovirus 2 Late Messenger RNAs." *Cell* 11, no. 3 (1977): 533–544.

Gilbert, Walter. "Genes-in-Pieces Revisited." *Science* 228 (1985): 823–825.

Gilbert, Walter. "Why Genes in Pieces?" *Nature* 271, no. 5645 (1978): 501.

Gillett, Alexander James. "Invention through Bricolage: Epistemic Engineering in Scientific Communities." *RT: A Journal on Research Policy and Evaluation* 6, no. 1 (2018).

Goldfarb, Mitchell, Kenji Shimizu, Manuel Perucho, and Michael Wigler. "Isolation and Preliminary Characterization of a Human Transforming Gene from T24 Bladder Carcinoma Cells." *Nature* 296, no. 5856 (1982): 404.

Gallo, Robert C., and Luc Montagnier. "The Chronology of AIDS Research." *Nature* 326, no. 6112 (1987): 435.

Gee, Julianne, Cindy Weinbaum, Lakshmi Sukumaran, and Lauri E. Markowitz. "Quadrivalent HPV Vaccine Safety Review and Safety Monitoring Plans for Nine-Valent HPV Vaccine in the United States." *Human Vaccines & Immunotherapeutics* 12, no. 6 (2016): 1406–1417.

Ghim, Shin-Je, A. Bennett Jenson, and Richard Schlegel. "HPV-1 L1 Protein Expressed in COS Cells Displays Conformational Epitopes Found on Intact Virions." *Virology* 190, no. 1 (1992): 548–552.

Giacomoni, Dario. "The Origin of DNA: RNA Hybridization." *Journal of the History of Biology* 26, no. 1 (1993): 89–107.

Goedert, James J., Robert J. Biggar, Deborah M. Winn, Mark H. Greene, Dean L. Mann, Robert C. Gallo, M. G. Sarngadharan, et al. "Determinants of Retrovirus (HTLV-III) Antibody and Immunodeficiency Conditions in Homosexual Men." *Lancet* 324, no. 8405 (1984): 711–716.

Goldenberg, Maya J. "Public Misunderstanding of Science? Reframing the Problem of Vaccine Hesitancy." *Perspectives on Science* 24, no. 5 (2016): 552–581.

Gottlieb, M. S., H. M. Schanker, P. T. Fan, A. Saxon, and J. D. Weisman. "Pneumocystis Pneumonia—Los Angeles." *Morbidity and Mortality Weekly Report* 30, no. 21 (1981): 250–252.

Gottlieb, Samantha D. *Not Quite a Cancer Vaccine: Selling HPV and Cervical Cancer*. Rutgers University Press, 2018.

Gradmann, Christoph. "A Spirit of Scientific Rigour: Koch's Postulates in Twentieth-Century Medicine." *Microbes and Infection* 16, no. 11 (2014): 885–892.

Graffi, A., H. Bielka, and F. Fey. "Leukämieerzeugung durch ein filtrierbares Agens aus malignen Tumoren." *Acta Haematologica* 15, no. 3 (1956): 145–174.

Graham, Frank L., and Alex J. Van Der Eb. "A New Technique for the Assay of Infectivity of Human Adenovirus 5 DNA." *Virology* 52, no. 2 (1973): 456–467.

Griesemer, James R., and Elihu M. Gerson. "Collaboration in the Museum of Vertebrate Zoology." *Journal of the History of Biology* (1993): 185–203.

Griffiths, Paul E. "Genetic Information: A Metaphor in Search of a Theory." *Philosophy of Science* 68, no. 3 (2001): 394–412.

Grodzicker, Terri, J. Williams, P. Sharp, and J. Sambrook. "Physical Mapping of Temperature-Sensitive Mutations of Adenoviruses." In *Cold Spring Harbor Symposia on Quantitative Biology*, vol. 39, 439–446. Cold Spring Harbor Laboratory Press, 1974.

Gross, Ludwik. "A Filterable Agent, Recovered from Ak Leukemic Extracts, Causing Salivary Gland Carcinomas in C3H Mice." *Proceedings of the Society for Experimental Biology and Medicine* 83 (1953): 414–421.

Gross, Ludwik. "The Fortuitous Isolation and Identification of the Polyoma Virus." *Cancer Research* 36, no. 11, part 1 (1976): 4195–4196.

Gross, Ludwik. "Induction of Parotid Carcinoma and/or Subcutaneous Sarcomas in C3H Mice with Normal C3H Organ Extracts." *Proceedings of the Society for Experimental Biology and Medicine* 88 (1955): 362–368.

Gross, Ludwik. *Ludzkość w walce o zdrowie* [Humanity fights for health]. Trzaska, Evert & Michalski, 1946.

Gross, Ludwik. "Neck Tumors, or Leukemia, Developing in Adult C3H Mice Following Inoculation, in Early Infancy, with Filtered (Berkefeld N), or Centrifugated (144,000 x g), Ak-Leukemic Extracts." *Cancer* 6 (1953b): 948–957.

Gross, Ludwik. *Oncogenic Viruses*. New York: Pergamon Press, 1961 (2nd ed. 1970; 3rd ed. 1983).

Gross, Ludwik. "Pathogenic Properties, and 'Vertical' Transmission of the Mouse Leukemia Agent." *Proceedings of the Society for Experimental Biology and Medicine* 78, no. 1 (1951): 342–348.

Gross, Ludwik. *Siewcy chorób i śmierci* [Sowers of death and disease]. Trzaska, Evert & Michalski, 1951.

Gross, Ludwik. "'Spontaneous' Leukemia Developing in G3H Mice Following Inoculation, in Infancy, with AK-Emkemic." *Proceedings of the Society for Experimental Biology and Medicine* 76, no. 1 (1951c): 27–32.

Gross, Ludwik. "Susceptibility of Newborn Mice of an Otherwise Apparently 'Resistant' Strain to Inoculation with Leukemia." *Proceedings of the Society for Experimental Biology and Medicine* 73, no. 2 (1950): 246–248.

Gross, Ludwik. "Viral Etiology of Cancer and Leukemia?" Guest editorial. *Journal of the American Medical Association* 162, no. 14 (1956): 1318–1319.

Grote, Mathias. *Membranes to Molecular Machines: Active Matter and the Remaking of Life*. University of Chicago Press, 2019.

Grote, Mathias, Lisa Onaga, Angela N. H. Creager, Soraya de Chadarevian, Daniel Liu,

Gina Surita, and Sarah E. Tracy. "The Molecular Vista: Current Perspectives on Molecules and Life in the Twentieth Century." *History and Philosophy of the Life Sciences* 43, no. 1 (2021): 1–18.

Guntaka, Ramareddy V. "Antecedents of a Nobel Prize (I)." *Nature* 343, no. 6256 (1990): 302.

Guntaka, Ramareddy V., B. W. J. Mahy, J. Michael Bishop, and Harold E. Varmus. "Ethidium Bromide Inhibits Appearance of Closed Circular Viral DNA and Integration of Virus-Specific DNA in Duck Cells Infected by Avian Sarcoma Virus." *Nature* 253, no. 5492 (1975): 507–511.

Gye, William Ewart. "The Aetiology of Malignant New Growths." *Lancet* 206, no. 5316 (1925): 109–117.

Hacking, Ian. *The Social Construction of What?* Harvard University Press, 1999.

Hajdu, Steven I. "A Note From History: Landmarks in History of Cancer, Part 1." *Cancer* 117, no. 5 (2011): 1097–1102.

Hager, Tom. *Linus Pauling: And the Chemistry of Life*. Oxford University Press, 2000.

Hammond, Andy. "JBS Haldane, Holism, and Synthesis in Evolution." *Transactions of the American Philosophical Society* 99, no. 1 (2009): 49–70.

Hanafusa, H., T. Hanafusa, and H. Rubin. "Analysis of the Defectiveness of Rous Sarcoma Virus, II: Specification of RSV Antigenicity by Helper Virus." *Proceedings of the National Academy of Sciences* 51, no. 1 (1964): 41–48.

Harden, Victoria A. *AIDS at 30: A History*. Potomac Books, 2012.

Harden, Victoria A. "Emerging Paradigm, Emerging Disease: Molecular Immunology and AIDS in the 1980s." In *Crafting Immunity: Working Histories of Clinical Immunology*. Eds. Kenton Kroker, Pauline Margaret Hodgson Mazumdar, and Jennifer E. Keelan. Ashgate, 2008.

Harris, C. C. "p53 Sweeps through Cancer Research." *Science* 262, no. 5142 (1993): 1958–1961.

Harris, Henry, O. J. Miller, George Klein, P. Worst, and T. Tachibana. "Suppression of Malignancy by Cell Fusion." *Nature* 223, no. 5204 (1969): 363.

Hayes, William. *The Genetics of Bacteria and Their Viruses. Studies in Basic Genetics and Molecular Biology*. Wiley, 1964.

Hejnar, Jiří. "Jan Svoboda (1934–2017): Sixty Years with Retroviruses." *Retrovirology* 14 (2017): 32.

Heller, J. R. "Conference Summary." In *Proceedings of the Third National Cancer Conference*, 7–8. Lippincott, 1957.

Hellman, A., M. N. Oxman, and R. Pollack. *Biohazards in Biological Research*. Cold Spring Harbor Laboratory Press, 1973.

Henle, Gertrude, Werner Henle, and Volker Diehl. "Relation of Burkitt's Tumor-Associated Herpes-Ytpe [*sic*] Virus to Infectious Mononucleosis." *Proceedings of the National Academy of Sciences* 59, no. 1 (1968): 94–101.

Hill, John. *Cautions against the Immoderate Use of Snuff: Founded on the Known Qualities of the Tobacco Plant; and the Effects It Must Produce When This Way Taken into the Body*. R. Baldwin; J. Jackson, 1761.

Hill, Miroslav, and Jana Hillova. "Recovery of the Temperature-Sensitive Mutant of Rous

Sarcoma Virus from Chicken Cells Exposed to DNA Extracted from Hamster Cells Transformed by the Mutant." *Virology* 49, no. 1 (1972): 309–313.

Hilleman, M. R., and Jacqueline H. Werner. "Recovery of New Agent from Patients with Acute Respiratory Illness." *Proceedings of the Society for Experimental Biology and Medicine* 85, no. 1 (1954): 183–188.

Holland, Mary, Kim Mack Rosenberg, and Eileen Iorio. *The HPV Vaccine on Trial: Seeking Justice for a Generation Betrayed*. Skyhorse, 2018.

Horsfall, F. L. "Oncogenic Viruses. Ludwik Gross. Pergamon, New York, 1961. Xi + 391 pp. Illus. $12." *Science* 135 (1962): 661–662.

Howley, Peter M., and David M. Livingston. "Small DNA Tumor Viruses: Large Contributors to Biomedical Sciences." *Virology* 384, no. 2 (2009): 256–259.

Huang, A. S., D. Baltimore, and M. Stampfer. "Ribonucleic Acid Synthesis of Vesicular Stomatitis Virus: III. Multiple Complementary Messenger RNA Molecules." *Virology* 42, no. 4 (1970): 946–957.

Huebner, Robert J., and George J. Todaro. "Oncogenes of RNA Tumor Viruses as Determinants of Cancer." *Proceedings of the National Academy of Sciences* 64, no. 3 (1969): 1087–1094.

Hughes, Sally Smith. "Making Dollars Out of DNA: The First Major Patent in Biotechnology and the Commercialization of Molecular Biology, 1974–1980." *Isis* 92, no. 3 (2001): 541–575.

Hughes, Sally Smith, *The Virus: A History of the Concept*. Heinemann Educational, 1977.

Hull, David L. *Science as a Process: An Evolutionary Account of the Social and Conceptual Development of Science*. University of Chicago Press, 2010.

Hunter, Tony, and Bartholomew M. Sefton. "Transforming Gene Product of Rous Sarcoma Virus Phosphorylates Tyrosine." *Proceedings of the National Academy of Sciences* 77, no. 3 (1980): 1311–1315.

Huzair, Farah, and Steve Sturdy. "Biotechnology and the Transformation of Vaccine Innovation: The Case of the Hepatitis B Vaccines 1968–2000." *Studies in History and Philosophy of Science Part C: Studies in History and Philosophy of Biological and Biomedical Sciences* 64 (2017): 11–21.

Jackson, Roscoe B., and C. C. Little. "The Existence of Non-chromosomal Influence in the Incidence of Mammary Tumors in Mice." *Science* 78, no. 2029 (1933): 465–466.

Jacobs, Charlotte DeCroes. *Jonas Salk: A Life*. Oxford University Press, 2015.

Jaenisch, Rudolf, Allen Mayer, and Arnold Levine. "Replicating SV40 Molecules Containing Closed Circular Template DNA Strands." *Nature New Biology* 233, no. 37 (1971): 72.

Jarrett, W. F. H., E. M. Crawford, W. B. Martin, and F. Davie. "A Virus-Like Particle Associated with Leukemia (Lymphosarcoma)." *Nature* 202 (1964): 567–569.

Jarrett, William, Oswald Jarrett, Lindsay Mackey, Helen Laird, William Hardy Jr., and Myron Essex. "Horizontal Transmission of Leukemia Virus and Leukemia in the Cat." *Journal of the National Cancer Institute* 51, no. 3 (1973): 833–841.

Javier, Ronald T., and Janet S. Butel. "The History of Tumor Virology." *Cancer Research* 68, no. 19 (2008): 7693–7706.

Jenkins, J. R., K. Rudge, and G. A. Currie. "Cellular Immortalization by a cDNA Clone

Encoding the Transformation-Associated Phosphoprotein p53." *Nature* 312, no. 5995 (1984): 651–654.

Joklik, Wolfgang K. Bill. "When Two Is Better Than One: Thoughts on Three Decades of Interaction between Virology and the *Journal of Virology*." *Journal of Virology* 73, no. 5 (1999): 3520–3523.

Judson, Horace Freeland. *The Eighth Day of Creation*. Jonathan Cape, 1979.

Judt, Tony. *Postwar: A History of Europe since 1945*. Penguin, 2006.

Kachelmeier, Inina. "Reflections on Image and Logic: Philosophy and the History of Experimentation on the Rous Sarcoma Virus in the Early to Mid 1900s." BS thesis, University of Oregon, 2016.

Kaiser, David. *Drawing Theories Apart: The Dispersion of Feynman Diagrams in Postwar Physics*. University of Chicago Press, 2009.

Kay, Lily E. *The Molecular Vision of Life: Caltech, the Rockefeller Foundation, and the Rise of the New Biology*. Oxford University Press, 1992.

Kay, Lily E. "W. M. Stanley's Crystallization of the Tobacco Mosaic Virus, 1930–1940." *Isis* 77, no. 3 (1986): 450–472.

Keating, Peter, and Alberto Cambrosio. *Cancer on Trial: Oncology as a New Style of Practice*. University of Chicago Press, 2011.

Keller, Evelyn Fox. *The Century of the Gene*. Harvard University Press, 2009.

Kendrew, John C. *The Thread of Life: An Introduction to Molecular Biology*. Harvard University Press, 1966.

Kennedy, Gregory, Jun Komano, and Bill Sugden. "Epstein-Barr Virus Provides a Survival Factor to Burkitt's Lymphomas." *Proceedings of the National Academy of Sciences* 100, no. 24 (2003): 14269–14274.

Kern, Scott E., Kenneth W. Kinzler, Arthur Bruskin, David Jarosz, Paula Friedman, Carol Prives, and Bert Vogelstein. "Identification of p53 as a Sequence-Specific DNA-Binding Protein." *Science* 252, no. 5013 (1991): 1708–1711.

Kessler, Irving I. "Human Cervical Cancer as a Venereal Disease." *Cancer Research* 36, no. 2, part 2 (1976): 783–791.

Kevles, D. J. "Pursuing the Unpopular: A History of Courage, Viruses, and Cancer." In *Hidden Histories of Science*. Eds. Robert B. Silvers, 69–112. New York Review, 1995.

Kevles, Daniel J. "Renato Dulbecco and the New Animal Virology: Medicine, Methods, and Molecules." *Journal of the History of Biology* 26, no. 3 (1993): 409–442.

Kevles, Daniel J., and Gerald L. Geison. "The Experimental Life Sciences in the Twentieth Century." *Osiris* 10 (1995): 97–121.

Khot, Sandeep, Buhm Soon Park, and W. T. Longstreth Jr. "The Vietnam War and Medical Research: Untold Legacy of the US Doctor Draft and the NIH 'Yellow Berets.'" *Academic Medicine* 86, no. 4 (2011): 502–508.

King, Madonna. *Ian Frazer: The Man Who Saved a Million Lives*. University of Queensland Press, 2013.

Kirnbauer, R., F. Booy, N. Cheng, D. R. Lowy, and J. T. Schiller. "Papillomavirus L1 Major Capsid Protein Self-Assembles into Virus-Like Particles That Are Highly Immunogenic." *Proceedings of the National Academy of Sciences* 89, no. 24 (1992): 12180–12184.

Kisselev, Lev L., Gary I. Abelev, and Feodor Kisseljov. "Lev Zilber, the Personality and the Scientist." *Advances in Cancer Research* 59 (1992): 1–40.

Klein, George. *The Atheist and the Holy City*. MIT Press, 1990.

Klein, George. "Summary of Papers Delivered at the Conference on Herpesvirus and Cervical Cancer (Key Biscayne, Florida)." *Cancer Research* 33, no. 6 (1973): 1557–1563.

Klein, Melissa K. "The Legacy of the 'Yellow Berets': The Vietnam War, the Doctor Draft, and the NIH Associate Training Program." Manuscript, 1998, NIH History Office, National Institutes of Health, Bethesda, MD.

Knight, Claude Arthur. *Molecular Virology*. McGraw-Hill, 1974.

Knudson, Alfred G. "Hereditary Cancer, Oncogenes, and Antioncogenes." *Cancer Research* 45, no. 4 (1985): 1437–1443.

Kolykhalov, Alexander A., Eugene V. Agapov, Keril J. Blight, Kathleen Mihalik, Stephen M. Feinstone, and Charles M. Rice. "Transmission of Hepatitis C by Intrahepatic Inoculation with Transcribed RNA." *Science* 277, no. 5325 (1997): 570–574.

Kontaratos, Nikolas. *Dissecting a Discovery: The Real Story of How the Race to Uncover the Cause of AIDS Turned Scientists against Disease, Politics against Science, Nation against Nation*. Xlibris, 2006.

Kontaratos, N., G. Sourvinos, and D. A. Spandidos. "Examining the Discovery of the Human Retrovirus." *JBUON* 15 (2010): 174–181.

Koshland, D. E., Jr. "Molecule of the Year." *Science* 262 (1993): 5142.

Kostyrka, Gladys, and Neeraja Sankaran. "From Obstacle to Lynchpin: The Evolution of the Role of Bacteriophage Lysogeny in Defining and Understanding Viruses." *Notes and Records* (2020). https://doi.org/10.1098/rsnr.2019.0033.

Kourany, Janet A., ed. *The Gender of Science*. Prentice Hall, 2002.

Koutsky, Laura A., Kevin A. Ault, Cosette M. Wheeler, Darron R. Brown, Eliav Barr, Frances B. Alvarez, Lisa M. Chiacchierini, and Kathrin U. Jansen. "A Controlled Trial of a Human Papillomavirus Type 16 Vaccine." *New England Journal of Medicine* 347, no. 21 (2002): 1645–1651.

Krementsov, Nikolai. "Lysenkoism in Europe: Export-Import of the Soviet Model." In *Academia in Upheaval: Origins, Transfers, and Transformations of the Communist Academic Regime in Russia and East Central Europe*, 179–202. Garland, 2000.

Krementsov, Nikolai. "A 'Second Front' in Soviet Genetics: The International Dimension of the Lysenko Controversy, 1944–1947." *Journal of the History of Biology* 29, no. 2 (1996): 229–250.

Krimsky, Sheldon. *Genetic Alchemy*. MIT Press, 1982.

Krontiris, Theodore G., and Geoffrey M. Cooper. "Transforming Activity of Human Tumor DNAs." *Proceedings of the National Academy of Sciences* 78, no. 2 (1981): 1181–1184.

Kuhn, Thomas S. *The Structure of Scientific Revolutions*. University of Chicago Press, 2012.

Kusin, Vladimir V. *The Intellectual Origins of the Prague Spring: The Development of Reformist Ideas in Czechoslovakia 1956–1967*. Vol. 5. Cambridge University Press, 2002.

Landecker, Hannah. *Culturing Life*. Harvard University Press, 2007.

Landecker, Hannah. "It Is What It Eats: Chemically Defined Media and the History of Surrounds." *Studies in History and Philosophy of Science Part C: Studies in History and Philosophy of Biological and Biomedical Sciences* 57 (2016): 148–160.

Lane, David P. "Such an Obsession." *Cancer Biology and Therapy* 5, no. 1 (2006): 120–123.

Lane, David P., and Lionel V. Crawford. "T antigen Is Bound to a Host Protein in SY40-Transformed Cells." *Nature* 278, no. 5701 (1979): 261–263.

Lane, D. P., and D. M. Silver. "Isolation of a Murine Liver-Specific Alloantigen, F antigen, and Examination of Its Immunogenic Properties by Radioimmunoassay." *European Journal of Immunology* 6, no. 7 (1976): 480–485.

Langård, Sverre. "Gregorius Agricola Memorial Lecture: Lung Cancer—a Work-Related Disease for 500 Years, as Predicted by Agricola." *Journal of Trace Elements in Medicine and Biology* 31 (2015): 214–218.

Latarjet, R., R. Cramer, and L. Montagnier. "Inactivation, by UV-, X-, and Γ-Radiations, of the Infecting and Transforming Capacities of Polyoma Virus." *Virology* 33, no. 1 (1967): 104–111.

Latour, Bruno, and Steve Woolgar. *Laboratory Life: The Construction of Scientific Facts.* Princeton University Press, 2013.

Lawler, Sean E., Maria-Carmela Speranza, Choi-Fong Cho, and E. Antonio Chiocca. "Oncolytic Viruses in Cancer Treatment: A Review." *JAMA Oncology* 3, no. 6 (2017): 841–849.

Lear, John. *Recombinant DNA: The Untold Story*. Crown, 1978.

Ledford, Heidi. "Cancer-Killing Viruses Show Promise—and Draw Billion-Dollar Investment." *Nature* 557, no. 7706 (2018): 150–152.

Leonelli, Sabina, and Rachel A. Ankeny. "What Makes a Model Organism?" *Endeavour* 37, no. 4 (2013): 209–212.

Levine, Arnold J., Cathy A. Finlay, and Philip W. Hinds. "p53 Is a Tumor Suppressor Gene." *Cell* 116 (2004): S67–S70.

Levy, Jay A. "Changing Dogmas in Retrovirology." In *International Symposium: Retroviruses and Human Pathology*, 35–59. Humana Press, 1985.

Levy, Jay A., Anthony D. Hoffman, Susan M. Kramer, Jill A. Landis, Joni M. Shimabukuro, and Lyndon S. Oshiro. "Isolation of Lymphocytopathic Retroviruses from San Francisco Patients with AIDS." *Science* 225, no. 4664 (1984): 840–842.

Lewis, Paul A., and Richard E. Shope. "The Blood in Hog Cholera." *Journal of Experimental Medicine* 50, no. 6 (1929): 719–737.

Lewontin, Richard, and Richard Levins. "The Problem of Lysenkoism." In *The Radicalisation of Science*, 32–64. Palgrave, 1976.

Levinson, Arthur D., Hermann Oppermann, Leon Levintow, Harold E. Varmus, and J. Michael Bishop. "Evidence That the Transforming Gene of Avian Sarcoma Virus Encodes a Protein Kinase Associated with a Phosphoprotein." *Cell* 15, no. 2 (1978): 561–572.

Linzer, Daniel I. H., and Arnold J. Levine. "Characterization of a 54K Dalton Cellular SV40 Tumor Antigen Present in SV40-Transformed Cells and Uninfected Embryonal Carcinoma Cells." *Cell* 17, no. 1 (1979): 43–52.

Little, C. C. "Evidence That Cancer Is Not a Simple Mendelian Recessive." *Faculty Research 1900–1939* (1928): 65.

Lloyd, Elisabeth A. "Feyerabend, Mill, and Pluralism." *Philosophy of Science* 64 (1997): S396–S407.

London, William T., Harvey J. Alter, Jerrold Lander, and Robert H. Purcell. "Serial Transmission in Rhesus Monkeys of an Agent Related to Hepatitis-Associated Antigen." *Journal of Infectious Diseases* 125, no. 4 (1972): 382–389.

Löwy, Ilana. *Between Bench and Bedside: Science, Healing, and Interleukin-2 in a Cancer Ward*. Harvard University Press, 1996.

Löwy, Ilana. "Epidemiology, Immunology, and Yellow Fever: The Rockefeller Foundation in Brazil, 1923–1939." *Journal of the History of Biology* 30, no. 3 (1997): 397–417.

Lu, Hua. "Legends of p53: Untold Four-Decade Stories." *Journal of Molecular Cell Biology* 11: (2019): 521–522.

Lucké, Balduin. "Carcinoma in the Leopard Frog: Its Probable Causation by a Virus." *Journal of Experimental Medicine* 68, no. 4 (1938): 457–468.

Luria, Salvador E. *General Virology*. John Wiley & Sons, 1953.

Luria, Salvador E. "Viruses, Cancer Cells, and the Genetic Concept of Virus Infection." *Cancer Research* 20, no. 5, part 1 (1960): 677–688.

Luria, S. E., and R. Dulbecco. "Lethal Mutations, and Inactivation of Individual Genetic Determinants in Bacteriophage." *Genetics* 33, no. 6 (1948): 618.

Lustig, Alice, and Arnold J. Levine. "One Hundred Years of Virology." *Journal of Virology* 66, no. 8 (1992): 4629.

Macphail, Theresa. *The Viral Network: A Pathography of the H1N1 Influenza Pandemic*. Cornell University Press, 2015.

Macpherson, Ian, and Luc Montagnier. "Agar Suspension Culture for the Selective Assay of Cells Transformed by Polyoma Virus." *Virology* 23, no. 2 (1964): 291–294.

Maienschein, Jane. "Why Collaborate?" *Journal of the History of Biology* 26, no. 2 (1993): 167–183.

Maienschein, Jane, Manfred Laubichler, and Andrea Loettgers. "How Can History of Science Matter to Scientists?" *Isis* 99, no. 2 (2008): 341–349.

Malinin, Theodore I. *Surgery and Life: The Extraordinary Career of Alexis Carrel*. New York: Harcourt Brace Jovanovich, 1979.

Mamo, Laura, and Steven Epstein. "The New Sexual Politics of Cancer: Oncoviruses, Disease Prevention, and Sexual Health Promotion." *Biosocieties* 12, no. 3 (2017): 367–391.

Manaker, Robert A., and Vincent Groupé. "Discrete Foci of Altered Chicken Embryo Cells Associated with Rous Sarcoma Virus in Tissue Culture." *Virology* 2, no. 6 (1956): 838–840.

Mandell, Joseph D., and Alfred Day Hershey. "A Fractionating Column for Analysis of Nucleic Acids." *Analytical Biochemistry* 1, no. 1 (1960): 66–77.

Maniatis, Tom, Edward F. Fritsch, and Joseph Sambrook. *Molecular Cloning: A Laboratory Manual*. Cold Spring Harbor Laboratory Press, 1982.

Mann, Ida. "Effect of Low Temperatures on Bittner Virus of Mouse Carcinoma." *British Medical Journal* 2, no. 4621 (1949): 251–253.

Manning, Patrick, and Mat Savelli, eds. *Global Transformations in the Life Sciences, 1945–1980*. University of Pittsburgh Press, 2018.

Manolov, George, and Yanka Manolova. "Marker Band in One Chromosome 14 from Burkitt Lymphomas." *Nature* 237, no. 5349 (1972): 33–34.

Marcum, James A. "From Heresy to Dogma in Accounts of Opposition to Howard Temin's DNA Provirus Hypothesis." *History and Philosophy of the Life Sciences* (2002): 165–192.

Marcus, Alan I. *Malignant Growth: Creating the Modern Cancer Research Establishment, 1875–1915*. University of Alabama Press, 2018.

Martin, G. Steven. "The Road to Src." *Oncogene* 23, no. 48 (2004): 7910–7917.

Martin, G. S. "Rous Sarcoma Virus: A Function Required for the Maintenance of the Transformed State." *Nature* 227 (1970): 1021–1023.

Matlashewski, Greg. "Human Papillomavirus Update." *International Journal of Cancer* 71, no. 5 (1997): 715–718.

McElheny, Victor K. *Watson and DNA: Making a Scientific Revolution*. Basic Books, 2004.

McIntyre, Peter. "Finding the Viral Link: The Story of Harald zur Hausen." *Cancer World* (2005): 32–37.

McKaughan, Daniel J. "The Influence of Niels Bohr on Max Delbrück: Revisiting the Hopes Inspired by 'Light and Life.'" *Isis* 96, no. 4 (2005): 507–529.

Medical World News. "'Epidemic' Takes Mystery Immune Deficiency beyond Gays." April 16, 1982.

Méthot, Pierre-Olivier. "Writing the History of Virology in the Twentieth Century: Discovery, Disciplines, and Conceptual Change." *Studies in History and Philosophy of Science Part C: Studies in History and Philosophy of Biological and Biomedical Sciences* 59 (2016): 145–153.

Miller, J. H. *Experiments in Molecular Genetics*. Cold Spring Laboratory Press.1972.

Miller, Janice M., Lyle D. Miller, Carl Olson, and Kenneth G. Gillette. "Virus-Like Particles in Phytohemagglutinin-Stimulated Lymphocyte Cultures with Reference to Bovine Lymphosarcoma 2." *Journal of the National Cancer Institute* 43, no. 6 (1969): 1297–1305.

Miller, J. M., and C. Olson. "Precipitating Antibody to an Internal Antigen of the C-Type Virus Associated with Bovine Lymphosarcoma." *Journal of the National Cancer Institute* 49, no. 5 (1972): 1459–1462.

Mitchell, Sandra D. "Through the Fractured Looking Glass." *Philosophy of Science* 87, no. 5 (2020): 771–792.

Mitsialis, S. Alex, Richard A. Katz, Jan Svoboda, and Ramareddy V. Guntaka. "Studies on the Structure and Organization of Avian Sarcoma Proviruses in the Rat XC Cell Line." *Journal of General Virology* 64, no. 9 (1983): 1885–1893.

Moloney, J. "Biological Studies on a Lymphoid Leukemia Virus Extracted from Sarcoma S. 37: I. Origin and Introductory Investigations." *Journal of the National Cancer Institute* 24 (1960): 933–951.

Moloney, J. "Preliminary Studies on a Mouse Lymphoid Leukemia Virus Extracted from Sarcoma 37." *Proceedings of the American Association for Cancer Research* 3 (1959): 44.

Montagnier, Luc. *Virus: The Co-discoverer of HIV Tracks Its Rampage and Charts the Future*. Norton, 2000.

Montagnier, Luc, Jamal Aissa, Stéphane Ferris, Jean-Luc Montagnier, and Claude Lavallée. "Electromagnetic Signals Are Produced by Aqueous Nanostructures Derived from Bacterial DNA Sequences." *Interdisciplinary Sciences: Computational Life Sciences* 1, no. 2 (2009): 81–90.

Montagnier, L., J. C. Chermann, F. Barre-Sinoussi, S. Chamaret, J. Gruest, M. T. Nugeyre,

F. Rey, et al. "A New Human T-Lymphotropic Retrovirus: Characterization and Possible Role in Lymphadenopathy and Acquired Immune Deficiency Syndromes." In *Human T-Cell Leukemia / Lymphoma Virus*, 363. Cold Spring Harbor Press, 1984.

Montagnier, L., and F. K. Sanders. "Replicative Form of Encephalomyocarditis Virus Ribonucleic Acid." *Nature* 199, no. 4894 (1963): 664–667.

Moore, Amanda R., Scott C. Rosenberg, Frank McCormick, and Shiva Malek. "RAS-Targeted Therapies." *Nature Reviews Drug Discovery* 19, no. 8 (2020): 533–552.

Morange, Michel. *The Black Box of Biology: A History of the Molecular Revolution*. Harvard University Press, 2020.

Morange, Michel. "The Discovery of Cellular Oncogenes." *History and Philosophy of the Life Sciences* (1993): 45–58.

Morange, Michel. "From the Regulatory Vision of Cancer to the Oncogene Paradigm, 1975–1985." *Journal of the History of Biology* 30, no. 1 (1997): 1–29.

Morange, Michel. *A History of Molecular Biology*. Harvard University Press, 1998.

Morange, Michel. "What History Tells Us." *Journal of Biosciences* 30, no. 3 (2005): 313–316.

Morgan, A. J., A. C. Allison, S. Finerty, F. T. Scullion, N. E. Byars, and M. A. Epstein. "Validation of a First-Generation Epstein-Barr Virus Vaccine Preparation Suitable for Human Use." *Journal of Medical Virology* 29, no. 1 (1989): 74–78.

Morgan, A. J., S. Finerty, K. Lovgren, F. T. Scullion, and B. Morein. "Prevention of Epstein-Barr (EB) Virus-Induced Lymphoma in Cottontop Tamarins by Vaccination with the EB Virus Envelope Glycoprotein gp340 Incorporated into Immune-Stimulating Complexes." *Journal of General Virology* 69, no. 8 (1988): 2093–2096.

Morgan, Doris Anne, Francis W. Ruscetti, and Robert Gallo. "Selective in Vitro Growth of T Lymphocytes from Normal Human Bone Marrows." *Science* 193, no. 4257 (1976): 1007–1008.

Morgan, Gregory J. "Early Theories of Virus Structure." In *Conformational Proteomics of Macromolecular Architecture: Approaching the Structure of Large Molecular Assemblies and Their Mechanisms of Action*, 3–40. Eds. R. Holland Cheng and Lena Hammar. World Scientific, 2004.

Morgan, Gregory J. "Historical Review: Viruses, Crystals and Geodesic Domes." *Trends in Biochemical Sciences* 28, no. 2 (2003): 86–90.

Morgan, Gregory J. "Virus Design, 1955–1962: Science Meets Art." *Phytopathology* 96, no. 11 (2006): 1287–1291.

Mowat, G. N., and W. G. Chapman. "Growth of Foot-and-Mouth Disease Virus in a Fibroblastic Cell Line Derived from Hamster Kidneys." *Nature* 194, no. 4825 (1962): 253.

Mukherjee, Siddhartha. *The Emperor of All Maladies: A Biography of Cancer*. Simon and Schuster, 2010.

Müller, Bruno, and Ueli Grossniklaus. "Model Organisms—a Historical Perspective." *Journal of Proteomics* 73, no. 11 (2010): 2054–2063.

Mullins, Nicholas C. "The Development of a Scientific Specialty: The Phage Group and the Origins of Molecular Biology." *Minerva* 10, no. 1 (1972): 51–82.

Munyon, William, Enzo Paoletti, and James T. Grace. "RNA Polymerase Activity in Purified Infectious Vaccinia Virus." *Proceedings of the National Academy of Sciences* 58, no. 6 (1967): 2280–2287.

Murphy, James B. "Experimental Approach to the Cancer Problem." *Bulletin of the Johns Hopkins Hospital* 56 (1935): 1–31.

Neiman, P. E., S. E. Wright, C. Mcmillin, and D. Macdonnell. "Nucleotide Sequence Relationships of Avian RNA Tumor Viruses: Measurement of the Deletion in a Transformation-Defective Mutant of Rous Sarcoma Virus." *Journal of Virology* 13, no. 4 (1974): 837–846.

Nemunaitis, John, Ian Ganly, Fadlo Khuri, James Arseneau, Joseph Kuhn, Todd Mccarty, Stephen Landers, et al. "Selective Replication and Oncolysis in p53 Mutant Tumors with ONYX-015, an E1B-55kd Gene-Deleted Adenovirus, in Patients with Advanced Head and Neck Cancer: A Phase II Trial." *Cancer Research* 60, no. 22 (2000): 6359–6366.

Novinsky, M. "O privivanii trakovikh novoobrazovanii [On the inoculation of cancerous neoplasms]." *Med. Vestnik* 16 (1876): 289–290.

Offit, Paul A. *Vaccinated: One Man's Quest to Defeat the World's Deadliest Diseases*. Smithsonian Books, 2007.

Okada, Yoshio. "The Fusion of Ehrlichs Tumor Cells Caused by HVJ Virus in Vitro." *Biken Journal* 1, no. 2 (1958): 103–110.

Olby, Robert Cecil. "Francis Crick, DNA, and the Central Dogma." *Daedalus* (1970): 938–987.

Olby, Robert Cecil. *Francis Crick: Hunter of Life's Secrets*. Cold Spring Harbor Laboratory Press, 2009.

Olby, Robert Cecil. *The Path to the Double Helix: The Discovery of DNA*. Dover, 1994.

Olitsky, Peter K. "The Action of Glycerol on the Virus of Experimental Typhus Fever and on Proteus Bacilli." *Proceedings of the Society for Experimental Biology and Medicine* 22, no. 7 (1925): 399–400.

Olson, James Stuart. *The History of Cancer: An Annotated Bibliography*. Greenwood Press, 1989.

Oppermann, Hermann, Arthur D. Levinson, Harold E. Varmus, Leon Levintow, and J. Michael Bishop. "Uninfected Vertebrate Cells Contain a Protein That Is Closely Related to the Product of the Avian Sarcoma Virus Transforming Gene (*src*)." *Proceedings of the National Academy of Sciences* 76, no. 4 (1979): 1804–1808.

Orth, Gérard, Stefania Jablonska, Maria Jarząbek-Chorzelska, Slavomir Obalek, Genowefa Rzesa, Michel Favre, and Odile Croissant. "Characteristics of the Lesions and Risk of Malignant Conversion Associated with the Type of Human Papillomavirus Involved in Epidermodysplasia Verruciformis." *Cancer Research* 39, no. 3 (1979): 1074–1082.

Oshinsky, David M. *Polio: An American Story*. Oxford University Press, 2005.

Parada, Luis F., Clifford J. Tabin, Chiaho Shih, and Robert A. Weinberg. "Human EJ Bladder Carcinoma Oncogene Is Homologue of Harvey Sarcoma Virus Gene." *Nature* 297, no. 5866 (1982): 474.

Pauling, Linus, Harvey A. Itano, Seymour J. Singer, and Ibert C. Wells. "Sickle Cell Anemia, a Molecular Disease." *Science* 110, no. 2865 (1949): 543–548.

Payne, L. N., and V. Nair. "The Long View: 40 Years of Avian Leukosis Research." *Avian Pathology* 41, no. 1 (2012): 11–19.

Pendlebury, D. "Citation Superstars of NIH: Most-Cited Scientists, 1981–88." *Scientist* (1990): 14.

Perlman, Robert L., and Ira Pastan. "Regulation of B-Galactosidase Synthesis in Escherichia Coli by Cyclic Adenosine 3′, 5′-Monophosphate." *Journal of Biological Chemistry* 243, no. 20 (1968): 5420–5427.

Pickering, Andrew, ed. *Science as Practice and Culture*. University of Chicago Press, 1992.

Pierrel, Jérôme. "An RNA Phage Lab: MS2 in Walter Fiers' Laboratory of Molecular Biology in Ghent, from Genetic Code to Gene and Genome, 1963–1976." *Journal of the History of Biology* 45, no. 1 (2012): 109–138.

Pieters, Toine. *Interferon: The Science and Selling of a Miracle Drug*. Routledge, 2005.

Pipas, James M. "DNA Tumor Viruses and Their Contributions to Molecular Biology." *Journal of Virology* 93, no. 9 (2019): E01524–18.

Pitt, Joseph C. "The Dilemma of Case Studies: Toward a Heraclitian Philosophy of Science." *Perspectives on Science* 9, no. 4 (2001): 373–382.

Plutynski, Anya. "Cancer and the Goals of Integration." *Studies in History and Philosophy of Science Part C: Studies in History and Philosophy of Biological and Biomedical Sciences* 44, no. 4 (2013): 466–476.

Poiesz, Bernard J., Francis W. Ruscetti, Adi F. Gazdar, Paul A. Bunn, John D. Minna, and Robert C. Gallo. "Detection and Isolation of Type C Retrovirus Particles from Fresh and Cultured Lymphocytes of a Patient with Cutaneous T-Cell Lymphoma." *Proceedings of the National Academy of Sciences* 77, no. 12 (1980): 7415–7419.

Porter, Keith R., and H. P. Thompson. "A Particulate Body Associated with Epithelial Cells Cultured from Mammary Carcinomas of Mice of a Milk-Factor Strain." *Journal of Experimental Medicine* 88, no. 1 (1948): 15–24.

Porter, Roy. *The Greatest Benefit to Mankind: A Medical History of Humanity*. The Norton History of Science. Norton, 1999.

Pott, Percival. "Chirurgical Observations Relative to the Cataract, the Polypus of the Nose, the Cancer of the Scrotum, the Different Kinds of Ruptures, and the Mortification of the Toes and Feet." *Weekly Entertainer and West of England Miscellany* 5, no. 129 (1776): 622–625.

Powell, Alexander, and John Dupré. "From Molecules to Systems: The Importance of Looking Both Ways." *Studies in History and Philosophy of Science Part C: Studies in History and Philosophy of Biological and Biomedical Sciences* 40, no. 1 (2009): 54–64.

Powers, Scott, and Robert E. Pollack. "Inducing Stable Reversion to Achieve Cancer Control." *Nature Reviews Cancer* 16, no. 4 (2016): 266.

Prince, Alfred M. *My Life with Viruses, Friends & Enemies*. Xlibris, 2008.

Prince, Alfred M. "Relation of Australia and SH antigens." *Lancet* 292, no. 7565 (1968): 462–463.

Proctor, Robert N., and Londa Schiebinger, eds. *Agnotology: The Making and Unmaking of Ignorance*. Stanford University Press, 2008.

Prusiner, Stanley B. "Discovering the Cause of AIDS." *Science* 298, no. 5599 (2002): 1726–1726.

Rader, K. A. *Making Mice: Standardizing Animals for American Biomedical Research, 1900–1955*. Princeton University Press, 2004.

Rader, Karen A. "The Origins of Mouse Genetics: Beyond the Bussey Institution I. Cold Spring Harbor: The Station for Experimental Evolution and the 'Mouse Club of America.'" *Mammalian Genome* 8, no. 7 (1997): 464–466.

Radetsky, Peter. *The Invisible Invaders: Viruses and the Scientists Who Pursue Them*. Little Brown, 1994.

Rajewsky, Klaus. "George Klein: 1925–2016." *Proceedings of the National Academy of Sciences* (2017): 201702501. doi:10.1073/pnas.1702501114.

Rasmussen, Nicolas. "DNA Technology: 'Moratorium' on Use and Asilomar Conference." *eLS* (2001): 1–5.

Rasmussen, Nicolas. *Gene Jockeys: Life Science and the Rise of Biotech Enterprise*. Johns Hopkins University Press, 2014.

Rasmussen, Nicolas. *Picture Control: The Electron Microscope and the Transformation of Biology in America, 1940–1960*. Stanford University Press, 1999.

Rauscher, Frank J. "A Virus-Induced Disease of Mice Characterized by Erythrocytopoiesis and Lymphoid Leukemia." *Journal of the National Cancer Institute* 29, no. 3 (1962): 515–543.

Raza, Azra. *The First Cell: And the Human Costs of Pursuing Cancer to the Last*. Basic Books, 2019.

Reece, E. Albert. "The Once and Future Robert C. Gallo." In *From Cause to Care*. AAAS Science Business Office, 2009.

Reynolds, Andrew S. *The Third Lens: Metaphor and the Creation of Modern Cell Biology*. University of Chicago Press, 2018.

Reynolds, Lois A., and E. M. Tansey. *Foot and Mouth Disease: The 1967 Outbreak and Its Aftermath*. Wellcome Trust Centre for the History of Medicine at UCL, 2003.

Reynolds, Lois A., and E. M. Tansey. *History of Cervical Cancer and the Role of the Human Papillomavirus, 1960–2000*. Vol. 38, no. 38. Wellcome Trust Centre for the History of Medicine at UCL, 2009.

Rheinberger, Hans-Jörg. "Patterns of the International and the National, the Global and the Local in the History of Molecular Biology." In *The Local Configuration of New Research Fields*, 193–204. Springer, 2016.

Rheinberger, Hans-Jörg. "Recent Orientations and Reorientations in the Life Sciences." In *Science in the Context of Application*, 161–168. Springer, 2011.

Rheinberger, Hans-Jörg. "What Happened to Molecular Biology?" *BioSocieties* 3, no. 3 (2008): 303–310.

Rigoni-Stern, D. "Fatti statistici relativi alle malattie cancerose."*Giornale Service Progr Pathol Terap Ser* 2 (1842): 507–517.

Rivers, Thomas M. "Viruses and Koch's Postulates." *Journal of Bacteriology* 33, no. 1 (1937): 1–12.

Rogers, Michael. *Biohazard*. Knopf, 1977.

Rose, R. C., William Bonnez, R. C. Reichman, and R. L. Garcea. "Expression of Human Papillomavirus Type 11 L1 Protein in Insect Cells: In Vivo and in Vitro Assembly of Viruslike Particles." *Journal of Virology* 67, no. 4 (1993): 1936–1944.

Rosenberg, Alexander. *Darwinian Reductionism: Or, How to Stop Worrying and Love Molecular Biology*. University of Chicago Press, 2008.

Rous, F. Peyton. "The Effect of Pilocarpine on the Output of Lymphocytes through the Thoracic Duct." *Journal of Experimental Medicine* 10, no. 3 (1908): 329–342.

Rous, F. Peyton. "An Experimental Comparison of Transplanted Tumor and a Transplanted Normal Tissue Capable of Growth." *Journal of Experimental Medicine* 12, no. 3 (1910): 344–366.

Rous, F. Peyton. "An Inquiry into Some Mechanical Factors in the Production of Lymphocytosis." *Journal of Experimental Medicine* 10, no. 2 (1908): 238–270.

Rous, P. Presentation of the Kober Medal to Richard Shope. *Transactions of the Association of American Physicians* 70 (1957): 29.

Rous, F. Peyton. "A Sarcoma of the Fowl Transmissible by an Agent Separable from the Tumor Cells." *Journal of Experimental Medicine* 13, no. 4 (1911): 397–411.

Rous, F. Peyton. "Some Differential Counts of the Cells in the Lymph of the Dog: Their Bearing on Problems in Haematology." *Journal of Experimental Medicine* 10, no. 4 (1908): 537.

Rous, Peyton. "A Transmissible Avian Neoplasm (Sarcoma of the Common Fowl)." *Journal of Experimental Medicine* 12, no. 5 (1910): 696–705.

Rous, F. Peyton. "The Virus Tumors and the Tumor Problem." *American Journal of Cancer* 28, no. 2 (1936): 233–272.

Rous, Peyton, and J. W. Beard. "The Progression to Carcinoma of Virus-Induced Rabbit Papillomas (Shope)." *Journal of Experimental Medicine* 62, no. 4 (1935): 523–548.

Rowe, Wallace P., Robert J. Huebner, Loretta K. Gilmore, Robert H. Parrott, and Thomas G. Ward. "Isolation of a Cytopathogenic Agent from Human Adenoids Undergoing Spontaneous Degeneration in Tissue Culture." *Proceedings of the Society for Experimental Biology and Medicine* 84, no. 3 (1953): 570–573.

Rowson, K. E., and B. W. Mahy. "Human Papova (Wart) Virus." *Bacteriological Reviews* 31, no. 2 (1967): 110.

Rubin, Andrew L., Adam Yao, and Harry Rubin. "Relation of Spontaneous Transformation in Cell Culture to Adaptive Growth and Clonal Heterogeneity." *Proceedings of the National Academy of Sciences* 87, no. 1 (1990): 482–486.

Rubin, Harry. "A Disease in Captive Egrets Caused by a Virus of the Psittacosis-Lymphogranuloma Venereum Group." *Journal of Infectious Diseases* 94, no. 1 (1954): 1–8.

Rubin, Harry. "The Early History of Tumor Virology: Rous, RIF, and RAV." *Proceedings of the National Academy of Sciences* 108, no. 35 (2011): 14389–14396.

Ruddy, Kathleen T. *Of Mice and Women: Unraveling the Mystery of the Breast Cancer Virus.* Breast Health & Healing Foundation, 2015.

Rusch, Harold P. *Something Attempted, Something Done: A Personal History of Cancer Research at the University of Wisconsin, 1934–1979.* Wisconsin Medical Alumni Association, 1984.

Ruse, Michael E. "Reduction, Replacement, and Molecular Biology." *Dialectica* (1971): 39–72.

Ryan, Frank. *Virolution.* Collins, 2009.

Rybicki, Edward P. "A Top Ten List for Economically Important Plant Viruses." *Archives of Virology* 160, no. 1 (2015): 17–20.

Sachs, Leo, and Dan Medina. "In Vitro Transformation of Normal Cells by Polyoma Virus." *Nature* 189, no. 4763 (1961): 457.

Sambrook, Joseph, Heiner Westphal, P. R. Srinivasan, and Renato Dulbecco. "The Integrated State of Viral DNA in SV40-Transformed Cells." *Proceedings of the National Academy of Sciences* 60, no. 4 (1968): 1288–1295.

Sambrook, Joseph, P. A. Sharp, and W. Keller. "Transcription of Simian Virus 40: Separation of the Strands of SV40 DNA and Hybridization of the Separated Strands to RNA Extracted from Lytically Infected and Transformed Cells." *Journal of Molecular Biology* 70 (1972): 57–71.

Sanford, K. K., W. R. Earle, and G. D. Likely. "The Growth in Vitro of Single Isolated Tissue Cells." *Journal of the National Cancer Institute* 9, no. 3 (1948): 229–246.

Sankaran, Neeraja. "On the Historical Significance of Beijerinck and His *Contagium Vivum Fluidum* for Modern Virology." *History and Philosophy of the Life Sciences* 40, no. 3 (2018): 41.

Sankaran, Neeraja. *A Tale of Two Viruses: Parallels in the Research Trajectories of Tumor and Bacterial Viruses*. University of Pittsburgh Press, 2021.

Sankaran, Neeraja. "When Viruses Were Not in Style: Parallels in the Histories of Chicken Sarcoma Viruses and Bacteriophages." *Studies in History and Philosophy of Science Part C: Studies in History and Philosophy of Biological and Biomedical Sciences* 48 (2014): 189–199.

Sankaran, Neeraja, and Ton Van Helvoort. "Andrewes's Christmas Fairy Tale: Atypical Thinking about Cancer Aetiology in 1935." *Notes Records: The Royal Society Journal of the History of Science* 70 no. 2 (2016): 175–201.

Sarkar, Sahotra. *Genetics and Reductionism*. Cambridge University Press, 1998.

Sarngadharan, M. G., Prem S. Sarin, Marvin S. Reitz, and Robert C. Gallo. "Reverse Transcriptase Activity of Human Acute Leukaemic Cells: Purification of the Enzyme, Response to AMV 70S RNA, and Characterization of the DNA Product." *Nature* 240, no. 98 (1972): 67–72.

Schaffer, F. L., A. J. Hackett, and M. E. Soergel. "Vesicular Stomatitis Virus RNA: Complementarity between Infected Cell RNA and RNA's from Infectious and Autointerfering Viral Fractions." *Biochemical and Biophysical Research Communications* 31, no. 5 (1968): 685–692.

Scheffler, Robin Wolfe. "Following Cancer Viruses through the Laboratory, Clinic, and Society." *Studies in History and Philosophy of Biological and Biomedical Sciences* 48 (2014): 185–188.

Scheffler, Robin Wolfe. "Managing the Future: The Special Virus Leukemia Program and the Acceleration of Biomedical Research." *Studies in History and Philosophy of Science Part C: Studies in History and Philosophy of Biological and Biomedical Sciences* 48 (2014): 231–249.

Scheffler, Robin Wolfe. "Protecting Children: The American Turn from Polio to Cancer Vaccines." *Canadian Medical Association Journal* 191, no. 26 (2019): E739–E741.

Scheffler, Robin Wolfe. *A Contagious Cause: The American Hunt for Cancer Viruses and the Rise of Molecular Medicine*. University of Chicago Press, 2019.

Schickore, Jutta. "More Thoughts on HPS: Another 20 Years Later." *Perspectives on Science* 19, no. 4 (2011): 453–481.

Scholthof, Karen-Beth G. "Making a Virus Visible: Francis O. Holmes and a Biological Assay for Tobacco Mosaic Virus." *Journal of the History of Biology* 47, no. 1 (2014): 107–145.

Scholthof, Karen-Beth G. "Tobacco Mosaic Virus: A Model System for Plant Biology." *Annual Review of Phytopathology* 42 (2004): 13–34.

Schwarz, Elisabeth, Ulrich Karl Freese, Lutz Gissmann, Wolfgang Mayer, Birgit Roggenbuck, Armin Stremlau, and Harald zur Hausen. "Structure and Transcription of Human Papillomavirus Sequences in Cervical Carcinoma Cells." *Nature* 314, no. 6006 (1985): 111–114.

Scolnick, Edward M. "A Vaccine to Prevent Cervical Cancer: Academic and Industrial Collaboration and a Lasker Award." *Clinical & Translational Immunology* 7, no. 1 (2018).

Scolnick, E. M., R. J. Goldberg, and W. P. Parks. "A Biochemical and Genetic Analysis of Mammalian RNA-Containing Sarcoma Viruses." In *Cold Spring Harbor Symposia on Quantitative Biology*, vol. 39, 885–895. Cold Spring Harbor Laboratory Press, 1974.

Scolnick, Edward M., Alex G. Papageorge, and Thomas Y. Shih. "Guanine Nucleotide-Binding Activity as an Assay for Src Protein of Rat-Derived Murine Sarcoma Viruses." *Proceedings of the National Academy of Sciences* 76, no. 10 (1979): 5355–5359.

Scolnick, Edward M., Elaine Rands, David Williams, and Wade P. Parks. "Studies on the Nucleic Acid Sequences of Kirsten Sarcoma Virus: A Model for Formation of a Mammalian RNA-Containing Sarcoma Virus." *Journal of Virology* 12, no. 3 (1973): 458–463.

Scolnick, E., R. Tompkins, T. Caskey, and M. Nirenberg. "Release Factors Differing in Specificity for Terminator Codons." *Proceedings of the National Academy of Sciences* 61, no. 2 (1968): 768–774.

Sedwick, Caitlin. "Joan Brugge: Running Rings around Cancer." *Journal of Cell Biology* 189, no. 6 (2010): 922–923.

Selya, Rena. "Primary Suspects: Reflections on Autobiography and Life Stories in the History of Molecular Biology." *History and Poetics of Scientific Biography* (2007): 199–206.

Shabad, L. M., and V. I. Ponomarkov. "Mstislav Novinsky, Pioneer of Tumour Transplantation." *Cancer Letters* 2, no. 1 (1976): 1–3.

Shatkin, A. J., and J. D. Sipe. "RNA Polymerase Activity in Purified Reoviruses." *Proceedings of the National Academy of Sciences* 61, no. 4 (1968): 1462–1469.

Sherlock, Sheila, S. P. Niazi, R. A. Fox, and P. J. Scheuer. "Chronic Liver Disease and Primary Liver-Cell Cancer with Hepatitis-Associated (Australia) Antigen in Serum." *Lancet* 295, no. 7659 (1970): 1243–1247.

Shih, Chiaho, L. C. Padhy, Mark Murray, and Robert A. Weinberg. "Transforming Genes of Carcinomas and Neuroblastomas Introduced into Mouse Fibroblasts." *Nature* 290, no. 5803 (1981): 261–264.

Shih, Chiaho, Ben-Zion Shilo, Mitchell P. Goldfarb, Ann Dannenberg, and Robert A. Weinberg. "Passage of Phenotypes of Chemically Transformed Cells via Transfection of DNA and Chromatin." *Proceedings of the National Academy of Sciences* 76, no. 11 (1979): 5714–5718.

Shilts, Randy. *And the Band Played On: Politics, People, and the AIDS Epidemic*. Souvenir Press, 2011.

Shimkin, Michael B. "As Memory Serves—an Informal History of the National Cancer Institute, 1937–57." *Journal of the National Cancer Institute* 59, no. 2, suppl. (1977): 559.

Shimkin, Michael B. *Some Classics of Experimental Oncology: 50 Selections, 1775–1965*. Washington, DC: National Institutes of Health, 1980.

Shope, Richard E. "A Filterable Virus Causing a Tumor-Like Condition in Rabbits and Its

Relationship to Virus Myxomatosum." *Journal of Experimental Medicine* 56 (1932): 803–822.

Shope, Richard E. "Immunization of Rabbits to Infectious Papillomatosis." *Journal of Experimental Medicine* 65 (1937): 219–231.

Shope, Richard E., and E. Weston Hurst. "Infectious Papillomatosis of Rabbits: With a Note on the Histopathology." *Journal of Experimental Medicine* 58, no. 5 (1933): 607–624.

Shope, Thomas, Douglas Dechairo, and George Miller. "Malignant Lymphoma in Cotton-top Marmosets after Inoculation with Epstein-Barr Virus." *Proceedings of the National Academy of Sciences* 70, no. 9 (1973): 2487–2491.

Silver, Donald M., and David P. Lane. "Dominant Nonresponsiveness in the Induction of Autoimmunity to Liver-Specific F antigen." *Journal of Experimental Medicine* 142, no. 6 (1975): 1455–1461.

Singer, Peter. *Galen: Selected Works*. Oxford University Press, 1997.

Skalka, Anna Marie. *Discovering Retroviruses: Beacons in the Biosphere*. Harvard University Press, 2018.

Skloot, Rebecca. *The Immortal Life of Henrietta Lacks*. Broadway Books, 2017.

Smith, Wilson, C. H. Andrewes, and P. P. Laidlaw. "A Virus Obtained from Influenza Patients." *Lancet* 222 (1933): 66–68.

Smith, Jane S. *Patenting the Sun: Polio and the Salk Vaccine*. William Morrow, 1990.

Smotkin, David, Alessandro M. Gianni, Shmuel Rozenblatt, and Robert A. Weinberg. "Infectious Viral DNA of Murine Leukemia Virus." *Proceedings of the National Academy of Sciences* 72, no. 12 (1975): 4910–4913.

Snell, George D. *Biology of the Laboratory Mouse*. Blakiston, 1941.

Snow, Charles Percy. *The Two Cultures*. Cambridge University Press, 2012.

Soussi, Thierry. "The History of p53: A Perfect Example of the Drawbacks of Scientific Paradigms." *EMBO Reports* 11, no. 11 (2010): 822–826.

Spurgeon, Megan E., and Paul F. Lambert. "Merkel Cell Polyomavirus: A Newly Discovered Human Virus with Oncogenic Potential." *Virology* 435, no. 1 (2013): 118–130.

Stark, Laura, and Nancy D. Campbell. "Stowaways in the History of Science: The Case of Simian Virus 40 and Clinical Research on Federal Prisoners at the US National Institutes of Health, 1960." *Studies in History and Philosophy of Science Part C: Studies in History and Philosophy of Biological and Biomedical Sciences* 48 (2014): 218–230.

Stehelin, Dominique, and Thomas Graf. "Avian Myelocytomatosis and Erythroblastosis Viruses Lack the Transforming Gene *src* of Avian Sarcoma Viruses." *Cell* 13, no. 4 (1978): 745–750.

Stehelin, Dominique, Ramareddy V. Guntaka, Harold E. Varmus, and J. Michael Bishop. "Purification of DNA Complementary to Nucleotide Sequences Required for Neoplastic Transformation of Fibroblasts by Avian Sarcoma Viruses." *Journal of Molecular Biology* 101, no. 3 (1976): 349–365.

Stehelin, Dominique, Harold E. Varmus, J. Michael Bishop, and Peter K. Vogt. "DNA Related to the Transforming Gene (S) of Avian Sarcoma Viruses Is Present in Normal Avian DNA." *Nature* 260, no. 5547 (1976): 170–173.

Sterner, Beckett, Joeri Witteveen, and Nico Franz. "Coordinating Dissent as an Alternative to Consensus Classification: Insights from Systematics for Bio-ontologies." *History and Philosophy of the Life Sciences* 42, no. 1 (2020): 8.

Stevens, Rosemary A. "Health Care in the Early 1960s." *Health Care Financing Review* 18, no. 2 (1996): 11.

Stewart, Sarah E. "Leukemia in Mice Produced by A Filterable Agent Present in AKR Leukemic Tissues with Notes on a Sarcoma Produced by the Same Agent." *Anatomical Record* 117, no. 3 (1953): 532–532.

Stewart, Sarah E., and Bernice E. Eddy. "Tumor Induction by SE Polyoma Virus and the Inhibition of Tumors by Specific Neutralizing Antibodies." *American Journal of Public Health and the Nation's Health* 49, no. 11 (1959): 1493–1496.

Stewart, Sarah E., Bernice E. Eddy, and Ninette G. Borgese. "Neoplasms in Mice Inoculated with a Tumor Agent Carried in Tissue Culture." *JNCI: Journal of the National Cancer Institute* 20, no. 6 (1958): 1223–1243.

Stewart, Sarah E., Bernice E. Eddy, Alice M. Gochenour, Ninette G. Borgese, and George E. Grubbs. "The Induction of Neoplasms with A Substance Released from Mouse Tumors by Tissue Culture." *Virology* 3, no. 2 (1957): 380–400.

Stewart, Sarah E., Elizabeth Lovelace, Jacqueline J. Whang, and V. Anomah Ngu. "Burkitt Tumor: Tissue Culture, Cytogenetic and Virus Studies." *Journal of the National Cancer Institute* 34, no. 2 (1965): 319–327.

Stoker, M. G. P. "Neoplastic Transformation by Polyoma Virus and Its Wider Implications." *British Medical Journal* 1, no. 5341 (1963): 1305.

Stoker, Michael. "Studies on the Oncogenic Activity of the Toronto Strain of Polyoma Virus." *British Journal of Cancer* 14, no. 4 (1960): 679.

Stoker, Michael, and Pamela Abel. "Conditions Affecting Transformation by Polyoma Virus." In *Cold Spring Harbor Symposia on Quantitative Biology*, vol. 27, 375–386. Cold Spring Harbor Laboratory Press, 1962.

Stotz, Karola, and Paul E. Griffiths. "Biohumanities: Rethinking the Relationship between Biosciences, Philosophy and History of Science, and Society." *Quarterly Review of Biology* 83, no. 1 (2008): 37–45.

Strasser, Bruno J. *Collecting Experiments: Making Big Data Biology*. University of Chicago Press, 2019.

Strasser, Bruno J. "Institutionalizing Molecular Biology in Post-war Europe: A Comparative Study." *Studies in History and Philosophy of Science Part C: Studies in History and Philosophy of Biological and Biomedical Sciences* 33, no. 3 (2002): 515–546.

Strasser, Bruno J. "Sickle Cell Anemia, a Molecular Disease." *Science* 286, no. 5444 (1999): 1488–1490.

Strasser, Bruno J. "A World in One Dimension: Linus Pauling, Francis Crick and the Central Dogma of Molecular Biology." *History and Philosophy of the Life Sciences* (2006): 491–512.

Strasser, Bruno J., and Bernardino Fantini. "Molecular Diseases and Diseased Molecules: Ontological and Epistemological Dimensions." *History and Philosophy of the Life Sciences* (1998): 189–214.

Straus, Eugene. *Rosalyn Yalow, Nobel Laureate: Her Life and Work in Medicine; A Biographical Memoir*. Basic Books, 1998.

Strong, Leonell C. "Obituary John Joseph Bittner 1904–1961." *Cancer Research* 22, no. 3 (1962): 393–393.

Summers, William C. *Felix d'Herelle and the Origins of Molecular Biology*. Yale University Press, 1999.

Summers, William C. "How Bacteriophage Came to Be Used by the Phage Group." *Journal of the History of Biology* 26, no. 2 (1993): 255–267.

Summers, William C. "Inventing Viruses." *Annual Review of Virology* 1 (2014): 25–35.

Summers, William C. "Microbe Hunters Revisited." *International Microbiology* 1 (1998): 65–68.

Suzich, Joann A., Shin-Je Ghim, Frances J. Palmer-Hill, Wendy I. White, James K. Tamura, Judith A. Bell, Joseph A. Newsome, A. Bennett Jenson, and Richard Schlegel. "Systemic Immunization with Papillomavirus L1 Protein Completely Prevents the Development of Viral Mucosal Papillomas." *Proceedings of the National Academy of Sciences* 92, no. 25 (1995): 11553–11557.

Svoboda, Jan. "Foundations in Cancer Research: The Turns of Life and Science." *Advances in Cancer Research* 99 (2008): 1–32.

Svoboda, Jan. "Further Findings on the Induction of Tumors by Rous Sarcoma in Rats and on the Rous Virus-Producing Capacity of One of the Induced Tumours (XC) in Chicks." *Folia Biologica* 8 (1962): 215–220.

Svoboda, Jan. "Presence of Chicken Tumour Virus in the Sarcoma of the Adult Rat Inoculated after Birth with Rous Sarcoma Tissue." *Nature* 186, no. 4729 (1960): 980–981.

Svoboda, J., P. Chyle, D. Simkovic, and I. Hilgert. "Demonstration of the Absence of Infectious Rous Virus in Rat Tumour XC, Whose Structurally Intact Cells Produce Rous Sarcoma when Transferred to Chicks." *Folia biologica* 9 (1963): 77–81.

Svoboda, J., and M. Hasek. "Influencing the Transplantability of the Virus of Rous Sarcoma by Immunological Approximation in Turkeys." *Folia Biologica-Praha* 2 (1956): 256–284.

Svoboda, J., I. Hložánek, O. Mach, and S. Zadražil. "Problems of RSV Rescue from Virogenic Mammalian Cells." In *Cold Spring Harbor Symposia on Quantitative Biology*, vol. 39, 1077–1083. Cold Spring Harbor Laboratory Press, 1974.

Svoboda, Jan, Vladimír Lhoták, Josef Geryk, Simon Saule, Marie Berthe Raes, and Dominique Stehelin. "Characterization of Exogenous Proviral Sequences in Hamster Tumor Cell Lines Transformed by Rous Sarcoma Virus Rescued from XC Cells." *Virology* 128, no. 1 (1983): 195–209.

Svoboda, Jan, Dusan Simkovic, and Hilary Koprowski. "Report on the Workshop on Virus Induction by Cell Association." *International Journal of Cancer* 3, no. 2 (1968): 317–322.

Sweet, Ben H., and Maurice R. Hilleman. "The Vacuolating Virus, SV 40." *Proceedings of the Society for Experimental Biology and Medicine* 105, no. 2 (1960): 420–427.

Szmuness, Wolf, Cladd E. Stevens, Edward J. Harley, Edith A. Zang, William R. Oleszko, Daniel C. William, Richard Sadovsky, et al. "Hepatitis B Vaccine: Demonstration of Efficacy in a Controlled Clinical Trial in a High-Risk Population in the United States." *New England Journal of Medicine* 303, no. 15 (1980): 833–841.

Tabin, Clifford J., Scott M. Bradley, Cornelia I. Bargmann, Robert A. Weinberg, Alex G.

Papageorge, Edward M. Scolnick, Ravi Dhar, et al. "Mechanism of Activation of a Human Oncogene." *Nature* 300, no. 5888 (1982): 143–149.

Tabor, Edward, Robert J. Gerety, Charles L. Vogel, Anne C. Bayley, Peter P. Anthony, Chao H. Chan, and Lewellys F. Barker. "Hepatitis B Virus Infection and Primary Hepatocellular Carcinoma." *Journal of the National Cancer Institute* 58, no. 5 (1977): 1197–1200.

Taub, R., I. Kirsch, C. Morton, G. Lenoir, D. Swan, S. Tronick, S. Aaronson, and P. Leder. "Translocation of the *c-myc* Gene into the Immunoglobulin Heavy Chain Locus in Human Burkitt Lymphoma and Murine Plasmacytoma Cells." *Proceedings of the National Academy of Sciences* 79, no. 24 (1982): 7837–7841.

Taylor, Milton W. *Viruses and Man: A History of Interactions*. Springer, 2014.

Temin, Howard. "Cancer and Viruses." *Engineering and Science* 23, no. 4 (1960): 21–24.

Temin, Howard M. "The Effects of Actinomycin D on Growth of Rous Sarcoma Virus in Vitro." *Virology* 20, no. 4 (1963): 577–582.

Temin, Howard M. "Homology between RNA from Rous Sarcoma Virus and DNA from Rous Sarcoma Virus–Infected Cells." *Proceedings of the National Academy of Sciences* 52, no. 2 (1964): 323–329.

Temin, Howard M. "Separation of Morphological Conversion and Virus Production in Rous Sarcoma Virus Infection." In *Cold Spring Harbor Symposia on Quantitative Biology*, vol. 27, 407–414. Cold Spring Harbor Laboratory Press, 1962.

Temin, Howard M., and S. Mizutami. "RNA-Dependent DNA Polymerase in Virions of Rous Sarcoma Virus." *Nature* 226 (1970): 1211–1213.

Thorley-Lawson, David, Kirk W. Deitsch, Karen A. Duca, and Charles Torgbor. "The Link Between Plasmodium Falciparum Malaria and Endemic Burkitt's Lymphoma—New Insight into a 50-Year-Old Enigma." *PLOS Pathogens* 12, no. 1 (2016): E1005331.

Thummel, Carl, Robert Tjian, and Terri Grodzicker. "Expression of SV40 T antigen under Control of Adenovirus Promoters." *Cell* 23, no. 3 (1981): 825–836.

Tonn, Jenna. "Gender." In *Encyclopedia of the History of Science* (March 2019). doi: 10.34758/vzgr-x490.

Tooze, J. *The Molecular Biology of Tumor Viruses*. Cold Spring Harbor Laboratory Press, 1973.

Tooze, J. "Period Piece Revised." *Nature* 227 (1970): 416–417.

Toyoshima, Kumao, and Peter K. Vogt. "Temperature Sensitive Mutants of an Avian Sarcoma Virus." *Virology* 39, no. 4 (1969): 930.

Uchiyama, Takashi, Junji Yodoi, Kimitaka Sagawa, Kiyoshi Takatsuki, and Haruto Uchino. "Adult T-Cell Leukemia: Clinical and Hematologic Features of 16 Cases." *Blood* 50, no. 3 (1977): 481–492.

van Helvoort, Ton. "A Century of Research into the Cause of Cancer: Is the New Oncogene Paradigm Revolutionary?" *History and Philosophy of the Life Sciences* (1999): 293–330.

van Helvoort, Ton. "The Construction of Bacteriophage as Bacterial Virus: Linking Endogenous and Exogenous Thought Styles." *Journal of the History of Biology* (1994): 91–139.

van Helvoort, Ton. "History of Virus Research in the Twentieth Century: The Problem of Conceptual Continuity." *History of Science* 32, no. 2 (1994): 185–235.

van Helvoort, Ton. "The Start of a Cancer Research Tradition: Peyton Rous, James Ewing, and Viruses as a Cause of Cancer." In *Creating a Tradition of Biomedical Research: Contributions to the History of the Rockefeller University*, 191–209. Rockefeller University Press, 2004.

van Helvoort, Ton, and Neeraja Sankaran. "How Seeing Became Knowing: The Role of the Electron Microscope in Shaping the Modern Definition of Viruses." *Journal of the History of Biology* (2018): 1–36.

Varmus, Harold. *The Art and Politics of Science*. Norton, 2009.

Varmus, Harold. "How Tumor Virology Evolved into Cancer Biology and Transformed Oncology." *Annual Review of Cancer Biology* 1 (2017): 1–18.

Varmus, Harold E., Ramareddy V. Guntaka, Warner J. W. Fan, Suzanne Heasley, and J. Michael Bishop. "Synthesis of Viral DNA in the Cytoplasm of Duck Embryo Fibroblasts and in Enucleated Cells after Infection by Avian Sarcoma Virus." *Proceedings of the National Academy of Sciences* 71, no. 10 (1974): 3874–3878.

Varmus, H. E., D. Stehelin, D. Spector, J. Tal, D. Fujita, T. Padgett, D. Roulland-Dussoix, et al. "Distribution and Function of Defined Regions of Avian Tumor Virus Genomes in Viruses and Uninfected Cells." In *Animal Virology*, 339–358. Cambridge University Press, 1976.

Villa, Luisa L., Ronaldo L. R. Costa, Carlos A. Petta, Rosires P. Andrade, Kevin A. Ault, Anna R. Giuliano, Cosette M. Wheeler, et al. "Prophylactic Quadrivalent Human Papillomavirus (Types 6, 11, 16, and 18) L1 Virus-Like Particle Vaccine in Young Women: A Randomised Double-Blind Placebo-Controlled Multicentre Phase II Efficacy Trial." *Lancet Oncology* 6, no. 5 (2005): 271–278.

Vilmer, E., C. Rouzioux, F. Vezinet Brun, A. Fischer, J. C. Chermann, F. Barre-Sinoussi, C. Gazengel, et al. "Isolation of New Lymphotropic Retrovirus from Two Siblings with Haemophilia B, One with AIDS." *Lancet* 323, no. 8380 (1984): 753–757.

Vogelstein, Bert, David Lane, and Arnold J. Levine. "Surfing the P53 Network." *Nature* 408, no. 6810 (2000): 307–310.

Vogt, Peter K. "A Humble Chicken Virus That Changed Biology and Medicine." *Lancet Oncology* 10, no. 1 (2009): 96.

Vogt, Peter K. Introduction to *Retroviruses*, edited by J. M. Coffin, S. H. Hughes, and H. E. Varmus. Cold Spring Harbor Laboratory Press, 1997.

Vogt, Peter K. "Oncogenes and the Revolution in Cancer Research: Homage to Hidesaburo Hanafusa (1929–2009)." *Genes & Cancer* 1 no. 1 (2010): 6–11.

Vogt, Marguerite. "Inhibitory Effects of the Corpora Cardiaca and of the Corpus Allatum in Drosophila." *Nature* 157, no. 3990 (1946): 512.

Vogt, Marguerite, and Renato Dulbecco. "Virus-Cell Interaction with a Tumor-Producing Virus." *Proceedings of the National Academy of Sciences of the United States of America* 46, no. 3 (1960): 365.

Wade, Nicholas. "Special Virus Cancer Program: Travails of a Biological Moonshot." *Science* 174, no. 4016 (1971): 1306–1311.

Wade, Nicolas. *The Ultimate Experiment: Man-Made Evolution*. Walker, 1979.

Walboomers, Jan M. M., Marcel V. Jacobs, M. Michele Manos, F. Xavier Bosch, J. Alain Kummer, Keerti V. Shah, et al. "Human Papillomavirus Is a Necessary Cause of Invasive Cervical Cancer Worldwide." *Journal of Pathology* 189, no. 1 (1999): 12–19.

Wapner, Jessica. *The Philadelphia Chromosome: A Mutant Gene and the Quest to Cure Cancer at the Genetic Level*. Experiment, 2013.

Waterson, A. P. *Introduction to Animal Virology*. Cambridge University Press, 1961.

Waterson, Anthony Peter, and Lise Wilkinson. *An Introduction to the History of Virology*. Cambridge University Press, 1978.

Watson, James D. *Avoid Boring People: Lessons from a Life in Science*. Vintage, 2009.

Watson, James D. *Father to Son: Truth, Reason, and Decency*. Cold Spring Harbor Laboratory Press, 2014.

Watson, James D. Foreword to *Cold Spring Harbor Symposia on Quantitative Biology*, vol. 39, xvii. Cold Spring Harbor Laboratory Press, 1974.

Watson, James D. *Molecular Biology of the Gene*. W. A. Benjamin, 1965.

Watson James D. *A Passion for DNA*. Oxford University Press, 2000.

Weinberg, Robert A. "Inadvertent Cancer Research." *Cancer Biology & Therapy* 3, no. 2 (2004): 238–239.

Weinberg, Robert. *Racing to the Beginning of the Road*. Harmony Books, 1996.

Weiss, Robin A. "Cancer, Infection and Immunity: A Personal Homage to Jan Svoboda." *Folia Biologica-Praha* 50, nos. 3/4 (2004): 78–86.

Weiss, Robin A. "Look Back at Viral Oncology." *Nature* 307 (1984): 665.

Weiss, Robin A. "On Viruses, Discovery, and Recognition." *Cell* 135, no. 6 (2008): 983–986.

Weiss, Robin A. "Remembering Jan Svoboda: A Personal Reflection." *Viruses* 10, no. 4 (2018): 203.

Weiss, Robin A., and Peter K. Vogt. "100 Years of Rous Sarcoma Virus." *Journal of Experimental Medicine* 208, no. 12 (2011): 2351–2355.

Wellerstein, Alex. "Manhattan Project." *Encyclopedia of the History of Science* 1, no. 3 (2019).

Weston, Kathleen M. *Blue Skies and Bench Space: Adventures in Cancer Research*. Cold Spring Harbor Laboratory Press, 2014.

Westphal, Heiner. "SV40 DNA Strand Selection by *Escherichia coli* RNA Polymerase." *Journal of Molecular Biology* 50, no. 2 (1970): 407–420.

Whewell, William. *The Philosophy of the Inductive Sciences: Founded upon Their History*. Vol. 1. J. W. Parker, 1840.

Whyte, Peter, Karen J. Buchkovich, Jonathan M. Horowitz, Stephen H. Friend, Margaret Raybuck, Robert A. Weinberg, and Ed Harlow. "Association between an Oncogene and an Anti-oncogene: The Adenovirus E1A Proteins Bind to the Retinoblastoma Gene Product." *Nature* 334, no. 6178 (1988): 124–129.

Wigler, Michael, Saul Silverstein, Lih-Syng Lee, Angel Pellicer, Yung-Chi Cheng, and Richard Axel. "Transfer of Purified Herpes Virus Thymidine Kinase Gene to Cultured Mouse Cells." *Cell* 11, no. 1 (1977): 223–232.

Wildy, P., M. G. P. Stoker, I. A. Macpherson, and R. W. Horne. "The Fine Structure of Polyoma Virus." *Virology* 11, no. 2 (1960): 444–457.

Wilson, Edward O. *Consilience: The Unity of Knowledge*. Vintage, 1999.

Williams, Greer. *Virus Hunters*. Hutchinson, 1960.

Williams, J. W. "The Development of the Ultracentrifuge and Its Contributions." *Annals of the New York Academy of Sciences* 325, no. 1 (1979): 77–94.

Winocour, Ernest. "Purification of Polyoma Virus." *Virology* 19, no. 2 (1963): 158–168.

Witkowski, Jan Anthony. *The Road to Discovery: A Short History of Cold Spring Harbor Laboratory*. Cold Spring Harbor Laboratory Press, 2016.

Witkowski, Jan, Alexander Gann, and Joe Sambrook. *Life Illuminated: Selected Papers from Cold Spring Harbor*. Vol. 2, *1972–1994*. Cold Spring Harbor Laboratory Press, 2008.

Witte, Owen N., Asim Dasgupta, and David Baltimore. "Abelson Murine Leukaemia Virus Protein Is Phosphorylated in Vitro to Form Phosphotyrosine." *Nature* 283, no. 5750 (1980): 826.

Wolfe, Audra J. *Freedom's Laboratory: The Cold War Struggle for the Soul of Science*. Johns Hopkins University Press, 2018.

Woolley, G. W., and Small, M. C. "Experiments on Cell-Free Transmission of Mouse Leukemia." *Cancer* 9 (1956): 1102–1106.

Wright, Susan. *Molecular Politics: Developing American and British Regulatory Policy for Genetic Engineering, 1972–1982*. University of Chicago Press, 1994.

Yi, Doogab. "Cancer, Viruses, and Mass Migration: Paul Berg's Venture into Eukaryotic Biology and the Advent of Recombinant DNA Research and Technology, 1967–1980." *Journal of the History of Biology* 41, no. 4 (2008): 589–636.

Yi, Doogab. *The Recombinant University: Genetic Engineering and the Emergence of Stanford Biotechnology*. University of Chicago Press, 2015.

Yi, Doogab. "The Scientific Commons in the Marketplace: The Industrialization of Biomedical Materials at the New England Enzyme Center, 1963–1980." *History and Technology* 25, no. 1 (2009): 69–87.

Zhao, Kong-Nan, Lifang Zhang, and Jia Qu. "Dr. Jian Zhou: The Great Inventor of Cervical Cancer Vaccine." *Protein & Cell* 8, no. 2 (2017): 79–82.

Zhelev, Nikolai. "Man of Science: Celebrating Professor Sir David Lane's 60th Anniversary." *Biodiscovery* 1 (2012): e8921.

Zhou, Jian, Xiao Yi Sun, Deborah J. Stenzel, and Ian H. Frazer. "Expression of Vaccinia Recombinant HPV 16 L1 and L2 ORF Proteins in Epithelial Cells Is Sufficient for Assembly of HPV Virion-Like Particles." *Virology* 185, no. 1 (1991): 251–257.

Zilber, L. A. "On the Interaction between Tumor Viruses and Cells: A Virogenetic Concept of Tumorigenesis 2." *Journal of the National Cancer Institute* 26, no. 6 (1961): 1311–1319.

Zimmer, C. *Rabbits with Horns and Other Astounding Viruses*. University of Chicago Press, 2012.

zur Hausen, H., et al. "EBV DNA in Biopsies of Burkitt Tumours and Anaplastic Carcinomas of the Nasopharynx." *Nature* 228 (1970): 1956–1958.

zur Hausen, H., and Heinrich Schulte-Holthausen. "Presence of EB Virus Nucleic Acid Homology in a 'Virus-Free' Line of Burkitt Tumour Cells." *Nature* 227, no. 5255 (1970): 245.fv

Index

Page numbers in *italics* refer to photographs.